# Peridynamic Theory and Its Applications

Erdogan Madenci • Erkan Oterkus

# Peridynamic Theory
# and Its Applications

 Springer

Erdogan Madenci
Department of Aerospace
  and Mechanical Engineering
University of Arizona
Tucson, AZ, USA

Erkan Oterkus
Naval Architecture
  and Marine Engineering
University of Strathclyde
Glasgow, UK

ISBN 978-1-4939-5322-6      ISBN 978-1-4614-8465-3 (eBook)
DOI 10.1007/978-1-4614-8465-3
Springer New York Heidelberg Dordrecht London

*We dedicate this book to*
**Dr. Stewart A. Silling**,
*the father of peridynamics*
*(Sandia National Laboratories,*
*Albuquerque, NM)*

# Preface

The effectiveness of computational techniques such as finite elements in modeling material failure has lagged far behind their capabilities in traditional stress analysis. This difficulty arises because the mathematical foundation on which all such methods are based assumes that the body remains continuous as it deforms. Existing computational methods for the modeling of fracture in a continuous body are based on the partial differential equations (PDEs) of classical continuum mechanics. These methods suffer from the inherent limitation that the spatial derivatives required by the PDEs do not, by definition, exist at crack tips or along crack surfaces. Therefore, the basic mathematical structure of the formulation breaks down whenever a crack appears in a body. Various special techniques have been developed in fracture mechanics to deal with this limitation. Generally, these techniques involve redefining a body in such a way as to exclude the crack and then applying conditions at the crack surfaces as boundary conditions. In addition, existing methods for fracture modeling suffer from the need of external crack growth criteria and, possibly, remeshing. This is a mathematical statement that prescribes how a crack evolves a priori based on local conditions. The requirement of tracking individual crack fronts, particularly in three dimensions, as well as the possibility of fractures moving between constituent materials, interfaces, and layers, makes it difficult to provide accurate crack growth criteria.

The difficulties encountered in the methods utilizing classical continuum mechanics can be overcome by performing molecular dynamics simulations or by using atomistic lattice models. Atomistic methods, although providing insight into the nature of fracture in certain materials, cannot be expected to provide a practical tool for the modeling of engineering structures. It is clear that the atomistic simulations are insufficient to model fracture processes in real-life structures.

The peridynamic theory provides the capability for improved modeling of progressive failure in materials and structures. Further, it paves the way for addressing multi-physics and multi-scale problems. Even though numerous journal articles and conference papers exist in the literature on the evolution and application of the peridynamic theory, it is still new to the technical community. Because it is based on concepts not commonly used in the past, the purpose of this book is to explain the peridynamic theory in a single framework. It presents not only the theoretical basis but also its numerical implementation. It starts with an overview of the peridynamic theory and the derivation of its governing equations. The relationship between peridynamics and classical continuum mechanics is established, and this leads to the ordinary state-based peridynamics formulations for both isotropic and composite materials. Numerical treatments of the peridynamic equations are presented in detail along with solutions to many benchmark and demonstration problems. In order to take advantage of salient features of peridynamics and the finite element method, a coupling technique is also described. Finally, an extension of the peridynamic theory for thermal diffusion and fully coupled thermomechanics is presented with applications.

Sample algorithms for the solutions of benchmark problems are available at the website http://extras.springer.com so that researchers and graduate students can modify these algorithms and develop their own solution algorithms for specific problems. The goal of this book is to provide students and researchers with a theoretical and practical knowledge of the peridynamic theory and the skills required to analyze engineering problems by developing their own algorithms.

## Acknowledgements

We appreciate the support provided by technical monitors Dr. Alex Tessler of NASA LaRC, Dr. David Stargel of AFOSR, and Dr. Abe Askari of the Boeing Company. Also, the first author had valuable discussions with Drs. Stewart Silling, Richard Lehoucq, Michael Parks, John Mitchell, and David Littlewood while at Sandia National Laboratories during his sabbatical. Last but not least, the first author also appreciates the encouragement and support of Dr. Nam Phan of NAVAIR.

We are greatly indebted to Ms. Connie Spencer for her invaluable efforts in typing, editing, and assisting with each detail associated with the completion of this book. Also, we appreciate the contributions made by Dr. Abigail Agwai, Dr. Atila Barut, Mr. Kyle Colavito, Dr. Ibrahim Guven, Dr. Bahattin Kilic, and Ms. Selda Oterkus at the University of Arizona in the development of various aspects of the theory and solutions of the example problems.

# Contents

# Chapter 1
# Introduction

## 1.1 Classical Local Theory

One of the underlying assumptions in the classical theory is its locality. The classical continuum theory assumes that a material point only interacts with its immediate neighbors; hence, it is a local theory. The interaction of material points is governed by the various balance laws. Therefore, in a local model a material point only exchanges mass, momentum, and energy with its closest neighbors. As a result, in classical mechanics the stress state at a point depends on the deformation at that point only. The validity of this assumption becomes questionable across different length scales. In general, at the macroscale this assumption is acceptable. However, the existence of long-range forces is evident from the atomic theory and as such the supposition of local interactions breaks down as the geometric length scale becomes smaller and approaches the atomic scale. Even at the macroscale there are situations when the validity of locality is questionable, for instance when small features and microstructures influence the entire macrostructure.

Despite the development of many important concepts to predict crack initiation and its growth in materials, it is still a major challenge within the framework of classical continuum mechanics. The main difficulty lies in the mathematical formulation, which assumes that a body remains continuous as it deforms. Therefore, the basic mathematical structure of the formulation breaks down whenever a discontinuity appears in a body. Mathematically, the classical theory is formulated using spatial partial differential equations, and these spatial derivatives are undefined at discontinuities. This introduces an inherent limitation to the classical theory, as the spatial derivatives in the governing equations, by definition, lose their meaning due to the presence of a discontinuity, such as a crack.

E. Madenci and E. Oterkus, *Peridynamic Theory and Its Applications*,
DOI 10.1007/978-1-4614-8465-3_1, © Springer Science+Business Media New York 2014

## 1.1.1   Shortcomings in Failure Prediction

The solutions within the realm of classical continuum mechanics result in infinite (singular) stresses at the crack tips, as derived in the pioneering study by Griffith (1921) that led to the concept of Linear Elastic Fracture Mechanics (LEFM). Within the realm of LEFM, a pre-existing crack in the material is necessary, and the stresses at the crack tip are mathematically singular. Therefore, the onset of crack initiation and crack growth are treated separately by introducing external criteria, such as critical energy release rate, that are not part of the governing equations of classical continuum mechanics. Furthermore, crack nucleation within LEFM still remains an unresolved issue.

Due to the presence of singular stresses, accurate calculation of the stress intensity factor or energy release rate can be highly challenging as these quantities are dependent on loading, geometry, and the numerical solution method. In addition to the requirement of an external criterion for the onset of crack growth, a criterion is also necessary for the direction of crack propagation. Understanding and prediction of the material failure process is rather complex due to the presence of a variety of mechanisms associated with grain boundaries, dislocations, microcracks, anisotropy, etc., each of which plays an important role at a specific length scale.

Many experiments indicated that materials with smaller cracks exhibit higher fracture resistance than those with larger cracks, and yet the solutions utilizing the classical continuum theory are independent of the crack size (Eringen et al. 1977). Furthermore, the classical continuum theory predicts no dispersion while experiments show otherwise for propagation of elastic plane waves with short wavelengths in elastic solids (Eringen 1972a). Within the realm of the classical (local) continuum theory, a material point in a continuum is influenced only by the other material points that are located within its immediate vicinity. Hence, there is no internal length parameter distinguishing different length scales.

Although the classical continuum theory is incapable of distinguishing among different scales, it can capture certain failure processes, and can be applied to a wide range of engineering problems, especially by employing the Finite Element Method (FEM). The FEM is robust in particular for determining stress fields, and it is also exceptionally suitable for modeling structures possessing complex geometries and different materials under general loading conditions. However, its governing equations are derived based on the classical continuum mechanics, and it also suffers from the presence of undefined spatial derivatives of displacements at crack tips or along crack surfaces.

When LEFM is adopted into the FEM, special elements are commonly needed in order to capture the correct singular behavior (mathematical artifact) at the crack tip. With traditional finite elements, the discontinuity in the displacement field that transpires as the crack propagates is remedied by redefining the body, i.e., defining the crack as a boundary.

The field of fracture mechanics is primarily concerned with the evolution of pre-existing cracks within a body, rather than the nucleation of new cracks. Even when addressing crack growth, the FEM with traditional elements suffers from the inherent limitation that it requires remeshing after each incremental crack growth. In addition to the need to remesh, existing methods for fracture modeling also suffer from the need to supply a kinetic relation for crack growth, a mathematical statement that prescribes how a crack evolves a priori based on local conditions. It guides the analysis as to when a crack should initiate; how fast it should grow and in what direction; whether it should turn, branch, oscillate, arrest, etc. Considering the difficulty in obtaining and generalizing experimental fracture data, providing such a kinetic relation for crack growth clearly presents a major obstacle to fracture modeling using conventional methods. Considering the presence of crack tip singularity, the need for external criteria, the inability to address crack initiation, and the requirement for redefining the body, it is clear that it is nearly impossible to solve a problem with multiple interacting cracks that propagate in a complex manner using traditional finite elements.

## 1.1.2  Remedies

Numerous studies were performed to improve the shortcomings of the FEM with traditional elements within the realm of LEFM. In particular, the cohesive zone concept introduced by Dugdale (1960) and Barenblatt (1962) has become prevalent among many other fracture criteria. However, the major breakthrough in computational fracture mechanics came with the introduction of Cohesive Zone Elements (CZE) by Hillerborg et al. (1976) for the Mode-I fracture mode and Xu and Needleman (1994) for a mixed-mode fracture. Materials and material interfaces are modeled through a traction-separation law for which the tractions are zero when the opening displacement (separation) reaches a critical value. Cohesive zone elements are usually surface elements that are placed along the element boundaries; hence, crack growth occurs only between traditional (regular) elements. Therefore, the material response exhibits characteristics of both regular and cohesive zone elements; the cohesive elements are only introduced to produce fracture behavior. The number of cohesive elements increases with decreasing mesh size, yet the size of the continuum region remains the same. Hence, softening of material properties can be observed with decreasing mesh size. Furthermore, mesh texture produces anisotropy, and it leads to mesh dependence. Crack paths are highly sensitive to mesh texture and alignment (Klein et al. 2001), and remeshing is required when crack paths are unknown a priori.

In an effort to resolve these difficulties, the concept of the eXtended Finite Element Method (XFEM) was introduced as a technique to model cracks and crack growth within the realm of finite elements without remeshing (Belytschko and Black 1999; Moes et al. 1999). It permits the cracks to propagate on any surface within an element, rather than only along element boundaries. Thus, it removes the limitation of CZE on the admissible direction of new fracture surfaces. XFEM is based on the partition of unity property of finite elements (Melenk and

Babuska 1996). Local enrichment functions, with additional degrees of freedom, are included in the standard finite element approximation. These functions are in the form of discontinuous displacement enrichment, in order to capture displacement discontinuity across a crack, and near-crack tip asymptotic displacement enrichment. Also, the additional number of degrees of freedom is minimized since the enrichment only includes the nodes that belong to the elements cut by cracks (Zi et al. 2007). According to Zi et al. (2007), the elements adjacent to the element in which the crack tip is positioned are partially enriched, and the partition of unity does not hold for them. Hence, the solution becomes inaccurate in the blending region. This prevents such methods from being applicable to problems in which multiple cracks grow and interact in complex patterns. The XFEM has been successfully employed to solve a number of fracture problems; however, it does require external criteria for injection of discontinuous displacement enrichment.

The difficulties encountered in the methods utilizing the classical continuum mechanics can be overcome by performing Molecular Dynamics Simulations (MDS) or atomistic lattice models. The atomistic simulation is certainly the most detailed and realistic one for predicting material fracture (Schlangen and van Mier 1992). The crack initiation and propagation can be simulated using the inter-atomic forces. However, atomistic studies focus on providing a fundamental understanding of the underlying basic physical processes of dynamic fracture, instead of being predictive (Cox et al. 2005). The reason for the limited focus stems from the availability of computational resources. In recent years, large-scale molecular dynamics simulations certainly have become possible with advancements in computer architectures. For instance, Kadau et al. (2006) performed simulations using 320 billion atoms, corresponding to a cubic piece of solid copper with an edge length of 1.56 $\mu$m. However, these length scales are still very small for real-life engineering structures. Furthermore, atomistic simulations suffer from the strict limitation on the total time since the time step is very small. Hence, most simulations are performed under very high loading rates, and it is not very clear if the fracture processes at artificially high rates resulting in high stresses are representative of events happening at lower rates.

Inspired from atomistic lattice models, lattice spring models eliminate the inadequacy of the atomistic simulations for large-scale structures by representing materials with discrete units interacting through springs, or, more generally, rheological elements (Ostoja-Starzewski 2002). The interactions among the lattice points can be short range by including nearest neighbors or long range (nonlocal) by containing neighbors beyond the nearest ones. Furthermore, lattice sites can be periodic or disordered and there are many different periodic lattices: triangular, square, honeycomb, etc. However, periodic lattices exhibit directional dependence on the elastic properties. Furthermore, the interaction force for one lattice type cannot easily be utilized for another type, and it is also not clear which lattice type is best suited for a specific problem.

Hence, it is clear that the atomistic simulations are insufficient to model fracture processes in real-life structures. Moreover, the experiments of physicists have revealed that cohesive forces reach finite distances among atoms, yet the classical continuum theory lacks an internal length parameter permitting modeling at different length scales because it is valid only for very long wavelengths (Eringen 1972a). Therefore, Eringen and Edelen (1972), Kroner (1967), and Kunin (1982) introduced the nonlocal continuum theory in an effort to account for the long-range effects.

## 1.2 Nonlocal Theories

The nonlocal theory of continuous media establishes the connection between classical continuum mechanics and molecular dynamics. In the case of the local (classical) continuum model, the state of a material point is influenced by the material points located in its immediate vicinity. In the case of the nonlocal continuum model, the state of a material point is influenced by material points located within a region of finite radius. As the radius becomes infinitely large, the nonlocal theory becomes the continuous version of the molecular dynamics model. Therefore, the nonlocal theory of continuous media establishes a connection between the classical (local) continuum mechanics and molecular dynamics models. The relationship between the local and nonlocal continuum models and the molecular dynamics model is illustrated in Fig. 1.1.

Any point $x$ interacts with other material points within a distance $\delta$. The material points within a distance $\delta$ of $x$ are called the family of $x$, $H_x$. The number of material points in a family of $x$, within the realm of classical continuum mechanics, is 3, 5, and 7 (including itself) for one-, two-, and three-dimensional analysis, respectively.

Various nonlocal theories, involving higher-order displacement gradients and spatial integrals, were introduced in the past. Early work by Eringen and Edelen (1972) and Eringen (1972a, b) resulted in a nonlocal continuum theory that accounted for nonlocality in the balance laws and thermodynamic statements. However, the resulting equations were rather complicated, and later work by these researchers simplified the theory by accounting for nonlocality in the

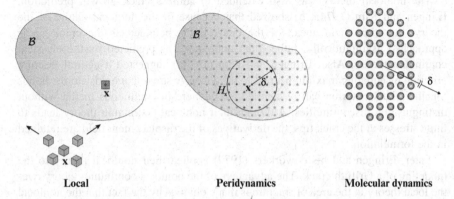

| Local | Peridynamics | Molecular dynamics |

**Fig. 1.1** Relationship between local and nonlocal continuum models

constitutive relation while keeping the equilibrium and kinematic equations in the local form (Eringen et al. 1977). Currently, most nonlocal theories account for nonlocality through the constitutive relation. Generally, integral-type nonlocal material models in continuum mechanics have a constitutive law that relates the forces (stresses) at a material point to some weighted average of deformation (strains) of other points that are some finite distance away. On the other hand, gradient-type nonlocal models include higher order derivatives to account for the field in the immediate vicinity of the point, such as the first derivative of the strain to the local constitutive law. Both types of nonlocal models have an associated characteristic length, which can be related to physical lengths such as grain size, fracture process zone size, or pore size.

The nonlocal continuum theory was noted for its ability to not only capture macroscale effects, but also the effects of molecular and atomic scales. Eringen (1972b) showed that a nonlocal model is capable of predicting a wide range of wavelengths. The nonlocal theory still assumes the media as a continuum, however it is computationally less demanding than the molecular dynamics while taking into account the long-range effects. Since the classical theory is the longwave limit of the atomic theory, they showed the ability of the nonlocal theory to capture deformation from the classical longwave limit to the atomic scale. According to Bazant and Jirasek (2002), there are many occasions when it is necessary to adopt a nonlocal approach in continuum mechanics. Such instances include capturing the effects of microstructure heterogeneity on small-scale continuum models. Nonlocality is also required in order to capture size effects—the dependence of nominal strength on structure size, observed in experiments and discrete modeling but not captured by local models. Nonlocality is also exhibited in the phenomena of microcracking. Distributed microcracking has been experimentally observed; however, it is challenging if not impossible to numerically simulate with local models because microcrack growth is not decided by local deformation or local stress. Evidence points to the fact that microcracking is not only dependent on the local deformation at the center of the microcrack, but is also dependent on the deformation that occurs within some neighborhood of the microcrack (Bazant 1991).

The nonlocal theory was also extended to address crack growth prediction. Eringen and Kim (1974a, b) showed that because of the nonlocal nature of the theory, the stress field ahead of the crack tip is bounded as the crack tip is approached asymptotically, rather than unbounded as predicted by the classical continuum theory. Also, Eringen and Kim (1974a) suggested a natural fracture criterion by equating maximum stress to the cohesive stress that holds atomic bonds together. This criterion can be applied everywhere in continuous media without distinguishing discontinuities. Although their nonlocal continuum theory leads to finite stresses at the crack tips, the derivatives of the displacement field are retained in the formulation.

Later, Eringen and his coworkers (1977) applied their nonlocal theory to the modeling of a Griffith crack. The advantage of the nonlocal continuum theory over the local theory in the area of fracture is made obvious by the fact that the nonlocal model predicts a physically meaningful finite stress field at the crack tip. This is opposed to the local theory that predicts infinite stresses at the crack tip, which is nonphysical because no real material can support infinite stresses. Also, Ari and Eringen (1983) showed that the analysis results of a Griffith crack using nonlocal elasticity are in agreement with the lattice model given by Elliott (1947). In spite of this, the governing equations of their model still lose meaning at the crack, as they are formulated in terms of spatial derivatives. In fact, most nonlocal models still break down in the presence of a discontinuity, such as a crack, because, similar to the classical local theory, spatial derivatives are included in their formulation. Typically, nonlocal models include nonlocality in the stress–strain relation through strain averaging (Eringen et al. 1977; Ozbolt and Bazant 1996) or through adding strain derivatives to the standard constitutive relation, hence retaining the spatial derivative.

Another type of nonlocal theory, introduced by Kunin (1982, 1983) and Rogula (1982), circumvents this difficulty because it uses displacement fields rather than their derivatives. However, it is only given for a one-dimensional medium by Kunin (1982) and Rogula (1982). Kunin (1983) derived a three-dimensional nonlocal model by approximating a continuous medium as a discrete lattice structure. More recently, Silling (2000) proposed a nonlocal theory that does not require spatial derivatives—the peridynamic (PD) theory. Compared to the previous non-local theory by Kunin (1982) and Rogula (1982), the PD theory is more general because it considers two- and three-dimensional media in addition to the one-dimensional medium. In contrast to the nonlocal theory by Kunin (1983), the PD theory provides nonlinear material response with respect to displacements. Furthermore, the material response includes damage in the PD theory.

## 1.2.1 Basics of Peridynamic Theory

In light of the inadequacies of local and nonlocal theories, the peridynamic theory, which is nonlocal, was introduced by Silling (2000) and Silling et al. (2007) in an attempt to deal with the discontinuities. Similar to the nonlocal theory formulated by Kunin (1982), the peridynamic theory employs displacements rather than dis-placement derivatives in its formulation. Basically, the peridynamic theory is a reformulation of the equation of motion in solid mechanics that is better suited for modeling bodies with discontinuities, such as cracks. The theory uses spatial integral equations that can be applied to a discontinuity. This stands in contrast to the partial differential equations used in the classical formulation, which are not defined at discontinuities. The peridynamic governing equations are defined at fracture surfaces; additionally, material damage is part of the peridynamic consti-tutive laws. These attributes permit fracture initiation and propagation to be modeled, with arbitrary paths, without the need for special crack growth treatment. Furthermore, interfaces between dissimilar materials have their own properties.

In the peridynamic theory, material points interact with each other directly through the prescribed response function, which contains all of the constitutive information associated with the material. The response function includes a length parameter called internal length (horizon), $\delta$. The locality of interactions depends on the horizon, and interactions become more local with a decreasing horizon. Hence, the classical theory of elasticity can be considered as a limiting case of the peridynamic theory as the internal length approaches zero. For instance, it has been shown that the peridynamic theory reduces to the linear theory of elasticity with the proper choice of response function (Silling et al. 2003; Weckner and Abeyaratne 2005). In another limiting case where the internal length approaches the inter-atomic distance, it was shown by Silling and Bobaru (2005) that van der Waals forces can be used as part of the response function to model nanoscale structures. Therefore, the peridynamic theory is capable of bridging the nano to macro length scales. With the PD theory, damage in the material is simulated in a much more realistic manner compared to the classical continuum-based methods. As the interactions between material points cease, cracks may initiate and align themselves along surfaces that form cracks, yet the integral equations continue to remain valid.

### *1.2.2   Attributes and Its Present State*

The main difference between the peridynamic theory and classical continuum mechanics is that the former is formulated using integral equations as opposed to derivatives of the displacement components. This feature allows damage initiation and propagation at multiple sites with arbitrary paths inside the material without resorting to special crack growth criteria. In the peridynamic theory, internal forces are expressed through nonlocal interactions between pairs of material points within a continuous body, and damage is part of the constitutive model. Interfaces between dissimilar materials have their own properties and damage can propagate when and where it is energetically favorable for it to do so. The PD theory provides the ability to link different length scales, and it can be viewed as the continuum version of MDS. It provides the ability to address multiphysics and multiscale failure prediction in a common framework.

The ability of the peridynamic theory to represent physical phenomena was demonstrated by Silling (2000). Silling investigated the propagation of linear stress waves and wave dispersion along with the shape of the crack tip within the realm of the peridynamic theory. The peridynamic linear elastic waves with long wavelengths were in agreement with those from the classical theory. At small scales, the peridynamic theory predicts nonlinear dispersion curves, which are found in real materials, unlike the curves predicted by the classical elasticity. In the crack tip study, the peridynamic theory predicts a cusp-like crack tip as opposed to the parabolic crack tip from LEFM. The parabolic crack tip in LEFM is associated with the unphysical unbounded stress at the crack tip.

The original peridynamic formulation by Silling (2000), later coined the "bond-based peridynamic theory," is based on the assumption of pairwise interactions of the same magnitude, thus resulting in a constraint on material properties, such as requiring the Poisson's ratio to be one-fourth for isotropic materials. Also, it does not distinguish between volumetric and distortional deformations; thus it is not suitable to capture the plastic incompressibility condition, or to utilize the existing material models.

In order to relax the constraint on material properties, Gerstle et al. (2007) introduced a "micropolar peridynamic model" by considering pairwise moments as well as forces in the "bond-based" peridynamics. Although this formulation overcomes the constraint limitation for isotropic materials, it is not clear if it can also capture the incompressibility condition. Therefore, Silling et al. (2007) introduced a more general formulation, coined the "state-based" peridynamic theory, which eliminates the limitations of the "bond-based" peridynamics. The "state-based" PD theory is based on the concept of peridynamic states that are infinite dimensional arrays containing information about peridynamic interactions. Silling (2010) also extended the "state-based" PD theory to account for the effects of indirect interactions between material points on other material points by introducing the "double state" concept. Recently, Lehoucq and Sears (2011) derived the energy and momentum conservation laws of the peridynamic theory by using the principles of classical statistical mechanics. They showed that the nonlocal interaction is intrinsic to continuum conservation laws. Recently, Silling (2011) also extended the use of PD theory for bridging different length scales by introducing a "coarse-graining method." According to this approach, the structural properties at a lower scale are reflected to its upper scale through a mathematically consistent technique.

The peridynamic theory does not concern the concept of stress and strain; however, it is possible to define a stress tensor within the PD framework. Lehoucq and Silling (2008) derived a PD stress tensor from nonlocal PD interactions. The stress tensor is obtained from the PD forces that pass through a material point volume. For sufficiently smooth motion, a constitutive model, and any existing nonhomogeneities, Silling and Lehoucq (2008) showed that the PD stress tensor converges to a Piola-Kirchhoff stress tensor in the limiting case where the horizon size converges to zero.

The integro-differential equation of peridynamic theory is difficult to solve analytically. However, a few analytical solutions exist in the literature. For instance, Silling et al. (2003) investigated the deformation of an infinite bar subjected to a self-equilibrated load distribution. The solution was achieved in the form of a linear Fredholm integral equation and solved by Fourier transformation. This solution revealed interesting results that cannot be captured by the classical theory, including decaying oscillations in the displacement field and progressively weakening discontinuities propagating outside of the loading region. Weckner et al. (2009) also used Laplace and Fourier transforms, and obtained an integral representation for the three-dimensional PD solution by utilizing Green's functions. This approach was independently pursued by Mikata (2012) to investigate peristatic and peridynamic solution of a one-dimensional infinite bar, and it was found that peridynamics can represent negative group velocities for certain wavenumbers, which can be used for modeling certain types of dispersive media with irregular dispersion.

Peridynamics permits not only linear elastic material behavior, but also nonlinear elastic (Silling and Bobaru 2005), plastic (Silling et al. 2007; Mitchell 2011a), viscoelastic (Kilic 2008; Taylor 2008; Mitchell 2011b), and viscoplastic (Taylor 2008; Foster et al. 2010) material behaviors. Dayal and Bhattacharya (2006) studied the kinetics of phase transformations in solids by using peridynamics. They derived a nucleation criterion by examining nucleation as a dynamic instability.

The solution of PD equations requires numerical integration both in time and space, for which explicit and Gaussian quadrature techniques can be adopted because of their simplicity. The description of these techniques and their application to peridynamics are presented by Silling and Askari (2005). They also provided the stability criterion for convergence of time integration, and discussed the order of accuracy for uniform discretization (grid) for spatial integration. Later, Emmrich and Weckner (2007) presented different spatial discretization schemes and tested them by considering a linear microelastic material of infinite length in one dimension. Recently, Bobaru et al. (2009) and Bobaru and Ha (2011) considered a nonuniform grid and nonuniform horizon sizes for spatial integration. In order to improve the accuracy and efficiency of numerical time integration, Polleschi (2010) proposed a mixed explicit-implicit time integration scheme; the integration through the time steps is explicit with an implicit cycle at every time step. In a similar way, Yu et al. (2011) proposed an adaptive trapezoidal integration scheme with a combined relative-absolute error control. Also, Mitchell (2011a, b) utilized an implicit time integration method.

Although the PD equation of motion includes the effects of inertia, it is possible to use it for quasi-static problems by appropriately allowing the inertia term to vanish through schemes, as demonstrated by Kilic and Madenci (2010a). Alternatively, Wang and Tian (2012) introduced a fast Galerkin method with efficient matrix assembly and storage.

The degree of nonlocality is defined by a PD parameter, referred to as the horizon; therefore, it is crucial to choose an appropriate size for it to obtain accurate results and represent the actual physical reality. In a recent study, Bobaru and Hu (2012) discuss the meaning, selection, and use of horizon in the PD theory and explain under what conditions the crack propagation speed depends on the horizon size and the role of incident waves on this speed. Influence function is another important parameter in PD theory; which determines the strength of interactions between material points. Seleson and Parks (2011) studied the effect of influence function by investigating wave propagation in simple one-dimensional models and brittle fracture in three-dimensional models.

The spatial integration of the PD equation is very suitable for parallel computing. However, the load distribution is a key issue to obtain the most efficient computational environment. An efficient load distribution scheme is described by Kilic (2008). Also, Liu and Hong (2012a) demonstrated the use of Graphics Processing Unit (GPU) architecture towards the same goal.

The PD theory permits crack initiation and growth. Silling et al. (2010) established a condition for the emergence of a discontinuity (crack nucleation) in an elastic body. For crack growth, it requires a critical material failure parameter. The original parameter for a brittle material is referred to as the "critical stretch," and it can be related to the critical energy release rate of the material, as explained in Silling and Askari (2005). Warren et al. (2009) demonstrated the capability of the nonordinary state-based PD theory for capturing failure based on either the critical equivalent strain (measure of shearing strain) or the averaged value of the volumetric strain (dilatation). Recently, Foster et al. (2011) proposed critical energy density as an alternative critical parameter and also related it to the critical energy release rate. As shown by Silling and Lehoucq (2010) and Hu et al. (2012b), the PD theory also permits the calculation of the J-integral value, which is an important parameter of fracture mechanics.

Silling (2003) considered a Kalthoff-Winkler experiment in which a plate having two parallel notches was hit by an impactor and peridynamic simulations successfully captured the angle of crack growth that was observed in the experiments. Silling and Askari (2004) also presented impact damage simulations including the Charpy V-notch test. Ha and Bobaru (2011) successfully captured various characteristics of dynamic fracture observed in experiments, including crack branching, crack-path instability, etc. Furthermore, Agwai et al. (2011) compared their PD analysis results against extended Finite Element Method (XFEM) and Cohesive Zone Model (CZM) predictions. Crack speeds computed from all approaches were found to be on the same order; however, the PD prediction of fracture paths are closer to the experimental observations, including both branching and microbranching behaviors.

The PD theory captures the interaction of local failure such as a crack growth with global failure due to structural stability. Kilic and Madenci (2009a) investigated the buckling characteristics of a rectangular column with a groove (crack initiation site) under compression and a constrained rectangular plate under uniform temperature load. They triggered lateral displacements using geometrical imperfection.

The PD theory also permits multiple load paths such as compression after impact. Demmie and Silling (2007) considered the extreme loadings on reinforced concrete structures by impacts from massive objects and explosive loading of concrete structures. This study was recently extended by Oterkus et al. (2012a) to predict the residual strength of impact damaged concrete structures.

Composite damage has also been modeled with the peridynamic theory. Within the PD framework, the simplest approach to model a composite layer with directional properties is achieved by assigning different material properties in the fiber and other (remaining) directions. The interactions between neighboring layers are defined by using inter-layer bonds. Askari et al. (2006) and Colavito et al. (2007a, b) predicted damage in laminated composites subjected to low-velocity impact and damage in woven composites subjected to static indentation. In addition, Xu et al. (2007) considered notched laminated composites under biaxial loads. Also,

Oterkus et al. (2010) demonstrated that PD analysis is capable of capturing bearing and shear-out failure modes in bolted composite lap joints.

Xu et al. (2008) analyzed the delamination and matrix damage process in composite laminates due to only low-velocity impact. Recently, Askari et al. (2011) considered the effect of both high- and low-energy hail impacts against a toughened-epoxy, intermediate-modulus, carbon-fiber composite. Also, Hu et al. (2012a) predicted the basic failure modes of fiber, matrix, and delamination in laminates with a pre-existing central crack under tension. The analytical derivation of the PD material parameters, including thermal loading conditions, was recently given by Oterkus and Madenci (2012). They also demonstrated the constraints on material constants due to the pairwise interaction assumption. The other approach to model composites was introduced by Kilic et al. (2009) by distinguishing fiber and matrix materials based on the volume fraction. Although this approach may have some advantages by taking into account the inhomogeneous structure, it is computationally more expensive than the homogenized technique. The other approach for modeling composites is the linking of micro- and macroscales as described by Alali and Lipton (2012). The method depends on a two-scale evolution equation. While the microscopic part of this equation governs the dynamics at the length scale of heterogeneities, the macroscopic part tracks the homogenized dynamics.

Since the numerical solution of peridynamic equations of motion is computationally more expensive than the local solutions, such as FEM, it may be advantageous to combine PD theory and local solutions. In a recent study, Seleson et al. (2013) proposed a force-based blended model that coupled PD theory and classical elasticity by using nonlocal weights composed of integrals of blending functions. They also generalized this approach to couple peridynamics and higher-order gradient models of any order. In another study, Lubineau et al. (2012) performed coupling of local and nonlocal solutions through a transition (morphing) that affects only constitutive parameters. The definition of the morphing functions in their approach relies on energy equivalence. In addition to these techniques, Kilic and Madenci (2010b) and Liu and Hong (2012b) coupled FEM and peridynamics. A more straightforward coupling procedure is given in Macek and Silling (2007), where the PD interactions are represented by truss elements. If only some part of the region is desired to be modeled by using peridynamics, then the other sections can be modeled by traditional finite elements. Another simple approach, demonstrated by Oterkus et al. (2012b) and Agwai et al. (2012), was first to solve the problem by using finite element analysis and obtain the displacement field. Then, by using the available information, the displacements can be applied as a boundary condition to the peridynamic model of a critical region.

The peridynamic theory is also suitable for thermal loading conditions; Kilic and Madenci (2010c) included the thermal term in the response function of peridynamic interactions. By utilizing this approach, Kilic and Madenci (2009b) predicted thermally driven crack propagation patterns in quenched glass plates containing single or multiple pre-existing cracks, and Kilic and Madenci (2010c) also predicted damage initiation and propagation in regions having dissimilar materials due to thermal loading.

Furthermore, the PD theory was extended to consider heat diffusion. Gerstle et al. (2008) developed a peridynamic model for electromigration that accounts for heat conduction in a one-dimensional body. Additionally, Bobaru and Duangpanya (2010, 2012) introduced a multidimensional peridynamic heat conduction equation, and considered domains with discontinuities such as insulated cracks. Both studies adopted the bond-based peridynamic approach. Later, Agwai (2011) derived the state-based peridynamic heat conduction equation. She also further extended it for fully coupled thermomechanics (Agwai 2011).

The peridynamic theory has been utilized successfully for damage prediction of many problems at different length scales from macro to nano. In order to take into account the effect of van der Waals interactions, Silling and Bobaru (2005) and Bobaru (2007) included an additional term to the peridynamic response function to represent van der Waals forces. This new formulation was used to investigate the mechanical behavior, strength, and toughness properties of three-dimensional nanofiber networks under imposed stretch deformation. It was found that the inclusion of van der Waals forces significantly changes the overall deformation behavior of the nanofiber network structure. In a recent study, Seleson et al. (2009) demonstrated that peridynamics can play the role of an upscale version of molecular dynamics and pointed out the extent where the molecular dynamics solutions can be recovered by peridynamics. Celik et al. (2011) utilized peridynamics to extract mechanical properties of nickel nanowires subjected to bending loads in a customized atomic force microscope (AFM) and scanning electron microscope (SEM). SEM images of fractured nanowires are also compared against peridynamic simulation results.

Even though numerous journal articles and conference papers exist in the literature on the evolution and application of the peridynamic theory, it is still new to the scientific community. Because it is based on concepts not commonly used in the past, the purpose of this book is to explain the peridynamic theory in a single framework. It presents not only the theoretical basis but also its numerical implementation.

It starts with an overview of the peridynamic theory and derivation of its governing equations. The relationship between peridynamics and classical continuum mechanics is established, and this leads to the ordinary state-based peridynamics formulations for both isotropic and composite materials. Numerical treatments of the peridynamic equations are presented in detail along with solutions to many benchmark and demonstration problems. In order to take advantage of salient features of the peridynamics and finite element methods, a coupling technique is also presented in detail. Finally, an extension of the peridynamic theory for thermal diffusion and fully coupled thermomechanics is presented with applications. [FORTRAN algorithms providing solutions to many of these benchmark problems can be found at http://extras.springer.com.]

# References

Agwai A (2011) A peridynamic approach for coupled fields. Dissertation, University of Arizona

Agwai A, Guven I, Madenci E (2011) Predicting crack propagation with peridynamics: a comparative study. Int J Fracture 171:65–78

Agwai A, Guven I, Madenci E (2012) Drop-shock failure prediction in electronic packages by using peridynamic theory. IEEE Trans Adv Pack 2(3):439–447

Alali B, Lipton R (2012) Multiscale dynamics of heterogeneous media in the peridynamic formulation. J Elast 106:71–103

Ari N, Eringen AC (1983) Nonlocal stress field at Griffith crack. Cryst Latt Defect Amorph Mater 10:33–38

Askari E, Xu J, Silling SA (2006) Peridynamic analysis of damage and failure in composites. Paper 2006–88 presented at the 44th AIAA aerospace sciences meeting and exhibit. Grand Sierra Resort Hotel, Reno, 9–12 Jan 2006

Askari A, Nelson K, Weckner O, Xu J, Silling S (2011) Hail impact characteristics of a hybrid material by advanced analysis techniques and testing. J Aerosp Eng 24:210–217

Barenblatt GI (1962) The mathematical theory of equilibrium cracks in brittle fracture. Adv Appl Mech 7:56–125

Bazant ZP (1991) Why continuum damage is nonlocal—micromechanics arguments. J Eng Mech 117:1070–1087

Bazant ZP, Jirasek M (2002) Nonlocal integral formulations of plasticity and damage: survey of progress. J Eng Mech 128(11):1119–1149

Belytschko T, Black T (1999) Elastic crack growth in finite elements with minimal remeshing. Int J Numer Meth Eng 45:601–620

Bobaru F (2007) Influence of Van Der Waals forces on increasing the strength and toughness in dynamic fracture of nanofiber networks: a peridynamic approach. Model Simul Mater Sci Eng 15:397–417

Bobaru F, Duangpanya M (2010) The peridynamic formulation for transient heat conduction. Int J Heat Mass Trans 53:4047–4059

Bobaru F, Duangpanya MA (2012) Peridynamic formulation for transient heat conduction in bodies with evolving discontinuities. J Comput Phys 231:2764–2785

Bobaru F, Ha YD (2011) Adaptive refinement and multiscale modeling in 2D peridynamics. Int J Multiscale Comput Eng 9(6):635–660

Bobaru F, Hu W (2012) The meaning, selection and use of the peridynamic horizon and its relation to crack branching in brittle materials. Int J Fract 176(2):215–222

Bobaru F, Yang M, Alves LF, Silling SA, Askari E, Xu J (2009) Convergence, adaptive refinement, and scaling in 1D peridynamics. Int J Numer Meth Eng 77:852–877

Celik E, Guven I, Madenci E (2011) Simulations of nanowire bend tests for extracting mechanical properties. Theor Appl Fract Mech 55:185–191

Colavito KW, Kilic B, Celik E, Madenci E, Askari E, Silling S (2007a) Effect of void content on stiffness and strength of composites by a peridynamic analysis and static indentation test. Paper 2007–2251 presented at the 48th AIAA/ASME/ASCE/AHS/ASC structures, structural dynamics, and materials conference, Waikiki, 23–26 Apr 2007

Colavito KW, Kilic B, Celik E, Madenci E, Askari E, Silling S (2007b) Effect of nano particles on stiffness and impact strength of composites. Paper 2007–2001 presented at the 48th AIAA/ASME/ASCE/AHS/ASC structures, structural dynamics, and materials conference, Waikiki, 23–26 Apr 2007

Cox BN, Gao H, Gross D, Rittel D (2005) Modern topics and challenges in dynamic fracture. J Mech Phys Solid 53:565–596

Dayal K, Bhattacharya K (2006) Kinetics of phase transformations in the peridynamic formulation of continuum mechanics. J Mech Phys Solids 54:1811–1842

Demmie PN, Silling SA (2007) An approach to modeling extreme loading of structures using peridynamics. J Mech Mater Struct 2(10):1921–1945

Dugdale DS (1960) Yielding of steel sheets containing slits. J Mech Phys Solids 8(2):100–104

Elliott HA (1947) An analysis of the conditions for rupture due to Griffith cracks. Proc Phys Soc 59:208–223

Emmrich E, Weckner O (2007) The peridynamic equation and its spatial discretization. J Math Model Anal 12(1):17–27

Eringen AC (1972a) Nonlocal polar elastic continua. Int J Eng Sci 10:1–16

Eringen AC (1972b) Linear theory of nonlocal elasticity and dispersion of plane waves. Int J Eng Sci 10:425–435

Eringen AC, Edelen DGB (1972) On nonlocal elasticity. Int J Eng Sci 10:233–248

Eringen AC, Kim BS (1974a) Stress concentration at the tip of crack. Mech Res Commun 1:233–237

Eringen AC, Kim BS (1974b) On the problem of crack tip in nonlocal elasticity. In: Thoft-Christensen P (ed) Continuum mechanics aspects of geodynamics and rock fracture mechanics. Proceedings of the NATO advanced study institute held in Reykjavik, 11–20 Aug 1974. D. Reidel, Dordrecht, pp 107–113

Eringen AC, Speziale CG, Kim BS (1977) Crack-tip problem in non-local elasticity. J Mech Phys Solids 25:339–355

Foster JT, Silling SA, Chen WW (2010) Viscoplasticity using peridynamics. Int J Numer Meth Eng 81:1242–1258

Foster JT, Silling SA, Chen W (2011) An energy based failure criterion for use with peridynamic states. Int J Multiscale Comput Eng 9(6):675–688

Gerstle W, Sau N, Silling S (2007) Peridynamic modeling of concrete structures. Nucl Eng Des 237(12–13):1250–1258

Gerstle W, Silling S, Read D, Tewary V, Lehoucq R (2008) Peridynamic simulation of electromigration. Comput Mater Continua 8(2):75–92

Griffith AA (1921) The phenomena of rupture and flow in solids. Philos Trans R Soc Lond A 221:163–198

Ha YD, Bobaru F (2011) Characteristics of dynamic brittle fracture captured with peridynamics. Eng Fract Mech 78:1156–1168

Hillerborg A, Modeer M, Petersson PE (1976) Analysis of crack formation and crack growth by means of fracture mechanics and finite elements. Cem Concr Res 6(6):773–781

Hu W, Ha YD, Bobaru F (2012a) Peridynamic model for dynamic fracture in unidirectional fiber-reinforced composites. Comput Meth Appl Mech Eng 217–220:247–261

Hu W, Ha YD, Bobaru F, Silling SA (2012b) The formulation and computation of the non-local J-integral in bond-based peridynamics. Int J Fract 176:195–206

Kadau K, Germann TC, Lomdahl PS (2006) Molecular dynamics comes of age: 320 billion atom simulation on BlueGene/L. Int J Mod Phys C 17:1755–1761

Kilic B (2008) Peridynamic theory for progressive failure prediction in homogeneous and heterogeneous materials. Dissertation, University of Arizona

Kilic B, Madenci E (2009a) Structural stability and failure analysis using peridynamic theory. Int J Nonlinear Mech 44:845–854

Kilic B, Madenci E (2009b) Prediction of crack paths in a quenched glass plate by using peridynamic theory. Int J Fract 156:165–177

Kilic B, Madenci E (2010a) An adaptive dynamic relaxation method for quasi-static simulations using the peridynamic theory. Theor Appl Fract Mech 53:194–201

Kilic B, Madenci E (2010b) Coupling of peridynamic theory and finite element method. J Mech Mater Struct 5:707–733

Kilic B, Madenci E (2010c) Peridynamic theory for thermomechanical analysis. IEEE Trans Adv Packag 33:97–105

Kilic B, Agwai A, Madenci E (2009) Peridynamic theory for progressive damage prediction in centre-cracked composite laminates. Compos Struct 90:141–151

Klein PA, Foulk JW, Chen EP, Wimmer SA, Gao H (2001) Physics-based modeling of brittle fracture: cohesive formulations and the application of meshfree methods. Theor Appl Fract Mech 37:99–166

Kroner E (1967) Elasticity theory of materials with long range cohesive forces. Int J Solids Struct 3:731–742

Kunin IA (1982) Elastic media with microstructure I: one dimensional models. Springer, Berlin

Kunin IA (1983) Elastic media with microstructure II: three-dimensional models. Springer, Berlin

Lehóucq RB, Sears MP (2011) Statistical mechanical foundation of the peridynamic nonlocal continuum theory: energy and momentum conservation laws. Phys Rev E 84:031112

Lehoucq RB, Silling SA (2008) Force flux and the peridynamic stress tensor. J Mech Phys Solids 56:1566–1577

Liu W, Hong J (2012a) Discretized peridynamics for brittle and ductile solids. Int J Numer Meth Eng 89(8):1028–1046

Liu W, Hong J (2012b) A coupling approach of discretized peridynamics with finite element method. Comput Meth Appl Mech Eng 245–246:163–175

Lubineau G, Azdoud Y, Han F, Rey C, Askari A (2012) A morphing strategy to couple non-local to local continuum mechanics. J Mech Phys Solids 60:1088–1102

Macek RW, Silling SA (2007) Peridynamics via finite element analysis. Finite Elem Anal Des 43 (15):1169–1178

Melenk JM, Babuska I (1996) The partition of unity finite element method: basic theory and applications. Comput Meth Appl Mech Eng 139:289–314

Mikata Y (2012) Analytical solutions of peristatic and peridynamic problems for a 1D infinite rod. Int J Solids Struct 49(21):2887–2897

Mitchell JA (2011a) A nonlocal, ordinary, state-based plasticity model for peridynamics. SAND2011-3166. Sandia National Laboratories, Albuquerque

Mitchell JA (2011b) A non-local, ordinary-state-based viscoelasticity model for peridynamics. SAND2011-8064. Sandia National Laboratories, Albuquerque

Moes N, Dolbow J, Belytschko T (1999) A finite element method for crack growth without remeshing. Int J Numer Meth Eng 46:131–150

Ostoja-Starzewski M (2002) Lattice models in micromechanics. Appl Mech Rev 55:35–60

Oterkus E, Madenci E (2012) Peridynamic analysis of fiber reinforced composite materials. J Mech Mater Struct 7(1):45–84

Oterkus E, Barut A, Madenci E (2010) Damage growth prediction from loaded composite fastener holes by using peridynamic theory. In: Proceedings of the 51st AIAA/ASME/ASCE/AHS/ASC structures, structural dynamics, and materials conference, April 2010. AIAA, Reston, Paper 2010–3026

Oterkus E, Guven I, Madenci E (2012a) Impact damage assessment by using peridynamic theory. Cent Eur J Eng 2(4):523–531

Oterkus E, Madenci E, Weckner O, Silling S, Bogert P, Tessler A (2012b) Combined finite element and peridynamic analyses for predicting failure in a stiffened composite curved panel with a central slot. Compos Struct 94:839–850

Ozbolt J, Bazant ZP (1996) Numerical smeared fracture analysis: nonlocal microcrack interaction approach. Int J Numer Meth Eng 39:635–661

Polleschi M (2010) Stability and applications of the peridynamic method. Thesis, Polytechnic University of Turin

Rogula D (1982) Nonlocal theory of material media. Springer, Berlin, pp 137–243

Schlangen E, van Mier JGM (1992) Simple lattice model for numerical simulation of fracture of concrete materials and structures. Mater Struct 25:534–542

Seleson P, Parks ML (2011) On the role of influence function in the peridynamic theory. Int J Multiscale Comput Eng 9(6):689–706

Seleson P, Parks ML, Gunzburger M, Lehocq RB (2009) Peridynamics as an upscaling of molecular dynamics. Multiscale Model Simul 8(1):204–227

Seleson P, Beneddine S, Prudhomme S (2013) A force-based coupling scheme for peridynamics and classical elasticity. Comput Mater Sci 66:34–49

Silling SA (2000) Reformulation of elasticity theory for discontinuities and long-range forces. J Mech Phys Solids 48:175–209

Silling SA (2003) Dynamic fracture modeling with a meshfree peridynamic code. In: Bathe KJ (ed) Computational fluid and solid mechanics. Elsevier, Amsterdam, pp 641–644

Silling SA (2010) Linearized theory of peridynamic states. J Elast 99:85–111

Silling SA (2011) A coarsening method for linear peridynamics. Int J Multiscale Comput Eng 9 (6):609–622

Silling SA, Askari A (2004) Peridynamic modeling of impact damage. In: Moody FJ (ed) PVP-vol. 489. American Society of Mechanical Engineers, New York, pp 197–205

Silling SA, Askari A (2005) A meshfree method based on the peridynamic model of solid mechanics. Comput Struct 83(17–18):1526–1535

Silling SA, Bobaru F (2005) Peridynamic modeling of membranes and fibers. Int J Nonlinear Mech 40:395–409

Silling SA, Lehoucq RB (2008) Convergence of peridynamics to classical elasticity theory. J Elast 93:13–37

Silling SA, Lehoucq RB (2010) Peridynamic theory of solid mechanics. Adv Appl Mech 44:73–168

Silling SA, Zimmermann M, Abeyaratne R (2003) Deformation of a peridynamic bar. J Elast 73:173–190

Silling SA, Epton M, Weckner O, Xu J, Askari A (2007) Peridynamics states and constitutive modeling. J Elast 88:151–184

Silling SA, Weckner O, Askari A, Bobaru F (2010) Crack nucleation in a peridynamic solid. Int J Fract 162:219–227

Taylor MJ (2008) Numerical simulation of thermo-elasticity, inelasticity and rupture in membrane theory. Dissertation, University of California, Berkeley

Wang H, Tian H (2012) A fast Galerkin method with efficient matrix assembly and storage for a peridynamic model. J Comput Phys 231:7730–7738

Warren TL, Silling SA, Askari A, Weckner O, Epton MA, Xu J (2009) A non-ordinary state-based peridynamic method to model solid material deformation and fracture. Int J Solids Struct 46:1186–1195

Weckner O, Abeyaratne R (2005) The effect of long-range forces on the dynamic bar. J Mech Phys Solids 53:705–728

Weckner O, Brunk G, Epton MA, Silling SA, Askari E (2009) Green's functions in non-local three-dimensional linear elasticity. Proc R Soc A 465:3463–3487

Xu XP, Needleman A (1994) Numerical simulations of fast crack growth in brittle solids. J Mech Phys Solids 42:1397–1434

Xu J, Askari A, Weckner O, Razi H, Silling S (2007) Damage and failure analysis of composite laminates under biaxial loads. In: Proceedings of the 48th AIAA/ASME/ASCE/AHS/ASC structures, structural dynamics, and materials conference, April 2007. AIAA, Reston. doi: 10.2514/6.2007-2315

Xu J, Askari A, Weckner O, Silling SA (2008) Peridynamic analysis of impact damage in composite laminates. J Aerosp Eng 21(3):187–194

Yu K, Xin XJ, Lease KB (2011) A new adaptive integration method for the peridynamic theory. Model Simul Mater Sci Eng 19:45003

Zi G, Rabczuk T, Wall W (2007) Extended meshfree methods without branch enrichment for cohesive cracks. Comput Mech 40:367–382

# Chapter 2
# Peridynamic Theory

## 2.1 Basics

At any instant of time, every point in the material denotes the location of a material particle, and these infinitely many material points (particles) constitute the continuum. In an undeformed state of the body, each material point is identified by its coordinates, $\mathbf{x}_{(k)}$ with $(k = 1, 2, \ldots, \infty)$, and is associated with an incremental volume, $V_{(k)}$, and a mass density of $\rho(\mathbf{x}_{(k)})$. Each material point can be subjected to prescribed body loads, displacement, or velocity, resulting in motion and deformation. With respect to a Cartesian coordinate system, the material point $\mathbf{x}_{(k)}$ experiences displacement, $\mathbf{u}_{(k)}$, and its location is described by the position vector $\mathbf{y}_{(k)}$ in the deformed state. The displacement and body load vectors at material point $\mathbf{x}_{(k)}$ are represented by $\mathbf{u}_{(k)}(\mathbf{x}_{(k)}, t)$ and $\mathbf{b}_{(k)}(\mathbf{x}_{(k)}, t)$, respectively. The motion of a material point conforms to the Lagrangian description.

According to the peridynamic (PD) theory introduced by Silling (2000), the motion of a body is analyzed by considering the interaction of a material point, $\mathbf{x}_{(k)}$, with the other, possibly infinitely many, material points, $\mathbf{x}_{(j)}$, with $(j = 1, 2, .., \infty)$, in the body. Therefore, an infinite number of interactions may exist between the material point at location $\mathbf{x}_{(k)}$ and other material points. However, the influence of the material points interacting with $\mathbf{x}_{(k)}$ is assumed to vanish beyond a local region (horizon), denoted by $H_{\mathbf{x}_{(k)}}$, shown in Fig. 2.1. Similarly, material point $\mathbf{x}_{(j)}$ interacts with material points in its own family, $H_{\mathbf{x}_{(j)}}$.

In other words, the PD theory is concerned with the physics of a material body at a point that interacts with all points within its range, as shown in Fig. 2.1. The range of material point $\mathbf{x}_{(k)}$ is defined by $\delta$, referred to as the "horizon." Also, the material points within a distance $\delta$ of $\mathbf{x}_{(k)}$ are called the family of $\mathbf{x}_{(k)}, H_{\mathbf{x}_{(k)}}$. The interaction of material points is prescribed through a micropotential that depends on the deformation and constitutive properties of the material. The locality of interactions depends on the horizon, and the interactions become more local with a decreasing horizon, $\delta$.

E. Madenci and E. Oterkus, *Peridynamic Theory and Its Applications*, 
DOI 10.1007/978-1-4614-8465-3_2, © Springer Science+Business Media New York 2014

**Fig. 2.1** Infinitely many
PD material points
and interaction of points at
$\mathbf{x}_{(k)}$ and $\mathbf{x}_{(j)}$

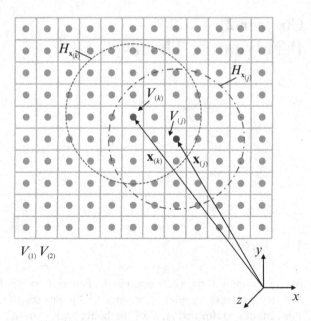

$V_{(1)}\ V_{(2)}$

Hence, the classical theory of elasticity can be considered a limiting case of the peridynamic theory as the horizon approaches zero (Silling and Lehoucq 2008).

## 2.2   Deformation

As shown in Fig. 2.2, material point $\mathbf{x}_{(k)}$ interacts with its family of material points, $H_{\mathbf{x}_{(k)}}$, and it is influenced by the collective deformation of all these material points. Similarly, material point $\mathbf{x}_{(j)}$ is influenced by deformation of the material points, $H_{\mathbf{x}_{(j)}}$, in its own family. In the deformed configuration, the material points $\mathbf{x}_{(k)}$ and $\mathbf{x}_{(j)}$ experience displacements, $\mathbf{u}_{(k)}$ and $\mathbf{u}_{(j)}$, respectively, as shown in Fig. 2.2. Their initial relative position vector $(\mathbf{x}_{(j)} - \mathbf{x}_{(k)})$ prior to deformation becomes $(\mathbf{y}_{(j)} - \mathbf{y}_{(k)})$ after deformation. The stretch between material points $\mathbf{x}_{(k)}$ and $\mathbf{x}_{(j)}$ is defined as

$$s_{(k)(j)} = \frac{\left(\left|\mathbf{y}_{(j)} - \mathbf{y}_{(k)}\right| - \left|\mathbf{x}_{(j)} - \mathbf{x}_{(k)}\right|\right)}{\left|\mathbf{x}_{(j)} - \mathbf{x}_{(k)}\right|}. \tag{2.1}$$

Associated with material point $\mathbf{x}_{(k)}$, all of the relative position vectors in the deformed configuration, $(\mathbf{y}_{(j)} - \mathbf{y}_{(k)})$ with $(j = 1, 2, .., \infty)$, can be stored in an infinite-dimensional array, or a deformation vector state, $\underline{\mathbf{Y}}$:

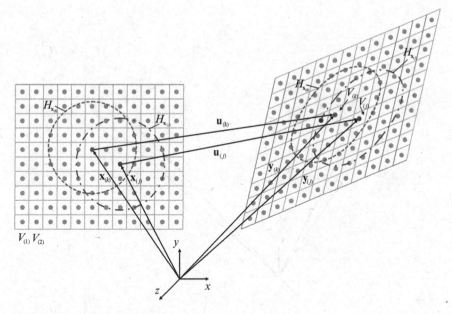

**Fig. 2.2** Kinematics of PD material points

$$\underline{\mathbf{Y}}(\mathbf{x}_{(k)}, t) = \left\{ \begin{array}{c} (\mathbf{y}_{(1)} - \mathbf{y}_{(k)}) \\ \vdots \\ (\mathbf{y}_{(\infty)} - \mathbf{y}_{(k)}) \end{array} \right\}. \tag{2.2}$$

The definitions and mathematical properties of vector states are presented by Silling et al. (2007). Their properties in relation to the derivation of PD equations are summarized in the Appendix.

## 2.3 Force Density

As illustrated in Fig. 2.3, the material point $\mathbf{x}_{(k)}$ interacts with its family of material points, $H_{\mathbf{x}_{(k)}}$, and it is influenced by the collective deformation of all these material points, thus resulting in a force density vector, $\mathbf{t}_{(k)(j)}$, acting at material point $\mathbf{x}_{(k)}$. It can be viewed as the force exerted by material point $\mathbf{x}_{(j)}$. Similarly, material point $\mathbf{x}_{(j)}$ is influenced by deformation of the material points, $H_{\mathbf{x}_{(j)}}$, in its own family, and the corresponding force density vector is $\mathbf{t}_{(j)(k)}$ at material point $\mathbf{x}_{(j)}$ and is exerted on by material point $\mathbf{x}_{(k)}$. These forces are determined jointly by the *collective* deformation of $H_{\mathbf{x}_{(k)}}$ and $H_{\mathbf{x}_{(j)}}$ through the material model.

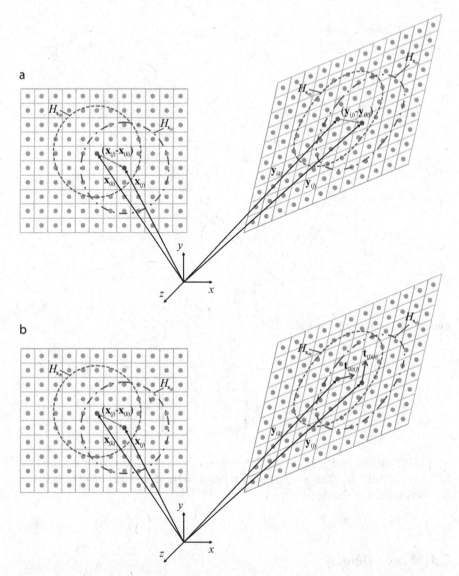

**Fig. 2.3** PD vector states: (**a**) deformation, $\underline{\mathbf{Y}}$, and (**b**) force, $\underline{\mathbf{T}}$

Associated with material point $\mathbf{x}_{(k)}$, all of the force density vectors, $\mathbf{t}_{(k)(j)}$ with $(j = 1, 2, .., \infty)$, can be stored in an infinite-dimensional array, or a force vector state, $\underline{\mathbf{T}}$:

$$\underline{\mathbf{T}}(\mathbf{x}_{(k)}, t) = \left\{ \begin{array}{c} \mathbf{t}_{(k)(1)} \\ \vdots \\ \mathbf{t}_{(k)(\infty)} \end{array} \right\}. \tag{2.3}$$

## 2.4  Peridynamic States

The PD theory mainly concerns the deformation state, $\underline{\mathbf{Y}}$, and the force state, $\underline{\mathbf{T}}$. As described in Fig. 2.3a, the relative position vector $(\mathbf{y}_{(j)} - \mathbf{y}_{(k)})$ can be obtained by operating the deformation state, $\underline{\mathbf{Y}}$, on the relative position vector $(\mathbf{x}_{(j)} - \mathbf{x}_{(k)})$ as

$$(\mathbf{y}_{(j)} - \mathbf{y}_{(k)}) = \underline{\mathbf{Y}}(\mathbf{x}_{(k)}, t)\langle \mathbf{x}_{(j)} - \mathbf{x}_{(k)} \rangle. \tag{2.4}$$

Similarly, the force density vector, $\mathbf{t}_{(k)(j)}$, shown in Fig. 2.3b, that the material point at location $\mathbf{x}_{(j)}$ exerts on the material point at location $\mathbf{x}_{(k)}$ can be expressed as

$$\mathbf{t}_{(k)(j)}\left(\mathbf{u}_{(j)} - \mathbf{u}_{(k)}, \mathbf{x}_{(j)} - \mathbf{x}_{(k)}, t\right) = \underline{\mathbf{T}}(\mathbf{x}_{(k)}, t)\langle \mathbf{x}_{(j)} - \mathbf{x}_{(k)} \rangle. \tag{2.5}$$

The difference between the force state and the deformation state is that the force state is dependent on the deformation state while the deformation state is independent. Therefore, the force state for material point $\mathbf{x}_{(k)}$ depends on the relative displacements between this material point and the other material points within its horizon. Hence, the force state can also be written as

$$\underline{\mathbf{T}}(\mathbf{x}_{(k)}, t) = \underline{\mathbf{T}}\left(\underline{\mathbf{Y}}(\mathbf{x}_{(k)}, t)\right). \tag{2.6}$$

## 2.5  Strain Energy Density

Due to the interaction between material points $\mathbf{x}_{(k)}$ and $\mathbf{x}_{(j)}$, a scalar-valued micropotential, $w_{(k)(j)}$, develops; it depends on the material properties as well as the stretch between point $\mathbf{x}_{(k)}$ and all other material points in its family. Note that the micropotential $w_{(j)(k)} \neq w_{(k)(j)}$, because $w_{(j)(k)}$ depends on the state of material points within the family of material point $\mathbf{x}_{(j)}$. These micropotentials can be expressed as

$$w_{(k)(j)} = w_{(k)(j)}\left(\mathbf{y}_{(1^k)} - \mathbf{y}_{(k)}, \ \mathbf{y}_{(2^k)} - \mathbf{y}_{(k)}, \cdots\right) \tag{2.7a}$$

and

$$w_{(j)(k)} = w_{(j)(k)}\left(\mathbf{y}_{(1^j)} - \mathbf{y}_{(j)}, \ \mathbf{y}_{(2^j)} - \mathbf{y}_{(j)}, \cdots\right), \tag{2.7b}$$

where $\mathbf{y}_{(k)}$ is the position vector of point $\mathbf{x}_{(k)}$ in the deformed configuration and $\mathbf{y}_{(1^k)}$ is the position vector of the first material point that interacts with point $\mathbf{x}_{(k)}$. Similarly, $\mathbf{y}_{(j)}$ is the position vector of point $\mathbf{x}_{(j)}$ in the deformed configuration and $\mathbf{y}_{(1^j)}$ is the position vector of the first material point that interacts with point $\mathbf{x}_{(j)}$.

The strain energy density, $W_{(k)}$, of material point $\mathbf{x}_{(k)}$ can be expressed as a summation of micropotentials, $w_{(k)(j)}$, arising from the interaction of material point $\mathbf{x}_{(k)}$ and the other material points, $\mathbf{x}_{(j)}$, within its horizon in the form

$$
\begin{aligned}
W_{(k)} = \frac{1}{2} \sum_{j=1}^{\infty} \frac{1}{2} & \left( w_{(k)(j)} \left( \mathbf{y}_{(1^k)} - \mathbf{y}_{(k)}, \ \mathbf{y}_{(2^k)} - \mathbf{y}_{(k)}, \cdots \right) \right. \\
& \left. + w_{(j)(k)} \left( \mathbf{y}_{(1^j)} - \mathbf{y}_{(j)}, \ \mathbf{y}_{(2^j)} - \mathbf{y}_{(j)}, \cdots \right) \right) V_{(j)},
\end{aligned}
\tag{2.8}
$$

in which $w_{(k)(j)} = 0$ for $k = j$.

## 2.6   Equations of Motion

The PD equations of motion at material point $\mathbf{x}_{(k)}$ can be derived by applying the principle of virtual work, i.e.,

$$
\delta \int_{t_0}^{t_1} (T - U) dt = 0,
\tag{2.9}
$$

where $T$ and $U$ represent the total kinetic and potential energies in the body. This principle is satisfied by solving for the Lagrange's equation

$$
\frac{d}{dt} \left( \frac{\partial L}{\partial \dot{\mathbf{u}}_{(k)}} \right) - \frac{\partial L}{\partial \mathbf{u}_{(k)}} = 0,
\tag{2.10}
$$

where the Lagrangian $L$ is defined as

$$
L = T - U.
\tag{2.11}
$$

The total kinetic and potential energies in the body can be obtained by summation of kinetic and potential energies of all material points, respectively,

$$
T = \sum_{i=1}^{\infty} \frac{1}{2} \rho_{(i)} \dot{\mathbf{u}}_{(i)} \cdot \dot{\mathbf{u}}_{(i)} V_{(i)}
\tag{2.12a}
$$

and

$$U = \sum_{i=1}^{\infty} W_{(i)} V_{(i)} - \sum_{i=1}^{\infty} \left( \mathbf{b}_{(i)} \cdot \mathbf{u}_{(i)} \right) V_{(i)}. \tag{2.12b}$$

Substituting for the strain energy density, $W_{(i)}$, of material point $\mathbf{x}_{(i)}$ from Eq. 2.8, the potential energy can be rewritten as

$$U = \sum_{i=1}^{\infty} \left\{ \frac{1}{2} \sum_{j=1}^{\infty} \frac{1}{2} \left[ \begin{array}{l} w_{(i)(j)}\left( \mathbf{y}_{(1^i)} - \mathbf{y}_{(i)}, \ \mathbf{y}_{(2^i)} - \mathbf{y}_{(i)}, \cdots \right) \\ + w_{(j)(i)}\left( \mathbf{y}_{(1^j)} - \mathbf{y}_{(j)}, \ \mathbf{y}_{(2^j)} - \mathbf{y}_{(j)}, \cdots \right) \end{array} \right] V_{(j)} - \left( \mathbf{b}_{(i)} \cdot \mathbf{u}_{(i)} \right) \right\} V_{(i)}. \tag{2.13}$$

By using Eq. 2.11, the Lagrangian can be written in an expanded form by showing only the terms associated with the material point $\mathbf{x}_{(k)}$:

$$\begin{aligned}
L = \ & \ldots + \frac{1}{2} \rho_{(k)} \, \dot{\mathbf{u}}_{(k)} \cdot \dot{\mathbf{u}}_{(k)} V_{(k)} + \cdots \\
& \cdots - \frac{1}{2} \sum_{j=1}^{\infty} \left\{ \frac{1}{2} \left[ w_{(k)(j)}\left( \mathbf{y}_{(1^k)} - \mathbf{y}_{(k)}, \ \mathbf{y}_{(2^k)} - \mathbf{y}_{(k)}, \cdots \right) \right. \right. \\
& \quad \left. \left. + w_{(j)(k)}\left( \mathbf{y}_{(1^j)} - \mathbf{y}_{(j)}, \ \mathbf{y}_{(2^j)} - \mathbf{y}_{(j)}, \cdots \right) \right] V_{(j)} \right\} V_{(k)} \cdots \\
& \cdots - \frac{1}{2} \sum_{i=1}^{\infty} \left\{ \frac{1}{2} \left[ w_{(i)(k)}\left( \mathbf{y}_{(1^i)} - \mathbf{y}_{(i)}, \ \mathbf{y}_{(2^i)} - \mathbf{y}_{(i)}, \cdots \right) \right. \right. \\
& \quad \left. \left. + w_{(k)(i)}\left( \mathbf{y}_{(1^k)} - \mathbf{y}_{(k)}, \ \mathbf{y}_{(2^k)} - \mathbf{y}_{(k)}, \cdots \right) \right] V_{(i)} \right\} V_{(k)} \cdots \\
& \cdots + \left( \mathbf{b}_{(k)} \cdot \mathbf{u}_{(k)} \right) V_{(k)} \cdots
\end{aligned} \tag{2.14a}$$

or

$$\begin{aligned}
L = \ & \cdots + \frac{1}{2} \rho_{(k)} \, \dot{\mathbf{u}}_{(k)} \cdot \dot{\mathbf{u}}_{(k)} V_{(k)} + \cdots \\
& \cdots - \frac{1}{2} \sum_{j=1}^{\infty} \left\{ w_{(k)(j)}\left( \mathbf{y}_{(1^k)} - \mathbf{y}_{(k)}, \ \mathbf{y}_{(2^k)} - \mathbf{y}_{(k)}, \cdots \right) V_{(j)} V_{(k)} \right\} \cdots \\
& \cdots - \frac{1}{2} \sum_{j=1}^{\infty} \left\{ w_{(j)(k)}\left( \mathbf{y}_{(1^j)} - \mathbf{y}_{(j)}, \ \mathbf{y}_{(2^j)} - \mathbf{y}_{(j)}, \cdots \right) V_{(j)} V_{(k)} \right\} \cdots \\
& \cdots + \left( \mathbf{b}_{(k)} \cdot \mathbf{u}_{(k)} \right) V_{(k)} \cdots \ .
\end{aligned} \tag{2.14b}$$

Substituting from Eq. 2.14b into Eq. 2.10 results in the Lagrange's equation of the material point $\mathbf{x}_{(k)}$ as

$$
\rho_{(k)}\ddot{\mathbf{u}}_{(k)}\,V_{(k)} + \left(\sum_{j=1}^{\infty}\frac{1}{2}\left(\sum_{i=1}^{\infty}\frac{\partial w_{(k)(j)}}{\partial\left(\mathbf{y}_{(j)}-\mathbf{y}_{(k)}\right)}V_{(i)}\right)\frac{\partial\left(\mathbf{y}_{(j)}-\mathbf{y}_{(k)}\right)}{\partial\mathbf{u}_{(k)}}\right.
$$
$$
\left. + \sum_{j=1}^{\infty}\frac{1}{2}\left(\sum_{i=1}^{\infty}\frac{\partial w_{(j)(k)}}{\partial\left(\mathbf{y}_{(k)}-\mathbf{y}_{(j)}\right)}V_{(i)}\right)\frac{\partial\left(\mathbf{y}_{(k)}-\mathbf{y}_{(j)}\right)}{\partial\mathbf{u}_{(k)}} - \mathbf{b}_{(k)}\right)V_{(k)} = 0
$$

$$(2.15a)$$

or

$$
\rho_{(k)}\ddot{\mathbf{u}}_{(k)} = \sum_{j=1}^{\infty}\frac{1}{2}\left(\sum_{i=1}^{\infty}\frac{\partial w_{(k)(i)}}{\partial\left(\mathbf{y}_{(j)}-\mathbf{y}_{(k)}\right)}V_{(i)}\right) - \sum_{j=1}^{\infty}\frac{1}{2}\left(\sum_{i=1}^{\infty}\frac{\partial w_{(i)(k)}}{\partial\left(\mathbf{y}_{(k)}-\mathbf{y}_{(j)}\right)}V_{(i)}\right) + \mathbf{b}_{(k)},
$$

$$(2.15b)$$

in which it is assumed that the interactions not involving material point $\mathbf{x}_{(k)}$ do not have any effect on material point $\mathbf{x}_{(k)}$. Based on the dimensional analysis of this equation, it is apparent that $\sum_{i=1}^{\infty}V_{(i)}\partial w_{(k)(i)}/\partial(\mathbf{y}_{(j)}-\mathbf{y}_{(k)})$ represents the force density that material point $\mathbf{x}_{(j)}$ exerts on material point $\mathbf{x}_{(k)}$ and $\sum_{i=1}^{\infty}V_{(i)}\partial w_{(i)(k)}/\partial(\mathbf{y}_{(k)}-\mathbf{y}_{(j)})$ represents the force density that material point $\mathbf{x}_{(k)}$ exerts on material point $\mathbf{x}_{(j)}$. With this interpretation, Eq. 2.15b can be rewritten as

$$
\rho_{(k)}\ddot{\mathbf{u}}_{(k)} = \sum_{j=1}^{\infty}\left[\mathbf{t}_{(k)(j)}\left(\mathbf{u}_{(j)}-\mathbf{u}_{(k)},\mathbf{x}_{(j)}-\mathbf{x}_{(k)},t\right)\right.
$$
$$
\left. - \mathbf{t}_{(j)(k)}\left(\mathbf{u}_{(k)}-\mathbf{u}_{(j)},\mathbf{x}_{(k)}-\mathbf{x}_{(j)},t\right)\right]V_{(j)} + \mathbf{b}_{(k)},
$$

$$(2.16)$$

where

$$
\mathbf{t}_{(k)(j)}\left(\mathbf{u}_{(j)}-\mathbf{u}_{(k)},\mathbf{x}_{(j)}-\mathbf{x}_{(k)},t\right) = \frac{1}{2}\frac{1}{V_{(j)}}\left(\sum_{i=1}^{\infty}\frac{\partial w_{(k)(i)}}{\partial\left(\mathbf{y}_{(j)}-\mathbf{y}_{(k)}\right)}V_{(i)}\right)
$$

$$(2.17a)$$

and

$$
\mathbf{t}_{(j)(k)}\left(\mathbf{u}_{(k)}-\mathbf{u}_{(j)},\mathbf{x}_{(k)}-\mathbf{x}_{(j)},t\right) = \frac{1}{2}\frac{1}{V_{(j)}}\left(\sum_{i=1}^{\infty}\frac{\partial w_{(i)(k)}}{\partial\left(\mathbf{y}_{(k)}-\mathbf{y}_{(j)}\right)}V_{(i)}\right).
$$

$$(2.17b)$$

By utilizing the state concept, the force densities $t_{(k)(j)}$ and $t_{(j)(k)}$ can be stored in force vector states that belong to material points $x_{(k)}$ and $x_{(j)}$, respectively, as

$$\underline{T}(x_{(k)}, t) = \left\{ \begin{array}{c} \vdots \\ t_{(k)(j)} \\ \vdots \end{array} \right\} \text{ and } \underline{T}(x_{(j)}, t) = \left\{ \begin{array}{c} \vdots \\ t_{(j)(k)} \\ \vdots \end{array} \right\}. \quad (2.18a,b)$$

The force densities $t_{(k)(j)}$ and $t_{(j)(k)}$ stored in vector states $\underline{T}(x_{(k)}, t)$ and $\underline{T}(x_{(j)}, t)$ can be extracted again by operating the force states on the corresponding initial relative position vectors

$$t_{(k)(j)} = \underline{T}(x_{(k)}, t) \langle x_{(j)} - x_{(k)} \rangle \quad (2.19a)$$

and

$$t_{(j)(k)} = \underline{T}(x_{(j)}, t) \langle x_{(k)} - x_{(j)} \rangle. \quad (2.19b)$$

By using Eqs. 2.19a and 2.19b, Lagrange's equation of the material point $x_{(k)}$ can be recast as

$$\rho_{(k)} \ddot{u}_{(k)} = \sum_{j=1}^{\infty} \left( \underline{T}(x_{(k)}, t) \langle x_{(j)} - x_{(k)} \rangle - \underline{T}(x_{(j)}, t) \langle x_{(k)} - x_{(j)} \rangle \right) V_{(j)} + b_{(k)}. \quad (2.20)$$

Because the volume of each material point $V_{(j)}$ is infinitesimally small, for the limiting case of $V_{(j)} \to 0$, the infinite summation can be expressed as integration while considering only the material points within the horizon,

$$\sum_{j=1}^{\infty} (\cdot) V_{(j)} \to \int_V (\cdot) \, dV' \to \int_H (\cdot) \, dH. \quad (2.21)$$

With this replacement, Eq. 2.20 can be written in integral equation form as

$$\rho(x)\ddot{u}(x, t) = \int_H \left( \underline{T}(x, t) \langle x' - x \rangle - \underline{T}(x', t) \langle x - x' \rangle \right) dH + b(x, t) \quad (2.22a)$$

or

$$\rho(x)\ddot{u}(x, t) = \int_H \left( t(u' - u, x' - x, t) - t'(u - u', x - x', t) \right) dH + b(x, t) . \quad (2.22b)$$

## 2.7    Initial and Constraint Conditions

The resulting PD equation of motion is a nonlinear integro-differential equation in time and space and is free of kinematic linearization, thus it is suitable for geometrically nonlinear analyses. It contains differentiation with respect to time and integration in the spatial domain. It does not contain any spatial derivatives of displacements. Thus, the PD equation of motion is valid everywhere whether or not displacement discontinuities exist in the material. Construction of its solution involves time and spatial integrations while being subject to constraints and/or loading conditions on the boundary, $\mathcal{B}$, of the material region, $\mathcal{R}$, and initial conditions on the displacement and velocity fields.

### 2.7.1    Initial Conditions

Time integration requires the application of initial displacement and velocity values at each material point in $\mathcal{R}$, and they can be specified as

$$\mathbf{u}(\mathbf{x}, t = 0) = \mathbf{u}^*(\mathbf{x}) \tag{2.23a}$$

and

$$\dot{\mathbf{u}}(\mathbf{x}, t = 0) = \mathbf{v}^*(\mathbf{x}). \tag{2.23b}$$

In addition to these required initial conditions, the initial conditions may also be necessary on the displacement and velocity gradients, $\mathbf{H}^*(\mathbf{x})$ and $\mathbf{L}^*(\mathbf{x})$, respectively. They can be specified as

$$\mathbf{H}(\mathbf{x}, t = 0) = \mathbf{H}^*(\mathbf{x}) \sim \frac{\partial u_i(x_k, 0)}{\partial x_j}, \quad \text{with } (i, j, k) = 1, 2, 3, \tag{2.24a}$$

and

$$\mathbf{L}(\mathbf{x}, t = 0) = \mathbf{L}^*(\mathbf{x}) \sim \frac{\partial \dot{u}_i(x_k, 0)}{\partial x_j}, \quad \text{with } (i, j, k) = 1, 2, 3 . \tag{2.24b}$$

The corresponding displacement and velocity fields are superimposed on the initial displacement and velocity fields as

$$\mathbf{u}(\mathbf{x}, t = 0) = \mathbf{u}^*(\mathbf{x}) + \mathbf{H}^*(\mathbf{x})(\mathbf{x} - \mathbf{x}_{ref}) \tag{2.25a}$$

and

**Fig. 2.4** Boundary regions
for constraint and external
load introduction

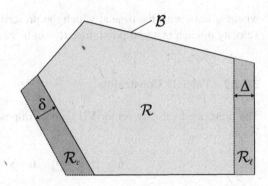

$$\dot{\mathbf{u}}(\mathbf{x}, t = 0) = \mathbf{v}^*(\mathbf{x}) + \mathbf{L}^*(\mathbf{x})(\mathbf{x} - \mathbf{x}_{ref}), \tag{2.25b}$$

where $\mathbf{x}_{ref}$ is a reference point (Silling 2004).

## 2.7.2   Constraint Conditions

The PD equation of motion does not contain any spatial derivatives; therefore, constraint conditions are, in general, not necessary for the solution of an integro-differential equation. However, such conditions can be imposed by prescribing constraints on displacement and velocity fields in a "fictitious material layer" along the boundary of a nonzero volume. Based on numerical experiments, Macek and Silling (2007) suggested that the extent of the fictitious boundary layer be equal to the horizon, $\delta$, in order to ensure that the imposed prescribed constraints are sufficiently reflected on the actual material region. Therefore, a fictitious boundary layer, $\mathcal{R}_c$, with depth $\delta$, is introduced along the boundary of the actual material region, $\mathcal{R}$, as shown in Fig. 2.4.

### 2.7.2.1   Displacement Constraints

The prescribed displacement vector $\mathbf{U}_0$ can be imposed through the material points in $\mathcal{R}_c$ as

$$\mathbf{u}(\mathbf{x}, t) = \mathbf{U}_0, \quad \text{for} \quad \mathbf{x} \in \mathcal{R}_c. \tag{2.26}$$

Also, in order to avoid abrupt constraint introduction, it can be applied as

$$\mathbf{u}(\mathbf{x}, t) = \begin{cases} \mathbf{U}_0 \dfrac{t}{t_0} & \text{for } 0 \leq t \leq t_0 \\ \mathbf{U}_0 & \text{for } t_0 \leq t \end{cases}, \tag{2.27}$$

where $t_0$ represents the time at which the prescribed displacement is reached. The velocity of each material point, $\dot{\mathbf{u}}(\mathbf{x}, t)$, can be calculated through differentiation.

#### 2.7.2.2  Velocity Constraints

The prescribed velocity vector $\mathbf{V}(t)$ can be imposed through the material points in $\mathcal{R}_c$ as

$$\dot{\mathbf{u}}(\mathbf{x}, t) = \mathbf{V}(t), \quad \text{for } \mathbf{x} \in \mathcal{R}_c. \tag{2.28}$$

Their displacement, $\mathbf{u}(\mathbf{x}, t)$, can be obtained from

$$\mathbf{u}(\mathbf{x}, t) = \int_0^t \mathbf{V}(t')dt'. \tag{2.29}$$

If $\mathbf{V}(t) = \mathbf{V}_0 H(t)$, with $\mathbf{V}_0$ containing constant constraint values, then $\mathbf{u}(\mathbf{x}, t) = \mathbf{V}_0 t$ for all material points in $\mathcal{R}_c$. The Heaviside step function is represented by $H(t)$. Also, in order to avoid abrupt velocity introduction, it can be applied as

$$\mathbf{V}(t) = \begin{cases} \mathbf{V}_0 \dfrac{t}{t_0} & \text{for } 0 \leq t \leq t_0 \\ \mathbf{V}_0 & \text{for } t_0 \leq t \end{cases}, \tag{2.30}$$

where $t_0$ represents the time at which the prescribed velocity is reached.

### 2.7.3  External Loads

Boundary traction does not directly appear in the PD equation of motion. Therefore, the application of external loads is also different from that of the classical continuum theory. The difference can be illustrated by considering a region, $\Omega$, that is subjected to external loads. If this region is fictitiously divided into two domains, $\Omega^-$ and $\Omega^+$, as shown in Fig. 2.5a, there must be a net force, $\mathbf{F}^+$, that is exerted to domain $\Omega^+$ by domain $\Omega^-$ so that force equilibrium is satisfied (Kilic 2008).

According to classical continuum mechanics, force $\mathbf{F}^+$ can be determined by integrating surface tractions over the cross-sectional area, $\partial\Omega$, of domains $\Omega^-$ and $\Omega^+$ as

$$\mathbf{F}^+ = \int_{\partial\Omega} \mathbf{T} dA, \tag{2.31}$$

in which $\mathbf{T}$ is the surface tractions (Fig. 2.5b).

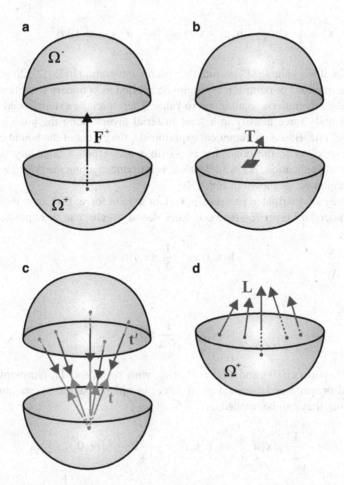

**Fig. 2.5** Boundary conditions: (**a**) domain of interest, (**b**) tractions in classical continuum mechanics, (**c**) interaction of a material point in domain $\Omega^+$ with other material points in domain $\Omega^-$, (**d**) force densities acting on domain $\Omega^+$ due to domain $\Omega^-$

In the case of the PD theory, the material points located in domain $\Omega^+$ interact with the other material points in domain $\Omega^-$ (Fig. 2.5c). Thus, the force $\mathbf{F}^+$ can be computed by volume integration of the force densities (Fig. 2.5d) over domain $\Omega^+$ as

$$\mathbf{F}^+ = \int_{\Omega^+} \mathbf{L}(\mathbf{x})dV, \qquad (2.32a)$$

in which $\mathbf{L}$, acting on a material point in domain $\Omega^+$, is determined by

$$\mathbf{L}(\mathbf{x}) = \int_{\Omega^-} [\mathbf{t}(\mathbf{u}' - \mathbf{u}, \mathbf{x}' - \mathbf{x}, t) - \mathbf{t}'(\mathbf{u} - \mathbf{u}', \mathbf{x} - \mathbf{x}', t)]dV . \tag{2.32b}$$

Note that if the volume $\Omega^-$ is void, the volume integration in Eq. 2.32b vanishes. Hence, the tractions or point forces cannot be applied as boundary conditions since their volume integrations result in a zero value. Therefore, the external loads can be applied as body force density in a "real material layer" along the boundary of a nonzero volume. Based on numerical experiments, the extent of the boundary layer should be as close to the boundary as possible. Therefore, a boundary layer for external load application, $\mathcal{R}_\ell$, with depth $\Delta$, is introduced along the boundary of the material region $\mathcal{R}$, as shown in Fig. 2.4.

In the case of distributed pressure, $p(\mathbf{x}, t)$, or a point force, $\mathbf{P}(t)$, over the surface $\mathcal{S}_\ell$ of the boundary layer $\mathcal{R}_\ell$, the body force density vector can be expressed as

$$\mathbf{b}(\mathbf{x}, t) = -\frac{1}{\Delta} p(\mathbf{x}, t)\mathbf{n} \tag{2.33a}$$

or

$$\mathbf{b}(\mathbf{x}, t) = \frac{1}{\mathcal{S}_\ell \Delta} \mathbf{P}(t) . \tag{2.33b}$$

If $p(\mathbf{x}, t) = p_0(\mathbf{x})H(t)$ and $\mathbf{P}(t) = \mathbf{P}_0 H(t)$, with $p_0(\mathbf{x})$ and $\mathbf{P}_0$ representing the distributed pressure and constant point force, in order to avoid abrupt constraint introduction, they can be applied as

$$\mathbf{b}(\mathbf{x}, t) = -\frac{1}{\Delta} p_0(\mathbf{x})\mathbf{n} \frac{t}{t_0} \text{ or } \mathbf{b}(\mathbf{x}, t) = \frac{1}{\mathcal{S}_\ell \Delta} \mathbf{P}_0 \frac{t}{t_0} \text{ for } 0 \le t \le t_0 \tag{2.34a}$$

and

$$\mathbf{b}(\mathbf{x}, t) = -\frac{1}{\Delta} p_0(\mathbf{x})\mathbf{n} \text{ or } \mathbf{b}(\mathbf{x}, t) = \frac{1}{\mathcal{S}_\ell \Delta} \mathbf{P}_0, \text{ for } t_0 \le t, \tag{2.34b}$$

where $t_0$ represents the time at which the prescribed external load is reached. The displacement and velocity of all points in the boundary layer $\mathcal{R}_\ell$ are calculated based on the equation of motion.

## 2.8   Balance Laws

The PD equation of motion must be further governed by the balance of linear momentum, angular momentum, and energy. These balance laws are viewed as having a primitive status in mechanics. The balance of linear momentum and energy are automatically satisfied, as the principle of virtual work, Eq. 2.9, represents their weak forms. However, the balance of angular momentum must be assured.

The linear momentum, **L**, and angular momentum (about the coordinate origin), $\mathbf{H}_0$, of a fixed set of particles at time $t$ in volume $V$ are given by

$$\mathbf{L} = \int_V \rho(\mathbf{x})\,\dot{\mathbf{u}}(\mathbf{x}, t)dV \tag{2.35a}$$

and

$$\mathbf{H}_0 = \int_V \mathbf{y}(\mathbf{x}, t) \times \rho(\mathbf{x})\,\dot{\mathbf{u}}(\mathbf{x}, t)dV, \tag{2.35b}$$

while the total force, **F**, and torque, $\mathbf{\Pi}_0$, about the origin are given by

$$\mathbf{F} = \int_V \mathbf{b}(\mathbf{x}, t)dV + \int_V\int_H \underline{\mathbf{T}}(\mathbf{x}, t)\langle \mathbf{x}' - \mathbf{x}\rangle\, dHdV - \int_V\int_H \underline{\mathbf{T}}(\mathbf{x}', t)\langle \mathbf{x} - \mathbf{x}'\rangle\, dHdV \tag{2.35c}$$

and

$$\mathbf{\Pi}_0 = \int_V \mathbf{y}(\mathbf{x}, t) \times \mathbf{b}(\mathbf{x}, t)dV + \int_V\int_H \mathbf{y}(\mathbf{x}, t) \times \underline{\mathbf{T}}(\mathbf{x}, t)\langle \mathbf{x}' - \mathbf{x}\rangle\, dHdV$$
$$- \int_V\int_H \mathbf{y}(\mathbf{x}, t) \times \underline{\mathbf{T}}(\mathbf{x}', t)\langle \mathbf{x} - \mathbf{x}'\rangle\, dHdV . \tag{2.35d}$$

Thus, the balance of linear momentum, $\dot{\mathbf{L}} = \mathbf{F}$, and angular momentum, $\dot{\mathbf{H}}_0 = \mathbf{\Pi}_0$, results in

$$\int_V \rho(\mathbf{x})\ddot{\mathbf{u}}(\mathbf{x}, t)\, dV = \int_V \mathbf{b}(\mathbf{x}, t)dV$$
$$+ \int_V\int_H \underline{\mathbf{T}}(\mathbf{x}, t)\langle \mathbf{x}' - \mathbf{x}\rangle\, dHdV$$
$$- \int_V\int_H \underline{\mathbf{T}}(\mathbf{x}', t)\langle \mathbf{x} - \mathbf{x}'\rangle\, dHdV \tag{2.36a}$$

and

$$\int_V \mathbf{y}(\mathbf{x},t) \times \rho(\mathbf{x})\ddot{\mathbf{u}}(\mathbf{x},t)\, dV = \int_V \mathbf{y}(\mathbf{x},t) \times \mathbf{b}(\mathbf{x},t) dV$$

$$+ \int_V \int_H \mathbf{y}(\mathbf{x},t) \times \underline{\mathbf{T}}(\mathbf{x},t)\langle \mathbf{x}' - \mathbf{x} \rangle\, dH dV \qquad (2.36b)$$

$$- \int_V \int_H \mathbf{y}(\mathbf{x},t) \times \underline{\mathbf{T}}(\mathbf{x}',t)\langle \mathbf{x} - \mathbf{x}' \rangle\, dH dV\;.$$

Because $\underline{\mathbf{T}}(\mathbf{x},t)\langle \mathbf{x}' - \mathbf{x} \rangle = \underline{\mathbf{T}}(\mathbf{x}',t)\langle \mathbf{x} - \mathbf{x}' \rangle = \mathbf{0}$ for $\mathbf{x}' \notin H$, these equations can be rewritten to include all of the material points in volume $V$ as

$$\int_V \rho(\mathbf{x})\ddot{\mathbf{u}}(\mathbf{x},t)\, dV = \int_V \mathbf{b}(\mathbf{x},t) dV$$

$$+ \int_V \int_V \underline{\mathbf{T}}(\mathbf{x},t)\langle \mathbf{x}' - \mathbf{x} \rangle\, dV' dV \qquad (2.37a)$$

$$- \int_V \int_V \underline{\mathbf{T}}(\mathbf{x}',t)\langle \mathbf{x} - \mathbf{x}' \rangle\, dV' dV$$

and

$$\int_V (\mathbf{y}(\mathbf{x},t) \times \rho(\mathbf{x})\ddot{\mathbf{u}}(\mathbf{x},t))\, dV = \int_V \mathbf{y}(\mathbf{x},t) \times \mathbf{b}(\mathbf{x},t) dV$$

$$+ \int_V \int_V \mathbf{y}(\mathbf{x},t) \times \underline{\mathbf{T}}(\mathbf{x},t)\langle \mathbf{x}' - \mathbf{x} \rangle\, dV' dV \qquad (2.37b)$$

$$- \int_V \int_V \mathbf{y}(\mathbf{x},t) \times \underline{\mathbf{T}}(\mathbf{x}',t)\langle \mathbf{x} - \mathbf{x}' \rangle\, dV' dV\;.$$

If the parameters $\mathbf{x}$ and $\mathbf{x}'$ in the third integrals on the right-hand side of Eqs. 2.37a, b are exchanged, the third integrals become

$$\int_V \int_V \underline{\mathbf{T}}(\mathbf{x}',t)\langle \mathbf{x} - \mathbf{x}' \rangle\, dV'\, dV = \int_V \int_V \underline{\mathbf{T}}(\mathbf{x},t)\langle \mathbf{x}' - \mathbf{x} \rangle\, dV\, dV' \qquad (2.38a)$$

and

$$\iint_{V\,V} (\mathbf{y}(\mathbf{x},t) \times \underline{\mathbf{T}}(\mathbf{x}',t)\langle \mathbf{x} - \mathbf{x}' \rangle)\, dV'\, dV$$

$$= \iint_{V\,V} (\mathbf{y}(\mathbf{x}',t) \times \underline{\mathbf{T}}(\mathbf{x},t)\langle \mathbf{x}' - \mathbf{x} \rangle)\, dV\, dV' \;. \tag{2.38b}$$

Therefore, Eqs. 2.37a, b can be rewritten as

$$\int_V (\rho(\mathbf{x})\ddot{\mathbf{u}}(\mathbf{x},t) - \mathbf{b}(\mathbf{x},t))\, dV = 0 \tag{2.39a}$$

and

$$\int_V (\mathbf{y}(\mathbf{x},t) \times \rho(\mathbf{x})\ddot{\mathbf{u}}(\mathbf{x},t))\, dV = \int_V \mathbf{y}(\mathbf{x},t) \times \mathbf{b}(\mathbf{x},t)\, dV$$

$$- \iint_{V\,V} ((\mathbf{y}(\mathbf{x}',t) - \mathbf{y}(\mathbf{x},t)) \times \underline{\mathbf{T}}(\mathbf{x},t)\langle \mathbf{x}' - \mathbf{x} \rangle)\, dV'\, dV. \tag{2.39b}$$

Hence, the balance of linear momentum, Eq. 2.39a, is automatically satisfied for arbitrary force density vectors $\underline{\mathbf{T}}(\mathbf{x},t)\langle \mathbf{x}' - \mathbf{x} \rangle$ and $\underline{\mathbf{T}}(\mathbf{x}',t)\langle \mathbf{x} - \mathbf{x}' \rangle$.

The difference between the locations of material points at $\mathbf{x}$ and $\mathbf{x}'$ in the deformed configuration can be written by using the state notation as

$$\mathbf{y}(\mathbf{x}',t) - \mathbf{y}(\mathbf{x},t) = (\mathbf{y}' - \mathbf{y}) = \underline{\mathbf{Y}}(\mathbf{x},t)\langle \mathbf{x}' - \mathbf{x} \rangle, \tag{2.40}$$

where $\mathbf{y}' = \mathbf{y}(\mathbf{x}',t) = \mathbf{x}' + \mathbf{u}'$ and $\mathbf{y} = \mathbf{y}(\mathbf{x},t) = \mathbf{x} + \mathbf{u}$. Considering only the material points within the horizon, substituting from Eq. 2.40 into Eq. 2.39b results in

$$\int_V \mathbf{y}(\mathbf{x},t) \times (\rho(\mathbf{x})\ddot{\mathbf{u}}(\mathbf{x},t) - \mathbf{b}(\mathbf{x},t))dV$$

$$= -\iint_{V\,H} (\underline{\mathbf{Y}}(\mathbf{x},t)\langle \mathbf{x}' - \mathbf{x} \rangle \times \underline{\mathbf{T}}(\mathbf{x},t)\langle \mathbf{x}' - \mathbf{x} \rangle)\, dH\, dV. \tag{2.41}$$

While invoking the requirement of a balance of linear momentum, Eq. 2.39a, in order to satisfy the balance of angular momentum, the integral on the right-hand side of Eq. 2.41 must be forced to vanish, i.e.,

$$\int_H (\underline{\mathbf{Y}}(\mathbf{x},t)\langle \mathbf{x}' - \mathbf{x} \rangle \times \underline{\mathbf{T}}(\mathbf{x},t)\langle \mathbf{x}' - \mathbf{x} \rangle)\, dH = 0 \tag{2.42a}$$

or

$$\int_H ((\mathbf{y}' - \mathbf{y}) \times \underline{\mathbf{T}}(\mathbf{x}, t)\langle \mathbf{x}' - \mathbf{x}\rangle)\, dH = 0 \; . \tag{2.42b}$$

It is apparent that this requirement is automatically satisfied if the force vectors $\mathbf{t}(\mathbf{u}' - \mathbf{u}, \mathbf{x}' - \mathbf{x}, t) = \underline{\mathbf{T}}(\mathbf{x}, t)\langle \mathbf{x}' - \mathbf{x}\rangle$ and $\mathbf{t}'(\mathbf{u} - \mathbf{u}', \mathbf{x} - \mathbf{x}', t) = \underline{\mathbf{T}}(\mathbf{x}', t)\langle \mathbf{x} - \mathbf{x}'\rangle$ are aligned with the relative position vector of the material points in the deformed state, $(\mathbf{y}' - \mathbf{y})$. However, their general form that satisfies the requirement of Eq. 2.42b can also be derived in terms of the deformation gradient and stress tensors of classical continuum mechanics.

## 2.9   Bond-Based Peridynamics

As a special case, the force density vectors can also be equal in magnitude as well as being parallel to the relative position vector in the deformed state, shown in Fig. 2.6, in order to satisfy the requirement for balance of angular momentum. Thus, they can be expressed in the form

$$\mathbf{t}(\mathbf{u}' - \mathbf{u}, \mathbf{x}' - \mathbf{x}, t) = \underline{\mathbf{T}}(\mathbf{x}, t)\langle \mathbf{x}' - \mathbf{x}\rangle = \frac{1}{2} C \frac{\mathbf{y}' - \mathbf{y}}{|\mathbf{y}' - \mathbf{y}|}$$
$$= \frac{1}{2}\mathbf{f}(\mathbf{u}' - \mathbf{u}, \mathbf{x}' - \mathbf{x}, t) \tag{2.43a}$$

and

$$\mathbf{t}'(\mathbf{u} - \mathbf{u}', \mathbf{x} - \mathbf{x}', t) = \underline{\mathbf{T}}(\mathbf{x}', t)\langle \mathbf{x} - \mathbf{x}'\rangle$$
$$= -\frac{1}{2} C \frac{\mathbf{y}' - \mathbf{y}}{|\mathbf{y}' - \mathbf{y}|} = -\frac{1}{2}\mathbf{f}(\mathbf{u}' - \mathbf{u}, \mathbf{x}' - \mathbf{x}, t), \tag{2.43b}$$

where $C$ is an unknown auxiliary parameter that depends on the engineering material constants, pairwise stretch between $\mathbf{x}'$ and $\mathbf{x}$, and the horizon. This particular form of the force vectors is referred to as "bond-based" peridynamics, as introduced by Silling (2000). As shown in Fig. 2.6, the bond-based peridynamic theory is concerned with pairwise interactions of material points.

Their substitution into Eq. 2.22b results in the bond-based PD equation of motion of the material point $\mathbf{x}$

$$\rho(\mathbf{x})\,\ddot{\mathbf{u}}(\mathbf{x}, t) = \int_H \mathbf{f}(\mathbf{u}' - \mathbf{u}, \mathbf{x}' - \mathbf{x}, t)\, dH + \mathbf{b}(\mathbf{x}, t), \tag{2.44}$$

**Fig. 2.6** Deformation of PD material points $\mathbf{x}$ and $\mathbf{x}'$, and developing equal and opposite pairwise force densities

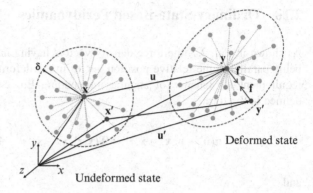

Deformed state

Undeformed state

in which the force density vector, $\mathbf{f}(\mathbf{u}' - \mathbf{u}, \mathbf{x}' - \mathbf{x})$ is referred to as the pairwise response function by Silling and Askari (2005). It is defined as the force vector per unit volume squared that the material point at $\mathbf{x}'$ exerts on the material point at $\mathbf{x}$. The force density vector can be assumed linearly dependent on the stretch between these material points in the form

$$\mathbf{f}(\mathbf{u}' - \mathbf{u}, \mathbf{x}' - \mathbf{x}) = [c_1 s(\mathbf{u}' - \mathbf{u}, \mathbf{x}' - \mathbf{x}) - c_2 T] \frac{\mathbf{y}' - \mathbf{y}}{|\mathbf{y}' - \mathbf{y}|}, \tag{2.45}$$

where the mean value of the temperatures at material points $\mathbf{x}'$ and $\mathbf{x}$ relative to the ambient temperature is denoted by $T$. The stretch $s(\mathbf{u}' - \mathbf{u}, \mathbf{x}' - \mathbf{x})$ can be interpreted as the strain in the classical continuum theory, and it is defined as

$$s(\mathbf{u}' - \mathbf{u}, \mathbf{x}' - \mathbf{x}) = \frac{|\mathbf{y}' - \mathbf{y}| - |\mathbf{x}' - \mathbf{x}|}{|\mathbf{x}' - \mathbf{x}|}. \tag{2.46}$$

For an isotropic material, the peridynamic material parameters $c_1$ and $c_2$ in Eq. 2.45 can be determined by considering an infinite homogeneous body under isotropic expansion, as suggested by Silling and Askari (2005). The body is also subjected to uniform temperature change, $T$. Equating the energy densities of peridynamic and classical continuum theory leads to the determination of $c_1$ and $c_2$ as

$$c_1 = c = \frac{18\kappa}{\pi \delta^4} \quad \text{and} \quad c_2 = c\alpha, \tag{2.47a, b}$$

in which $\kappa$ is the bulk modulus and $\alpha$ is the coefficient of thermal expansion of the material. The PD material parameter $c$ is referred to as the bond-constant. In this case, the PD theory limits the number of independent material constants to one for isotropic materials with a constraint on the Poisson's ratio. It permits only total deformation without distinguishing the distortional and volumetric deformations. Furthermore, it does not allow plastic incompressibility.

## 2.10  Ordinary State-Based Peridynamics

As shown in Fig. 2.7, the force density vectors having unequal magnitudes while being parallel to the relative position vector in the deformed state also satisfy the requirement for balance of angular momentum, Eq. 2.42b. Thus, they can be defined in the form

$$\mathbf{t}(\mathbf{u}' - \mathbf{u}, \mathbf{x}' - \mathbf{x}, t) = \underline{\mathbf{T}}(\mathbf{x}, t)\langle \mathbf{x}' - \mathbf{x}\rangle = \frac{1}{2}A\frac{\mathbf{y}' - \mathbf{y}}{|\mathbf{y}' - \mathbf{y}|} \qquad (2.48a)$$

and

$$\mathbf{t}'(\mathbf{u} - \mathbf{u}', \mathbf{x} - \mathbf{x}', t) = \underline{\mathbf{T}}(\mathbf{x}', t)\langle \mathbf{x} - \mathbf{x}'\rangle = -\frac{1}{2}B\frac{\mathbf{y}' - \mathbf{y}}{|\mathbf{y}' - \mathbf{y}|}, \qquad (2.48b)$$

where $A$ and $B$ are auxiliary parameters that are dependent on engineering material constants, deformation field, and the horizon. As coined by Silling et al. (2007), the choice of the force density vectors in this form is referred to as "ordinary state-based" peridynamics. It permits decoupled distortional and volumetric deformations. Also, it enables the enforcement of plastic incompressibility.

In light of the definition of the strain energy density function, Eq. 2.8, and the expressions for force density vectors in terms of micropotentials, Eqs. 2.17a, b, while considering the requirement on their direction, Eqs. 2.48a, b, the force density vectors can be related to the strain energy density function, $W$, as

$$\mathbf{t}(\mathbf{u}' - \mathbf{u}, \mathbf{x}' - \mathbf{x}, t) \sim \frac{\partial W(\mathbf{x})}{\partial(|\mathbf{y}' - \mathbf{y}|)}\frac{\mathbf{y}' - \mathbf{y}}{|\mathbf{y}' - \mathbf{y}|}, \qquad (2.49a)$$

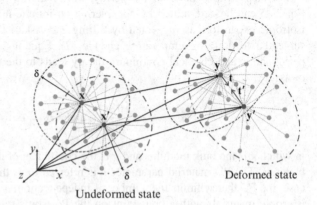

**Fig. 2.7** Deformation of PD material points $\mathbf{x}$ and $\mathbf{x}'$, and developing unequal pairwise force densities

Undeformed state

Deformed state

or

$$\mathbf{t}'(\mathbf{u} - \mathbf{u}', \mathbf{x} - \mathbf{x}', t) \sim \frac{\partial W(\mathbf{x}')}{\partial(|\mathbf{y} - \mathbf{y}'|)} \frac{\mathbf{y}' - \mathbf{y}}{|\mathbf{y}' - \mathbf{y}|}. \tag{2.49b}$$

These relations permit the determination of the auxiliary parameters $A$ and $B$ in Eq. 2.48, and thus the peridynamic constitutive parameters that describe the material behavior. The explicit forms of the expressions for these parameters are derived in Chap. 4 for isotropic and in Chap. 5 for fiber-reinforced composite materials.

## 2.11 Nonordinary State-Based Peridynamics

As shown in Fig. 2.8, a general form of a force density vector that satisfies the requirement of Eq. 2.42b necessary for balance of angular momentum can be derived by applying the principle of virtual displacements to Eq. 2.22a as

$$\rho(\mathbf{x})\ddot{\mathbf{u}}(\mathbf{x}, t) \cdot \Delta\mathbf{u} = \int_H (\underline{\mathbf{T}}(\mathbf{x}, t)\langle\mathbf{x}' - \mathbf{x}\rangle \tag{2.50}$$
$$-\underline{\mathbf{T}}(\mathbf{x}', t)\langle\mathbf{x} - \mathbf{x}'\rangle) \cdot \Delta\mathbf{u}\, dH + \mathbf{b}(\mathbf{x}, t) \cdot \Delta\mathbf{u},$$

where $\Delta\mathbf{u}$ represents the virtual displacement vector applied to the PD material point at $\mathbf{x}$. This equation can also be written in matrix notation as

$$\rho(\mathbf{x})\ddot{\mathbf{u}}^T(\mathbf{x}, t)\Delta\mathbf{u} = \int_H (\underline{\mathbf{T}}(\mathbf{x}, t)\langle\mathbf{x}' - \mathbf{x}\rangle \tag{2.51}$$
$$-\underline{\mathbf{T}}(\mathbf{x}', t)\langle\mathbf{x} - \mathbf{x}'\rangle)^T \Delta\mathbf{u}\, dH + \mathbf{b}^T(\mathbf{x}, t)\Delta\mathbf{u}.$$

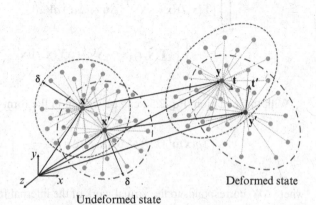

**Fig. 2.8** Deformation of PD material points x and x′, and developing force densities in arbitrary directions

Undeformed state

Deformed state

Noting that $\underline{\mathbf{T}}(\mathbf{x}, t)\langle \mathbf{x}' - \mathbf{x}\rangle = \underline{\mathbf{T}}(\mathbf{x}', t)\langle \mathbf{x} - \mathbf{x}'\rangle = \mathbf{0}$ for $\mathbf{x}' \notin H$ and integrating Eq. 2.51 throughout the body result in

$$\int_V \left(\rho(\mathbf{x})\ddot{\mathbf{u}}^T(\mathbf{x}, t) - \mathbf{b}^T(\mathbf{x}, t)\right)\Delta \mathbf{u} \, dV = \int_V \int_V (\underline{\mathbf{T}}(\mathbf{x}, t)\langle \mathbf{x}' - \mathbf{x}\rangle)^T \Delta \mathbf{u} \, dV' dV$$

$$- \int_V \int_V (\underline{\mathbf{T}}(\mathbf{x}', t)\langle \mathbf{x} - \mathbf{x}'\rangle)^T \Delta \mathbf{u} \, dV' dV. \qquad (2.52)$$

Exchanging the parameters $\mathbf{x}$ and $\mathbf{x}'$ in the second integral on the right-hand side of Eq. 2.52 leads to

$$\int_V \int_V (\underline{\mathbf{T}}(\mathbf{x}', t)\langle \mathbf{x} - \mathbf{x}'\rangle)^T \Delta \mathbf{u} \, dV' dV = \int_V \int_V (\underline{\mathbf{T}}(\mathbf{x}, t)\langle \mathbf{x}' - \mathbf{x}\rangle)^T \Delta \mathbf{u}' \, dV dV'. \qquad (2.53)$$

This relationship permits the right-hand side of Eq. 2.52 to be rewritten as

$$\int_V \int_V (\underline{\mathbf{T}}(\mathbf{x}, t)\langle \mathbf{x}' - \mathbf{x}\rangle)^T \Delta \mathbf{u} \, dV' dV - \int_V \int_V (\underline{\mathbf{T}}(\mathbf{x}', t)\langle \mathbf{x} - \mathbf{x}'\rangle)^T \Delta \mathbf{u} \, dV' dV$$

$$= \int_V \int_V (\underline{\mathbf{T}}(\mathbf{x}, t)\langle \mathbf{x}' - \mathbf{x}\rangle)^T (\Delta \mathbf{u} - \Delta \mathbf{u}') \, dV' dV. \qquad (2.54)$$

The difference in virtual displacements of material points at locations $\mathbf{x}$ and $\mathbf{x}'$ can be written in state form as

$$\Delta \mathbf{u}' - \Delta \mathbf{u} = \Delta \underline{\mathbf{Y}}(\mathbf{x}, t)\langle \mathbf{x}' - \mathbf{x}\rangle. \qquad (2.55)$$

Therefore, Eq. 2.54 can be rewritten as

$$\int_V \int_V (\underline{\mathbf{T}}(\mathbf{x}, t)\langle \mathbf{x}' - \mathbf{x}\rangle)^T (\Delta \mathbf{u} - \Delta \mathbf{u}') \, dV' dV$$

$$= -\int_V \int_V (\underline{\mathbf{T}}(\mathbf{x}, t)\langle \mathbf{x}' - \mathbf{x}\rangle)^T (\Delta \underline{\mathbf{Y}}(\mathbf{x}, t)\langle \mathbf{x}' - \mathbf{x}\rangle) \, dV' dV. \qquad (2.56)$$

With this equation, Eq. 2.52 can be written in the form

$$\int_V \left(\rho(\mathbf{x})\ddot{\mathbf{u}}^T(\mathbf{x}, t) - \mathbf{b}^T(\mathbf{x}, t)\right)\Delta \mathbf{u} \, dV = -\int_V \Delta W_I dV, \qquad (2.57)$$

where $\Delta W_I$ corresponds to the virtual work of the internal forces at location $\mathbf{x}$ due to its interactions with all other material points:

$$\Delta W_I = \int_V (\underline{\mathbf{T}}(\mathbf{x}, t)\langle \mathbf{x}' - \mathbf{x}\rangle)^T (\Delta \underline{\mathbf{Y}}(\mathbf{x}, t)\langle \mathbf{x}' - \mathbf{x}\rangle) \, dV'. \tag{2.58}$$

Considering only the material points within the horizon, Eq. 2.58 can be rewritten as

$$\Delta W_I = \int_H (\underline{\mathbf{T}}(\mathbf{x}, t)\langle \mathbf{x}' - \mathbf{x}\rangle)^T (\Delta \underline{\mathbf{Y}}(\mathbf{x}, t)\langle \mathbf{x}' - \mathbf{x}\rangle) \, dH. \tag{2.59}$$

The corresponding internal virtual work at location $\mathbf{x}$ in classical continuum mechanics can be expressed as

$$\Delta \hat{W}_I = \mathrm{tr}(\mathbf{S}^T \, \Delta \mathbf{E}) \tag{2.60}$$

where $\mathbf{S} = \mathbf{S}^T$ is the second Piola-Kirchhoff (Kirchhoff) stress tensor, and the Green-Lagrange strain tensor, $\mathbf{E} = \mathbf{E}^T$, can be related to the deformation gradient tensor, $\mathbf{F}$,

$$\mathbf{E} = \frac{1}{2}\left(\mathbf{F}^T\mathbf{F} - \mathbf{I}\right). \tag{2.61}$$

Using Eq. 2.61, the virtual form of the Green-Lagrange strain tensor can be written as

$$\Delta \mathbf{E} = \frac{1}{2}\left(\Delta\mathbf{F}^T\mathbf{F} + \mathbf{F}^T\Delta\mathbf{F}\right). \tag{2.62}$$

After substituting from Eq. 2.62 into Eq. 2.60, the internal virtual work expression in classical continuum mechanics takes the form

$$\Delta \hat{W}_I = \mathrm{tr}(\mathbf{S}^T \, \mathbf{F}^T \Delta \mathbf{F}) = \mathrm{tr}(\mathbf{P} \, \Delta \mathbf{F}), \tag{2.63}$$

where $\mathbf{P} = (\mathbf{S}^T \, \mathbf{F}^T)$ is the first Piola-Kirchhoff (Lagrangian) stress tensor.

By using the vector state reduction to a second-order tensor, given in Eq. A.8, the deformation gradient tensor, which corresponds to the deformation state in PD theory, can be obtained as

$$\mathbf{F} = (\underline{\mathbf{Y}} * \underline{\mathbf{X}})\mathbf{K}^{-1}, \tag{2.64}$$

whose virtual form can be written as

$$\Delta \mathbf{F} = (\Delta\underline{\mathbf{Y}} * \underline{\mathbf{X}})\mathbf{K}^{-1}, \tag{2.65}$$

in which the explicit form of the shape tensor, $\mathbf{K}$, serving as a volume-averaging quantity, is derived in the Appendix; it is symmetric and diagonal. The symbol $*$ denotes the convolution of vector states, also defined in the Appendix.

Substituting from Eq. 2.65 into the internal virtual work expression of classical continuum mechanics, Eq. 2.63, in conjunction with Eq. A.7, results in

$$\Delta \hat{W}_I = \text{tr}\left( \mathbf{P} \left( \int_H \underline{w}\langle \mathbf{x}' - \mathbf{x}\rangle \Delta \underline{\mathbf{Y}}\langle \mathbf{x}' - \mathbf{x}\rangle \otimes \underline{\mathbf{X}}\langle \mathbf{x}' - \mathbf{x}\rangle \, dH \right) \mathbf{K}^{-1} \right), \qquad (2.66)$$

where the influence (weight) function, $\underline{w}$, is a scalar state, and $\otimes$ denotes the dyadic product of two vectors, i.e., $\mathbf{C} = \mathbf{a} \otimes \mathbf{b}$ or $C_{ij} = a_i\, b_j$. The scalar state influence function provides a means to control the influence of PD points away from the current point.

Using Eqs. A.4 and 2.55, this equation can be expressed in indicial form as

$$\Delta \hat{W}_I = P_{ij}\left( \int_H \underline{w}\langle \mathbf{x}' - \mathbf{x}\rangle (\Delta u'_i - \Delta u_i)(x'_k - x_k)\, dH \right) K_{kj}^{-1}, \text{ with } (i,j,k) = 1,2,3$$

$$(2.67)$$

Because the shape tensor is symmetric, this equation can be rearranged in the form

$$\Delta \hat{W}_I = \int_H \underline{w}\langle \mathbf{x}' - \mathbf{x}\rangle P_{ij}\, K_{jk}^{-1}(x'_k - x_k)(\Delta u'_i - \Delta u_i)\, dH, \text{ with } (i,j,k) = 1,2,3$$

$$(2.68a)$$

or, in matrix form,

$$\Delta \hat{W}_I = \int_H \left( \underline{w}\langle \mathbf{x}' - \mathbf{x}\rangle \mathbf{P}\, \mathbf{K}^{-1}(\mathbf{x}' - \mathbf{x}) \right)^T (\Delta \mathbf{u}' - \Delta \mathbf{u})\, dH. \qquad (2.68b)$$

After invoking Eq. 2.55 into Eq. 2.68b, equating the virtual work expressions from the PD theory, Eq. 2.59, and classical continuum mechanics, Eq. 2.68b, results in

$$\int_H (\underline{\mathbf{T}}(\mathbf{x},t)\langle \mathbf{x}' - \mathbf{x}\rangle)^T (\Delta \underline{\mathbf{Y}}(\mathbf{x},t)\langle \mathbf{x}' - \mathbf{x}\rangle)\, dH$$

$$(2.69)$$

$$\equiv \int_H \left( \underline{w}\langle \mathbf{x}' - \mathbf{x}\rangle \mathbf{P}\, \mathbf{K}^{-1}(\mathbf{x}' - \mathbf{x}) \right)^T (\Delta \underline{\mathbf{Y}}(\mathbf{x},t)\langle \mathbf{x}' - \mathbf{x}\rangle)\, dH.$$

This requirement leads to the relation between the force vector state and the deformation gradient and stress tensors of classical continuum mechanics as

$$\mathbf{t}(\mathbf{u}' - \mathbf{u}, \mathbf{x}' - \mathbf{x}, t) = \underline{\mathbf{T}}(\mathbf{x}, t)\langle \mathbf{x}' - \mathbf{x} \rangle \equiv \underline{w}\langle \mathbf{x}' - \mathbf{x} \rangle \mathbf{P}\,\mathbf{K}^{-1}(\mathbf{x}' - \mathbf{x}) \qquad (2.70)$$

Although this expression for the force density vector, Eq. 2.70, is identical to that derived by Silling et al. (2007), this derivation based on the principle of virtual displacements proves that the force density vector is valid for any material model provided that the Piola-Kirchhoff stress tensor can be obtained directly or by using incremental procedures. Therefore, this equation also forms the basis for implementing any material behavior in the PD theory.

# References

Kilic B (2008) Peridynamic theory for progressive failure prediction in homogeneous and heterogeneous materials. Dissertation, University of Arizona

Macek RW, Silling SA (2007) Peridynamics via finite element analysis. Finite Elem Anal Des 43:1169–1178

Silling SA (2000) Reformulation of elasticity theory for discontinuities and long-range forces. J Mech Phys Solids 48:175–209

Silling SA (2004) EMU user's manual, Code Ver. 2.6d. Sandia National Laboratories, Albuquerque

Silling SA, Askari E (2005) A meshfree method based on the peridynamic model of solid mechanics. Comput Struct 83:1526–1535

Silling SA, Lehoucq RB (2008) Convergence of peridynamics to classical elasticity theory. J Elast 93:13–37

Silling SA, Epton M, Weckner O, Xu J, Askari A (2007) Peridynamics states and constitutive modeling. J Elast 88:151–184

# Chapter 3
# Peridynamics for Local Interactions

## 3.1 Equations of Motion

Within classical continuum mechanics, a material point can only interact with other material points in its nearest neighborhood. As depicted in Fig. 3.1, the material point $k$ at location $\mathbf{x}_{(k)}$ can only have interactions with the material points labeled as $(k-1)$, $(k+1)$, $(k-m)$, $(k+m)$, $(k-n)$, and $(k+n)$. These interactions are represented by "*internal traction vectors*." For the material point $k$ that is located on a surface whose unit normal is $\mathbf{n}^T = (n_x, n_y, n_z)$, the components of a traction vector, $\mathbf{T}^T = (T_x, T_y, T_z)$, are related to the Cauchy stress components as

$$
\begin{Bmatrix} T_x \\ T_y \\ T_z \end{Bmatrix} = \begin{bmatrix} \sigma_{xx(k)} & \sigma_{xy(k)} & \sigma_{xz(k)} \\ \sigma_{xy(k)} & \sigma_{yy(k)} & \sigma_{yz(k)} \\ \sigma_{xz(k)} & \sigma_{yz(k)} & \sigma_{zz(k)} \end{bmatrix} \begin{Bmatrix} n_x \\ n_y \\ n_z \end{Bmatrix}, \tag{3.1}
$$

in which $(\sigma_{xx(k)}, \sigma_{yy(k)}, \sigma_{zz(k)})$ and $(\sigma_{xy(k)}, \sigma_{xz(k)}, \sigma_{yz(k)})$ are the normal and shear stress components, respectively.

Associated with material point $k$, the traction vectors acting on surfaces with unit normal vectors $\mathbf{n} = \pm\mathbf{e}_x, \pm\mathbf{e}_y, \pm\mathbf{e}_z$ can be expressed in a slightly different form

$$
\mathbf{T}_{(k)(j)} = T_{x(k)(j)}\mathbf{e}_x + T_{y(k)(j)}\mathbf{e}_y + T_{z(k)(j)}\mathbf{e}_z, \tag{3.2}
$$

with $j = (k+1), (k-1), (k+m), (k-m), (k+n), (k-n)$ and the traction vector, $\mathbf{T}_{(k)(j)}$, representing the force exerted by material point $j$ on $k$.

The equations of motion for the classical continuum (local) theory can be derived in a manner similar to the derivation of equations of motion for the nonlocal PD theory. The only difference in the derivation is that the expression for the strain energy density, $W_{(k)}$, of material point $k$ is expressed as a summation of micropotential, $w_{(k)(j)}$, arising from the interaction of material point $k$ and the other six material points denoted as $(k-1)$, $(k+1)$, $(k-m)$, $(k+m)$, $(k-n)$, and

E. Madenci and E. Oterkus, *Peridynamic Theory and Its Applications*,
DOI 10.1007/978-1-4614-8465-3_3, © Springer Science+Business Media New York 2014

**Fig. 3.1** Material point interacting with others in its immediate vicinity

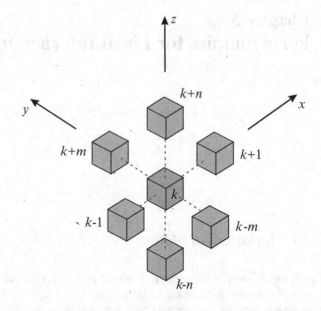

$(k + n)$, as shown in Fig. 3.1. Therefore, the strain energy expression, Eq. 2.8, for the nonlocal theory is modified as

$$W_{(k)} = \frac{1}{2} \sum_{j=k-1,k+1,k-m,k+m,k-n,k+n} \frac{1}{2}\left( w_{(k)(j)}\left( \mathbf{y}_{(1^k)} - \mathbf{y}_{(k)},\ \mathbf{y}_{(2^k)} - \mathbf{y}_{(k)}, \cdots \right)\right.$$
$$\left. + w_{(j)(k)}\left( \mathbf{y}_{(1^j)} - \mathbf{y}_{(j)},\ \mathbf{y}_{(2^j)} - \mathbf{y}_{(j)}, \cdots \right) \right) V_{(j)} .$$
(3.3)

Following the derivation of PD equations of motion (nonlocal), the equations of motion for the material point $k$ in the context of local theory can be obtained as

$$\rho_{(k)} \ddot{\mathbf{u}}_{(k)} = \sum_{j=k-1,k+1,k-m,k+m,k-n,k+n} \left( \mathbf{t}_{(k)(j)} - \mathbf{t}_{(j)(k)} \right) V_{(j)} + \mathbf{b}_{(k)},$$
(3.4a)

where

$$\mathbf{t}_{(k)(j)} = \frac{1}{2} \frac{\partial w_{(k)(j)}}{\partial \left( \mathbf{y}_{(j)} - \mathbf{y}_{(k)} \right)} \quad \text{and} \quad \mathbf{t}_{(j)(k)} = \frac{1}{2} \frac{\partial w_{(j)(k)}}{\partial \left( \mathbf{y}_{(k)} - \mathbf{y}_{(j)} \right)},$$
(3.4b)

with the interpretation that $\mathbf{t}_{(k)(j)}$ represents the force density that material point $\mathbf{x}_{(j)}$ exerts on material point $\mathbf{x}_{(k)}$, and $\mathbf{t}_{(j)(k)}$ represents the force density that material point $\mathbf{x}_{(k)}$ exerts on material point $\mathbf{x}_{(j)}$.

## 3.2 Relationship Between Cauchy Stresses and Peridynamic Forces

For material point $\mathbf{x}_{(k)}$, the equations of motion based on classical continuum mechanics can also be expressed in terms of stress components, $\sigma_{\alpha\beta(k)}$, in the form

$$\rho_{(k)}\ddot{u}_{\alpha(k)} = \sigma_{\alpha x,x(k)} + \sigma_{\alpha y,y(k)} + \sigma_{\alpha z,z(k)} + b_{\alpha(k)}, \tag{3.5}$$

with $\alpha = (x, y, z)$. Invoking the finite difference approximation, using both forward and backward formulae, these equations can be expressed as

$$\begin{aligned}
\rho_{(k)}\ddot{u}_{x(k)} = {} & \frac{1}{2}\frac{\left(\sigma_{xx(k)} - \sigma_{xx(k-1)}\right)}{\Delta x} + \frac{1}{2}\frac{\left(\sigma_{xx(k+1)} - \sigma_{xx(k)}\right)}{\Delta x} \\
& + \frac{1}{2}\frac{\left(\sigma_{xy(k)} - \sigma_{xy(k-m)}\right)}{\Delta y} + \frac{1}{2}\frac{\left(\sigma_{xy(k+m)} - \sigma_{xy(k)}\right)}{\Delta y} \\
& + \frac{1}{2}\frac{\left(\sigma_{xz(k)} - \sigma_{xz(k-n)}\right)}{\Delta z} + \frac{1}{2}\frac{\left(\sigma_{xz(k+n)} - \sigma_{xz(k)}\right)}{\Delta z} + b_{x(k)},
\end{aligned} \tag{3.6a}$$

$$\begin{aligned}
\rho_{(k)}\ddot{u}_{y(k)} = {} & \frac{1}{2}\frac{\left(\sigma_{xy(k)} - \sigma_{xy(k-1)}\right)}{\Delta x} + \frac{1}{2}\frac{\left(\sigma_{xy(k+1)} - \sigma_{xy(k)}\right)}{\Delta x} \\
& + \frac{1}{2}\frac{\left(\sigma_{yy(k)} - \sigma_{yy(k-m)}\right)}{\Delta y} + \frac{1}{2}\frac{\left(\sigma_{yy(k+m)} - \sigma_{yy(k)}\right)}{\Delta y} \\
& + \frac{1}{2}\frac{\left(\sigma_{yz(k)} - \sigma_{yz(k-n)}\right)}{\Delta z} + \frac{1}{2}\frac{\left(\sigma_{yz(k+n)} - \sigma_{yz(k)}\right)}{\Delta z} + b_{y(k)},
\end{aligned} \tag{3.6b}$$

and

$$\begin{aligned}
\rho_{(k)}\ddot{u}_{z(k)} = {} & \frac{1}{2}\frac{\left(\sigma_{xz(k)} - \sigma_{xz(k-1)}\right)}{\Delta x} + \frac{1}{2}\frac{\left(\sigma_{xz(k+1)} - \sigma_{xz(k)}\right)}{\Delta x} \\
& + \frac{1}{2}\frac{\left(\sigma_{yz(k)} - \sigma_{yz(k-m)}\right)}{\Delta y} + \frac{1}{2}\frac{\left(\sigma_{yz(k+m)} - \sigma_{yz(k)}\right)}{\Delta y} \\
& + \frac{1}{2}\frac{\left(\sigma_{zz(k)} - \sigma_{zz(k-n)}\right)}{\Delta z} + \frac{1}{2}\frac{\left(\sigma_{zz(k+n)} - \sigma_{zz(k)}\right)}{\Delta z} + b_{z(k)}.
\end{aligned} \tag{3.6c}$$

Equating each term of Eqs. 3.6a, 3.6b, 3.6c to those of Eq. 3.4a leads to the relations between stress and PD force densities as

$$\sigma_{\alpha\beta(k)} = 2t_{\beta(k)(q_a)}\Delta\alpha V_{(q_a)},$$
$$\text{with } q_x = (k+1), q_y = (k+m), q_z = (k+n),$$

(3.7a)

$$\sigma_{\alpha\beta(k)} = -2t_{\beta(k)(q_a)}\Delta\alpha V_{(q_a)},$$
$$\text{with } q_x = (k-1), q_y = (k-m), q_z = (k-n),$$

(3.7b)

with $\alpha, \beta = x, y, z$. The normal stresses of the Cauchy stress tensor can also be written as

$$\sigma_{\alpha\alpha(k)} = 2t_{(k)(q_a)} \cdot \left(\mathbf{x}_{(q_a)} - \mathbf{x}_{(k)}\right) V_{(q_a)},$$

(3.8a)

with

$$\mathbf{t}_{(k)(q_a)} = t_{x(k)(q_a)}\mathbf{e}_x + t_{y(k)(q_a)}\mathbf{e}_y + t_{z(k)(q_a)}\mathbf{e}_z,$$

(3.8b)

and

$$\mathbf{x}_{(q_a)} - \mathbf{x}_{(k)} = \Delta\alpha\mathbf{e}_\alpha,$$

(3.8c)

for $\alpha = x, y, z$.

The base vectors of the Cartesian coordinate system $(x, y, z)$ are denoted as $\mathbf{e}_\alpha$. It is also worth noting the following expression involving the normal and shear stress components can be expressed in terms of the PD force densities:

$$\sum_{\beta=x,y,z} \sigma_{\alpha\beta(k)}^2 = \sum_{\beta=x,y,z} 4t_{\beta(k)(q_a)}^2 (\Delta\alpha)^2 V_{(q_a)}^2$$

(3.9a)

or

$$\sum_{\beta=x,y,z} \sigma_{\alpha\beta(k)}^2 = 4\left(\mathbf{t}_{(k)(q_a)}\left|\mathbf{x}_{(q_a)} - \mathbf{x}_{(k)}\right|V_{(q_a)}\right) \cdot \left(\mathbf{t}_{(k)(q_a)}\left|\mathbf{x}_{(q_a)} - \mathbf{x}_{(k)}\right|V_{(q_a)}\right).$$

(3.9b)

## 3.3 Strain Energy Density

Based on classical continuum mechanics, the strain energy density at material point $k$ is expressed as

$$W_{(k)} = \frac{\kappa}{2}\left(\theta_{(k)} - 3\alpha T_{(k)}\right)^2 + \left[\frac{1}{4\mu}\left(\sigma_{xx(k)}^2 + \sigma_{yy(k)}^2 + \sigma_{zz(k)}^2\right)\right.$$
$$\left. + \frac{1}{2\mu}\left(\sigma_{xy(k)}^2 + \sigma_{xz(k)}^2 + \sigma_{yz(k)}^2\right) - \frac{3\kappa^2}{4\mu}\theta_{(k)}^2\right],$$

(3.10)

in which $T_{(k)}$ is the temperature change, and $\alpha$ represents the coefficient of thermal expansion for an isotropic material whose bulk modulus is $\kappa$. The first and second terms on the right-hand side represent the dilatational and distortional energy densities, respectively. This equation can also be rewritten in a slightly different form as

$$
\begin{aligned}
W_{(k)} = {} & \frac{\kappa}{2}\left(\theta_{(k)} - 3\,\alpha\,T_{(k)}\right)^2 - \frac{3\kappa^2}{4\mu}\theta_{(k)}^2 \\
& + \frac{1}{8\mu}\left[\left(\sigma_{xx(k)}^2 + \sigma_{xy(k)}^2 + \sigma_{xz(k)}^2\right) + \left(\sigma_{xx(k)}^2 + \sigma_{xy(k)}^2 + \sigma_{xz(k)}^2\right)\right] \\
& + \frac{1}{8\mu}\left[\left(\sigma_{yy(k)}^2 + \sigma_{xy(k)}^2 + \sigma_{yz(k)}^2\right) + \left(\sigma_{yy(k)}^2 + \sigma_{xy(k)}^2 + \sigma_{yz(k)}^2\right)\right] \\
& + \frac{1}{8\mu}\left[\left(\sigma_{zz(k)}^2 + \sigma_{xz(k)}^2 + \sigma_{yz(k)}^2\right) + \left(\sigma_{zz(k)}^2 + \sigma_{xz(k)}^2 + \sigma_{yz(k)}^2\right)\right],
\end{aligned} \tag{3.11}
$$

in which each term involving the stress components corresponds to the contribution of PD forces exerted by material points $(k+1)$, $(k-1)$, $(k+m)$, $(k-m)$, $(k+n)$, and $(k-n)$ on material point $k$.

Utilizing the expressions given by Eq. 3.9b, the strain energy density can be rewritten in terms of PD force densities as

$$
\begin{aligned}
W_{(k)} = {} & \frac{\kappa}{2}\left(\theta_{(k)} - 3\,\alpha\,T_{(k)}\right)^2 - \frac{3\,\kappa^2}{4\mu}\theta_{(k)}^2 \\
& + \frac{1}{2\mu}\sum_{\substack{j=k-1,k+1,\\ k-m,k+m,\\ k-n,k+n}}\left(\mathbf{t}_{(k)(j)}\left|\mathbf{x}_{(j)} - \mathbf{x}_{(k)}\right|V_{(j)}\right)\cdot\left(\mathbf{t}_{(k)(j)}\left|\mathbf{x}_{(j)} - \mathbf{x}_{(k)}\right|V_{(j)}\right).
\end{aligned} \tag{3.12}
$$

In accordance with Eq. 2.43a, for a pairwise interaction of material point $k$ with the other six material points denoted as $(k-1)$, $(k+1)$, $(k-m)$, $(k+m)$, $(k-n)$, and $(k+n)$, the PD force density vector $\mathbf{t}_{(k)(j)}$ can be replaced with $\mathbf{f}_{(k)(j)}$, leading to

$$
\begin{aligned}
W_{(k)} = {} & \frac{\kappa}{2}\left(\theta_{(k)} - 3\,\alpha\,T_{(k)}\right)^2 - \frac{3\,\kappa^2}{4\mu}\theta_{(k)}^2 \\
& + \frac{1}{8\mu}\sum_{\substack{j=k-1,k+1,\\ k-m,k+m,\\ k-n,k+n}}\left(\mathbf{f}_{(k)(j)}\left|\mathbf{x}_{(j)} - \mathbf{x}_{(k)}\right|V_{(j)}\right)\cdot\left(\mathbf{f}_{(k)(j)}\left|\mathbf{x}_{(j)} - \mathbf{x}_{(k)}\right|V_{(j)}\right).
\end{aligned} \tag{3.13}
$$

Substituting for the pairwise force density from Eq. 2.45 in conjunction with Eq. 2.46 results in

$$W_{(k)} = \frac{\kappa}{2}\left(\theta_{(k)} - 3\,\alpha\,T_{(k)}\right)^2 - \frac{3\,\kappa^2}{4\mu}\,\theta_{(k)}^2$$

$$+ \frac{c^2}{8\mu}\sum_{\substack{j=k-1,k+1, \\ k-m,k+m, \\ k-n,k+n}}\left(s_{(k)(j)} - \alpha\,T_{(k)}\right)^2\left|x_{(j)} - x_{(k)}\right|^2 V_{(j)}^2. \tag{3.14}$$

A general form of this expression that it is suitable for both bond-based and ordinary state-based peridynamics can be written as

$$W_{(k)} = a\,\theta_{(k)}^2 - a_2\,\theta_{(k)}\,T_{(k)} + a_3\,T_{(k)}^2$$

$$+ \sum_{\substack{j=k-1,k+1, \\ k-m,k+m, \\ k-n,k+n}}b\left(s_{(k)(j)} - \alpha\,T_{(k)}\right)^2\left|\mathbf{x}_{(j)} - \mathbf{x}_{(k)}\right|^2 V_{(j)} \tag{3.15a}$$

or

$$W_{(k)} = a\theta_{(k)}^2 - a_2\,\theta_{(k)}\,T_{(k)} + a_3\,T_{(k)}^2$$

$$+ \sum_{\substack{j=k-1,k+1, \\ k-m,k+m, \\ k-n,k+n}}b\left(\left(\left(\left|\mathbf{y}_{(j)} - \mathbf{y}_{(k)}\right| - \left|\mathbf{x}_{(j)} - \mathbf{x}_{(k)}\right|\right) - \alpha T_{(k)}\,\left|\mathbf{x}_{(j)} - \mathbf{x}_{(k)}\right|\right)^2 V_{(j)}\right),$$

$$\tag{3.15b}$$

where $a$, $a_2$, $a_3$, and $b$ are the peridynamic parameters.

The dilatation, $\theta_{(k)}$, term at material point $k$ is defined in classical continuum mechanics as

$$\theta_{(k)} = \left(\varepsilon_{xx(k)} + \varepsilon_{yy(k)} + \varepsilon_{zz(k)}\right) = \frac{\left(\sigma_{xx(k)} + \sigma_{yy(k)} + \sigma_{zz(k)}\right)}{3\,\kappa} + 3\,\alpha\,T_{(k)}, \tag{3.16}$$

in which the normal strain components are $\left(\varepsilon_{xx(k)}, \varepsilon_{yy(k)}, \varepsilon_{zz(k)}\right)$. This expression is rewritten in a slightly different form as

$$\theta_{(k)} = \frac{1}{3\kappa}\left(\frac{1}{2}\sigma_{xx(k)} + \frac{1}{2}\sigma_{xx(k)} + \frac{1}{2}\sigma_{yy(k)} + \frac{1}{2}\sigma_{yy(k)} + \frac{1}{2}\sigma_{zz(k)} + \frac{1}{2}\sigma_{zz(k)}\right)$$

$$+ 3\,\alpha\,T_{(k)}, \tag{3.17}$$

in which each term corresponds to peridynamic forces exerted by material points $(k+1), (k-1), (k+m), (k-m), (k+n)$, and $(k-n)$ on material point $k$. Utilizing the expressions given by Eq. 3.8a, the dilatation can be rewritten in terms of peridynamic force densities,

$$\theta_{(k)} = \frac{1}{3\kappa}\left(\sum_{j=k-1,k+1,k-m,k+m,k-n,k+n} \left(\mathbf{t}_{(k)(j)} \cdot \left(\mathbf{x}_{(j)} - \mathbf{x}_{(k)}\right)\right)V_{(j)}\right) + 3\alpha T_{(k)}. \quad (3.18)$$

As in the strain energy density expression, this expression can be written for a pairwise interaction of material point $k$ with the other six material points, leading to

$$\theta_{(k)} = \frac{1}{6\kappa}\left(\sum_{j=k-1,k+1,k-m,k+m,k-n,k+n} \left(\mathbf{f}_{(k)(j)} \cdot \left(\mathbf{x}_{(j)} - \mathbf{x}_{(k)}\right)\right)V_{(j)}\right)$$
$$+ 3\alpha T_{(k)}. \quad (3.19)$$

Substituting for the pairwise force density from Eq. 2.45 in conjunction with Eq. 2.46 results in

$$\theta_{(k)} = \frac{c}{6\kappa}\sum_{\substack{j=k-1\\k+1\\k-m\\k+m\\k-n\\k+n}} \left(s_{(k)(j)} - \alpha T_{(k)}\right)\frac{\left(\mathbf{y}_{(j)} - \mathbf{y}_{(k)}\right)}{\left|\mathbf{y}_{(j)} - \mathbf{y}_{(k)}\right|} \cdot \left(\mathbf{x}_{(j)} - \mathbf{x}_{(k)}\right)V_{(j)} + 3\alpha T_{(k)}.$$
$$(3.20)$$

A general form of this expression can be written as

$$\theta_{(k)} = d\sum_{\substack{j=k-1\\k+1\\k-m\\k+m\\k-n\\k+n}} \left(s_{(k)(j)} - \alpha T_{(k)}\right)\frac{\left(\mathbf{y}_{(j)} - \mathbf{y}_{(k)}\right)}{\left|\mathbf{y}_{(j)} - \mathbf{y}_{(k)}\right|} \cdot \left(\mathbf{x}_{(j)} - \mathbf{x}_{(k)}\right)V_{(j)} + 3\alpha T_{(k)}, \quad (3.21)$$

where $d$ is a peridynamic parameter. The expressions for dilatation, $\theta_{(k)}$, and strain energy density, $W_{(k)}$, at material point $k$ will take their general form within the ordinary state-based peridynamic framework, where the number of interactions are not limited to the immediate vicinity of material points as in classical continuum mechanics.

# Chapter 4
# Peridynamics for Isotropic Materials

## 4.1 Material Parameters

The auxiliary parameters, $C$ in Eq. 2.43 and $A$ and $B$ in Eq. 2.48, can be determined by using the relationship between the force density vector and the strain energy density, $W_{(k)}$, at material point $k$ given by Eq. 2.49 in the form,

$$\mathbf{t}_{(k)(j)}\left(\mathbf{u}_{(j)} - \mathbf{u}_{(k)}, \mathbf{x}_{(j)} - \mathbf{x}_{(k)}, t\right) = \frac{1}{V_{(j)}} \frac{\partial W_{(k)}}{\partial\left(\left|\mathbf{y}_{(j)} - \mathbf{y}_{(k)}\right|\right)} \frac{\mathbf{y}_{(j)} - \mathbf{y}_{(k)}}{\left|\mathbf{y}_{(j)} - \mathbf{y}_{(k)}\right|}, \tag{4.1}$$

in which $V_{(j)}$ represents the volume of material point $j$, and the direction of the force density vector is aligned with the relative position vector in the deformed configuration. The material point $j$ exerts the force density $\mathbf{t}_{(k)(j)}$ on material point $k$. Determination of the auxiliary parameters requires an explicit form of the strain energy density function.

For an isotropic and elastic material, the explicit form of the strain energy density, $W_{(k)}$, at material point $\mathbf{x}_{(k)}$ can be obtained by generalizing the expression given by Eq. 3.15 as

$$W_{(k)} = a\theta_{(k)}^2 - a_2\,\theta_{(k)}\,T_{(k)} + a_3\,T_{(k)}^2$$
$$+ b\sum_{j=1}^{N} w_{(k)(j)}\left(\left(\left|\mathbf{y}_{(j)} - \mathbf{y}_{(k)}\right| - \left|\mathbf{x}_{(j)} - \mathbf{x}_{(k)}\right|\right) - \alpha T_{(k)}\left|\mathbf{x}_{(j)} - \mathbf{x}_{(k)}\right|\right)^2 V_{(j)},$$

$$\tag{4.2}$$

where $N$ represents the number of material points within the family of $\mathbf{x}_{(k)}$. The nondimensional influence function, $w_{(k)(j)} = w(\left|\mathbf{x}_{(j)} - \mathbf{x}_{(k)}\right|)$, provides a means to control the influence of material points away from the current material point at $\mathbf{x}_{(k)}$. The temperature change at material point $k$ is $T_{(k)}$, with $\alpha$ representing the coefficient

E. Madenci and E. Oterkus, *Peridynamic Theory and Its Applications*,
DOI 10.1007/978-1-4614-8465-3_4, © Springer Science+Business Media New York 2014

of thermal expansion. Similarly, the explicit expression for $\theta_{(k)}$ can be obtained from Eq. 3.21 in a general form as

$$\theta_{(k)} = d \sum_{j=1}^{N} w_{(k)(j)} \left( s_{(k)(j)} - \alpha T_{(k)} \right) \frac{\mathbf{y}_{(j)} - \mathbf{y}_{(k)}}{\left| \mathbf{y}_{(j)} - \mathbf{y}_{(k)} \right|} \cdot \left( \mathbf{x}_{(j)} - \mathbf{x}_{(k)} \right) V_{(j)} + 3\alpha T_{(k)}, \quad (4.3)$$

in which the PD parameter $d$ ensures that $\theta_{(k)}$ remains nondimensional. The PD material parameters, $a$, $a_2$, $a_3$, and $b$, in Eq. 4.2 can be related to the engineering material constants of shear modulus, $\mu$, bulk modulus, $\kappa$, and thermal expansion coefficient, $\alpha$, of classical continuum mechanics by considering simple loading conditions.

After substituting for $\theta_{(k)}$ from Eq. 4.3 in the expression for $W_{(k)}$, given by Eq. 4.2, and performing differentiation, the force density vector $\mathbf{t}_{(k)(j)} \left( \mathbf{u}_{(j)} - \mathbf{u}_{(k)}, \mathbf{x}_{(j)} - \mathbf{x}_{(k)}, t \right)$ can be rewritten in terms of PD material parameters as

$$\mathbf{t}_{(k)(j)} \left( \mathbf{u}_{(j)} - \mathbf{u}_{(k)}, \mathbf{x}_{(j)} - \mathbf{x}_{(k)}, t \right) = \frac{1}{2} A \frac{\mathbf{y}_{(j)} - \mathbf{y}_{(k)}}{\left| \mathbf{y}_{(j)} - \mathbf{y}_{(k)} \right|}, \quad (4.4a)$$

with

$$A = 4w_{(k)(j)} \left\{ d \frac{\mathbf{y}_{(j)} - \mathbf{y}_{(k)}}{\left| \mathbf{y}_{(j)} - \mathbf{y}_{(k)} \right|} \cdot \frac{\mathbf{x}_{(j)} - \mathbf{x}_{(k)}}{\left| \mathbf{x}_{(j)} - \mathbf{x}_{(k)} \right|} \left( a\theta_{(k)} - \frac{1}{2} a_2 T_{(k)} \right) \right.$$
$$\left. + b \left( \left( \left| \mathbf{y}_{(j)} - \mathbf{y}_{(k)} \right| - \left| \mathbf{x}_{(j)} - \mathbf{x}_{(k)} \right| \right) - \alpha T_{(k)} \left| \mathbf{x}_{(j)} - \mathbf{x}_{(k)} \right| \right) \right\} . \quad (4.4b)$$

Similarly, the force density vector $\mathbf{t}_{(j)(k)} \left( \mathbf{u}_{(k)} - \mathbf{u}_{(j)}, \mathbf{x}_{(k)} - \mathbf{x}_{(j)}, t \right)$ can be expressed as

$$\mathbf{t}_{(j)(k)} \left( \mathbf{u}_{(k)} - \mathbf{u}_{(j)}, \mathbf{x}_{(k)} - \mathbf{x}_{(j)}, t \right) = -\frac{1}{2} B \frac{\mathbf{y}_{(j)} - \mathbf{y}_{(k)}}{\left| \mathbf{y}_{(j)} - \mathbf{y}_{(k)} \right|}, \quad (4.5a)$$

with

$$B = 4w_{(j)(k)} \left\{ d \frac{\mathbf{y}_{(k)} - \mathbf{y}_{(j)}}{\left| \mathbf{y}_{(k)} - \mathbf{y}_{(j)} \right|} \cdot \frac{\mathbf{x}_{(k)} - \mathbf{x}_{(j)}}{\left| \mathbf{x}_{(k)} - \mathbf{x}_{(j)} \right|} \left( a\theta_{(j)} - \frac{1}{2} a_2 T_{(j)} \right) \right.$$
$$\left. + b \left( \left( \left| \mathbf{y}_{(k)} - \mathbf{y}_{(j)} \right| - \left| \mathbf{x}_{(k)} - \mathbf{x}_{(j)} \right| \right) - \alpha T_{(j)} \left| \mathbf{x}_{(k)} - \mathbf{x}_{(j)} \right| \right) \right\} . \quad (4.5b)$$

Although Eqs. 4.4b and 4.5b appear to be similar, they are different because the values of $(\theta_{(k)}, T_{(k)})$ and $(\theta_{(j)}, T_{(j)})$ for the material points at $\mathbf{x}_{(k)}$ and $\mathbf{x}_{(j)}$, respectively, are not necessarily equal to each other. However, $A$ and $B$ must be equal to each other for the bond-based PD theory. Therefore, the terms associated with $\theta_{(k)}$ and $\theta_{(j)}$ in Eqs. 4.4b and 4.5b must disappear, thus requiring that

$$ad = 0. \tag{4.6}$$

Thus, the parameter $C$ in Eq. 2.43 becomes

$$C = 4bw_{(k)(j)} \left( \left( \left| \mathbf{y}_{(j)} - \mathbf{y}_{(k)} \right| - \left| \mathbf{x}_{(j)} - \mathbf{x}_{(k)} \right| \right) - \alpha T_{(k)} \left| \mathbf{x}_{(j)} - \mathbf{x}_{(k)} \right| \right). \tag{4.7}$$

The force density vector can be rewritten as

$$\mathbf{t}_{(k)(j)} = 2bw_{(k)(j)} \left( \left( \left| \mathbf{y}_{(j)} - \mathbf{y}_{(k)} \right| - \left| \mathbf{x}_{(j)} - \mathbf{x}_{(k)} \right| \right) \right.$$
$$\left. - \alpha T_{(k)} \left| \mathbf{x}_{(j)} - \mathbf{x}_{(k)} \right| \right) \frac{\mathbf{y}_{(j)} - \mathbf{y}_{(k)}}{\left| \mathbf{y}_{(j)} - \mathbf{y}_{(k)} \right|} . \tag{4.8}$$

Based on Eq. 2.43, the bond-based force density vector between the material points at $\mathbf{x}_{(k)}$ and $\mathbf{x}_{(j)}$ can be obtained as

$$\mathbf{f}_{(k)(j)} = 4bw_{(k)(j)} \left| \mathbf{x}_{(j)} - \mathbf{x}_{(k)} \right| \left( s_{(k)(j)} - \alpha T_{(k)} \right) \frac{\mathbf{y}_{(j)} - \mathbf{y}_{(k)}}{\left| \mathbf{y}_{(j)} - \mathbf{y}_{(k)} \right|} . \tag{4.9}$$

Comparing this expression with the bond-based definition of the force density vector, Eq. 2.45 leads to the explicit form of the influence function as

$$w_{(k)(j)} = \frac{c}{4b} \frac{1}{\left| \mathbf{x}_{(j)} - \mathbf{x}_{(k)} \right|}. \tag{4.10}$$

Performing dimensional analysis on Eq. 4.2 requires that parameter $b$ have dimensions $Force/(Length)^7$ whereas the parameter $c = c_1$ in Eq. 2.45 has dimensions $Force/(Length)^6$. Therefore, the ratio of $c/b$ has a dimension of $Length$, rendering the influence function to be nondimensional. The horizon, $\delta$, can be taken as the $Length$ dimension to include the influence of other material points within a family. Thus, the influence (weight) function for the state-based peridynamics becomes

$$w_{(k)(j)} = \frac{\delta}{\left| \mathbf{x}_{(j)} - \mathbf{x}_{(k)} \right|}. \tag{4.11}$$

Thus, the ratio of $c/b$ is established as

$$\frac{c}{b} = 4\delta. \tag{4.12}$$

Substituting for the influence function results in the final form of the expressions for the force density vectors

$$\mathbf{t}_{(k)(j)} = 2\delta \left\{ d \frac{\Lambda_{(k)(j)}}{|\mathbf{x}_{(j)} - \mathbf{x}_{(k)}|} \left( a\theta_{(k)} - \frac{1}{2} a_2 T_{(k)} \right) + b \left( s_{(k)(j)} - \alpha T_{(k)} \right) \right\}$$
$$\times \frac{\mathbf{y}_{(j)} - \mathbf{y}_{(k)}}{|\mathbf{y}_{(j)} - \mathbf{y}_{(k)}|}, \tag{4.13}$$

where the parameter, $\Lambda_{(k)(j)}$, is defined as

$$\Lambda_{(k)(j)} = \left( \frac{\mathbf{y}_{(j)} - \mathbf{y}_{(k)}}{|\mathbf{y}_{(j)} - \mathbf{y}_{(k)}|} \right) \cdot \left( \frac{\mathbf{x}_{(j)} - \mathbf{x}_{(k)}}{|\mathbf{x}_{(j)} - \mathbf{x}_{(k)}|} \right). \tag{4.14}$$

For the bond-based PD theory, the dilatation term $\theta_{(k)}$ must disappear, resulting in

$$\mathbf{t}_{(k)(j)} = 2\delta b \left( s_{(k)(j)} - \alpha T_{(k)} \right) \frac{\mathbf{y}_{(j)} - \mathbf{y}_{(k)}}{|\mathbf{y}_{(j)} - \mathbf{y}_{(k)}|}. \tag{4.15}$$

Based on Eq. 2.43 in conjunction with Eq. 4.12, the bond-based force density vector, $\mathbf{f}_{(k)(j)}$, in Eq. 2.44, becomes

$$\mathbf{f}_{(k)(j)} = c \left( s_{(k)(j)} - \alpha T_{(k)(j)} \right) \frac{\mathbf{y}_{(j)} - \mathbf{y}_{(k)}}{|\mathbf{y}_{(j)} - \mathbf{y}_{(k)}|}, \tag{4.16}$$

where $T_{(k)(j)} = (T_{(j)} + T_{(k)})/2$. This expression is the same as that given by Silling and Askari (2005), who coined the term "bond-constant" for the parameter $c$ for bond-based peridynamics.

Although all structures are three dimensional in nature, they can be idealized under certain assumptions as one dimensional or two dimensional in order to simplify the computational effort. For instance, long bars can be treated as one-dimensional structures. Similarly, thin plates can be treated as two-dimensional structures. The PD material constants must reflect these idealizations. A two-dimensional plate can be discretized with a single layer of material points in the thickness direction. The spherical domain of integral $H$ becomes a disk with radius $\delta$ and thickness $h$.

A one-dimensional bar can be discretized with a single row of material points. The spherical domain of integral $H$ becomes a line with a length $2\delta$ and cross-sectional area of $A$.

## 4.1.1  Three-Dimensional Structures

For three-dimensional analysis, the strain energy density based on classical continuum mechanics can be obtained from

$$W_{(k)} = \frac{1}{2}\boldsymbol{\sigma}_{(k)}^T\,\boldsymbol{\varepsilon}_{(k)}, \tag{4.17}$$

in which the stress and strain vectors $\boldsymbol{\sigma}_{(k)}$ and $\boldsymbol{\varepsilon}_{(k)}$ are defined as

$$\boldsymbol{\sigma}_{(k)}^T = \left\{ \sigma_{xx(k)} \quad \sigma_{yy(k)} \quad \sigma_{zz(k)} \quad \sigma_{yz(k)} \quad \sigma_{xz(k)} \quad \sigma_{xy(k)} \right\} \tag{4.18a}$$

and

$$\boldsymbol{\varepsilon}_{(k)}^T = \left\{ \varepsilon_{xx(k)} \quad \varepsilon_{yy(k)} \quad \varepsilon_{zz(k)} \quad \gamma_{yz(k)} \quad \gamma_{xz(k)} \quad \gamma_{xy(k)} \right\}. \tag{4.18b}$$

For an isotropic material with bulk modulus, $\kappa$, and shear modulus, $\mu$, the stress and strain components are related through the constitutive relation as

$$\boldsymbol{\sigma}_{(k)} = \mathbf{C}\,\boldsymbol{\varepsilon}_{(k)}, \tag{4.19}$$

where the material property matrix $\mathbf{C}$ is defined as

$$\mathbf{C} = \begin{bmatrix} \kappa+(4\mu/3) & \kappa-(2\mu/3) & \kappa-(2\mu/3) & 0 & 0 & 0 \\ \kappa-(2\mu/3) & \kappa+(4\mu/3) & \kappa-(2\mu/3) & 0 & 0 & 0 \\ \kappa-(2\mu/3) & \kappa-(2\mu/3) & \kappa+(4\mu/3) & 0 & 0 & 0 \\ 0 & 0 & 0 & \mu & 0 & 0 \\ 0 & 0 & 0 & 0 & \mu & 0 \\ 0 & 0 & 0 & 0 & 0 & \mu \end{bmatrix}, \tag{4.20a}$$

with

$$\kappa = \frac{E}{3(1-2\nu)} \text{ and } \mu = \frac{E}{2(1+\nu)}. \tag{4.20b}$$

Two different loading cases resulting in *isotropic expansion* and *simple shear* can be considered to determine the peridynamic parameters $a$, $a_2$, $a_3$, $b$, and $d$ in terms of engineering material constants of classical continuum mechanics.

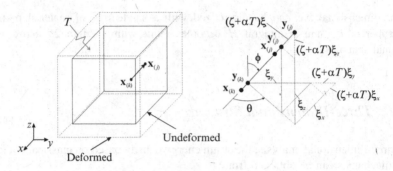

**Fig. 4.1** A three-dimensional body subjected to isotropic expansion

As illustrated in Fig. 4.1, a loading case of *isotropic expansion* can be achieved by applying a normal strain of $\zeta$ in all directions and a uniform temperature change of $T$. Thus, the strain components in the body are

$$\varepsilon_{xx(k)} = \varepsilon_{yy(k)} = \varepsilon_{zz(k)} = \zeta + \alpha T \tag{4.21a}$$

and

$$\gamma_{xy(k)} = \gamma_{xz(k)} = \gamma_{yz(k)} = 0, \tag{4.21b}$$

for which the dilatation, $\theta_{(k)}$, and the strain energy density, $W_{(k)}$, within the realm of classical continuum mechanics become

$$\theta_{(k)} = \varepsilon_{xx(k)} + \varepsilon_{yy(k)} + \varepsilon_{zz(k)} = 3\zeta + 3\alpha T \tag{4.22a}$$

and

$$W_{(k)} = \frac{9}{2}\kappa\zeta^2. \tag{4.22b}$$

The relative position vector between the material points at $\mathbf{x}_{(j)}$ and $\mathbf{x}_{(k)}$ in the deformed configuration becomes

$$\left|\mathbf{y}_{(j)} - \mathbf{y}_{(k)}\right| = \left(1 + \zeta + \alpha T_{(k)}\right)\left|\mathbf{x}_{(j)} - \mathbf{x}_{(k)}\right|, \tag{4.23}$$

in which $T_{(k)} = T$.

Defining $\boldsymbol{\xi} = \mathbf{x}_{(j)} - \mathbf{x}_{(k)}$, with $\xi = |\boldsymbol{\xi}|$, and substituting for $w_{(k)(j)}$ from Eq. 4.11 and the relative position vector from Eq. 4.23, the strain energy density, $W_{(k)}$, at material point $\mathbf{x}_{(k)}$ that interacts with other material points within a sphere of radius, $\delta$, from Eq. 4.2 can be evaluated as

$$W_{(k)} = a\,\theta_{(k)}^2 - a_2\,\theta_{(k)}\,T_{(k)} + a_3\,T_{(k)}^2 + b \int_0^\delta \int_0^{2\pi} \int_0^\pi \frac{\delta}{\xi} \left( \left[ \left( 1 + \zeta + \alpha T_{(k)} \right) \xi - \xi \right] \right.$$

$$\left. - \alpha T_{(k)}\,\xi \right)^2 \xi^2\,\sin(\phi)\,d\phi d\theta d\xi,$$

$$(4.24)$$

in which $(\xi, \theta, \phi)$ serve as spherical coordinates. After invoking from Eq. 4.22a, its evaluation leads to

$$W_{(k)} = a\left(3\zeta + 3\,\alpha T_{(k)}\right)^2 - a_2\left(3\zeta + 3\,\alpha T_{(k)}\right) T_{(k)} + a_3\,T_{(k)}^2 + \pi b\zeta^2\,\delta^5. \quad (4.25)$$

Equating the expressions for strain energy density from Eqs. 4.22b and 4.25 provides the relationships between the PD parameters and engineering material constants as

$$9\,a + \pi\,b\,\delta^5 = \frac{9}{2}\kappa, \qquad (4.26a)$$

$$a_2 = 6\,\alpha\,a, \qquad (4.26b)$$

$$a_3 = 9\,\alpha^2\,a. \qquad (4.26c)$$

Similarly, the expression for $\theta_{(k)}$ from Eq. 4.3 can be recast as

$$\theta_{(k)} = d \int_0^\delta \int_0^{2\pi} \int_0^\pi \frac{\delta}{\xi} \left( \left[ \left( 1 + \zeta + \alpha T_{(k)} \right) \xi - \xi \right] - \alpha T_{(k)}\,\xi \right)$$

$$\times \left( \frac{\xi}{\xi} \cdot \frac{\xi}{\xi} \right) \xi^2\,\sin(\phi) d\phi d\theta d\xi + 3\alpha T_{(k)},$$

$$(4.27)$$

whose explicit evaluation leads to

$$\theta_{(k)} = \frac{4\,\pi\,d\,\delta^4}{3}\zeta + 3\alpha T_{(k)}. \qquad (4.28)$$

Equating the expressions for dilatation from Eqs. 4.22a and 4.28 permits the determination of the peridynamic parameter $d$ as

$$d = \frac{9}{4\,\pi\,\delta^4}. \qquad (4.29)$$

As illustrated in Fig. 4.2, a loading case of *simple shear* can be achieved by applying

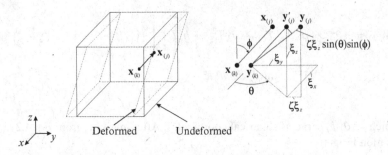

**Fig. 4.2** A three-dimensional body subjected to simple shear

$$\gamma_{xy(k)} = \zeta \text{ and } \varepsilon_{xx(k)} = \varepsilon_{yy(k)} = \varepsilon_{zz(k)} = \gamma_{xz(k)} = \gamma_{yz(k)} = T_{(k)} = 0, \qquad (4.30)$$

for which the dilatation, $\theta_{(k)}$, and the strain energy density, $W_{(k)}$, within the realm of classical continuum mechanics become

$$\theta_{(k)} = 0 \qquad (4.31a)$$

and

$$W_{(k)} = \frac{1}{2}\mu\zeta^2. \qquad (4.31b)$$

The relative position vector in the deformed state becomes

$$\left| \mathbf{y}_{(j)} - \mathbf{y}_{(k)} \right| = \left[ 1 + \frac{\zeta \sin(2\phi) \sin(\theta)}{2} \right] \left| \mathbf{x}_{(j)} - \mathbf{x}_{(k)} \right|. \qquad (4.32)$$

Therefore, the strain energy density, $W_{(k)}$, from Eq. 4.2 can be evaluated as

$$W_{(k)} = b \int_0^\delta \int_0^{2\pi} \int_0^\pi \frac{\delta}{\xi} \left( \left[ 1 + \frac{\zeta \sin(2\phi) \sin(\theta)}{2} \right] \xi - \xi \right)^2 \xi^2 \sin(\phi) d\phi d\theta d\xi, \qquad (4.33a)$$

whose evaluation leads to

$$W_{(k)} = \frac{b\pi\delta^5\zeta^2}{15}. \qquad (4.33b)$$

Equating the strain energy density expressions of Eqs. 4.31b and 4.33b obtained from classical continuum mechanics and the PD theory gives the relationship between the peridynamic parameter $b$ and shear modulus, $\mu$, as

$$b = \frac{15\mu}{2\pi\delta^5}. \qquad (4.34)$$

Substituting from Eq. 4.34 into Eq. 4.26a results in the evaluation of the peridynamic parameter $a$ in terms of bulk modulus, $\kappa$, and shear modulus, $\mu$, as

$$a = \frac{1}{2}\left(\kappa - \frac{5\mu}{3}\right). \tag{4.35}$$

In summary, the PD parameters for a three-dimensional analysis can be expressed as

$$a = \frac{1}{2}\left(\kappa - \frac{5\mu}{3}\right), \quad a_2 = 6\alpha a, \tag{4.36a,b}$$

$$a_3 = 9\alpha^2 a, \quad b = \frac{15\mu}{2\pi\delta^5}, \quad d = \frac{9}{4\pi\delta^4}. \tag{4.36c-e}$$

In view of Eqs. 4.6 and 4.12, a constraint condition of $\kappa = 5\mu/3$ or $\nu = 1/4$ emerges for bond-based peridynamics with a bond constant of $c = 30\mu/\pi\delta^4$ or $c = 18\kappa/\pi\delta^4$.

## 4.1.2 Two-Dimensional Structures

Under two-dimensional idealization, the stress and strain vectors $\boldsymbol{\sigma}_{(k)}$ and $\boldsymbol{\varepsilon}_{(k)}$ are defined as

$$\boldsymbol{\sigma}_{(k)}^T = \left\{ \sigma_{xx(k)} \quad \sigma_{yy(k)} \quad \sigma_{xy(k)} \right\} \tag{4.37a}$$

and

$$\boldsymbol{\varepsilon}_{(k)}^T = \left\{ \varepsilon_{xx(k)} \quad \varepsilon_{yy(k)} \quad \gamma_{xy(k)} \right\}. \tag{4.37b}$$

The material property matrix $\mathbf{C}$ in Eq. 4.19 is reduced to

$$\mathbf{C} = \begin{bmatrix} \kappa + \mu & \kappa - \mu & 0 \\ \kappa - \mu & \kappa + \mu & 0 \\ 0 & 0 & \mu \end{bmatrix}. \tag{4.38}$$

Due to two-dimensional idealization, the expression for bulk modulus differs from that given in Eq. 4.20b and is given by

$$\kappa = \frac{E}{2(1-\nu)}. \tag{4.39}$$

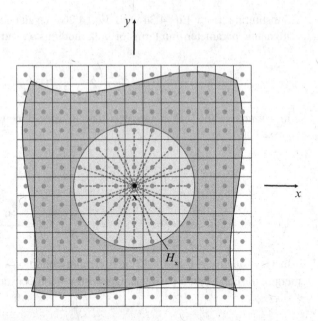

As shown in Fig. 4.3, a two-dimensional plate is discretized with a single layer of
material points in the thickness direction. The domain of integral $H$ in Eq. 2.22a
becomes a disk with radius $\delta$ and thickness $h$. As in the previous case, two different
loading cases to achieve *isotropic expansion* and *simple shear* are considered to
determine the peridynamic parameters.

As illustrated in Fig. 4.4, *isotropic expansion* can be achieved by applying an
equal normal strain of $\zeta$ in all directions and a uniform temperature change of $T$.
Thus, the strain components in the body are

$$\varepsilon_{xx(k)} = \varepsilon_{yy(k)} = \zeta + \alpha T \quad \text{and} \quad \gamma_{xy(k)} = 0, \tag{4.40}$$

for which the dilatation, $\theta_{(k)}$, and the strain energy density, $W_{(k)}$, within the realm of
classical continuum mechanics become

$$\theta_{(k)} = \varepsilon_{xx(k)} + \varepsilon_{yy(k)} = 2\zeta + 2\alpha T_{(k)} \tag{4.41a}$$

and

$$W_{(k)} = 2\kappa \zeta^2. \tag{4.41b}$$

The relative position vector between the material points at $\mathbf{x}_{(j)}$ and $\mathbf{x}_{(k)}$ in the
deformed configuration becomes

$$\left| \mathbf{y}_{(j)} - \mathbf{y}_{(k)} \right| = \left( 1 + \zeta + \alpha T_{(k)} \right) \left| \mathbf{x}_{(j)} - \mathbf{x}_{(k)} \right|, \tag{4.42}$$

in which $T_{(k)} = T$.

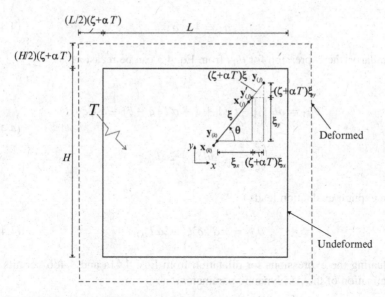

**Fig. 4.4** A two-dimensional plate subjected to isotropic expansion

The strain energy density, $W_{(k)}$, at material point $\mathbf{x}_{(k)}$ that interacts with other material points within a disk of radius $\delta$ and thickness $h$ from Eq. 4.2 can be evaluated as

$$
W_{(k)} = a\,\theta_{(k)}^2 - a_2\,\theta_{(k)}\,T_{(k)} + a_3\,T_{(k)}^2
$$
$$
+ b\,h \int\limits_0^\delta \int\limits_0^{2\pi} \frac{\delta}{\xi} \left( \left[ (1 + \zeta + \alpha T_{(k)})\,\xi - \xi \right] - \alpha T_{(k)}\,\xi \right)^2 \xi\,d\theta\,d\xi , \tag{4.43}
$$

in which $(\xi, \theta)$ serve as polar coordinates. While invoking from Eq. 4.41a, its evaluation leads to

$$
W_{(k)} = a\left(2\zeta + 2\,\alpha\,T_{(k)}\right)^2 - a_2\left(2\zeta + 2\,\alpha\,T_{(k)}\right)T_{(k)} + a_3\,T_{(k)}^2 + \frac{2}{3}\pi\,b\,h\delta^4\zeta^2 . \tag{4.44}
$$

Equating the expressions for strain energy density from Eqs. 4.41b and 4.44 provides the relationships between the PD parameters and engineering material constants as

$$
4a + \frac{2}{3}\pi\,bh\,\delta^4 = 2\kappa, \tag{4.45a}
$$

$$
a_2 = 4\,\alpha\,a, \tag{4.45b}
$$

$$a_3 = 4\alpha^2 a.\tag{4.45c}$$

Similarly, the expression for $\theta_{(k)}$ from Eq. 4.3 can be recast as

$$\theta_{(k)} = dh \int_0^\delta \int_0^{2\pi} \frac{\delta}{\xi} (((1 + \zeta + \alpha T)\,\xi - \xi) - \alpha T\,\xi)$$

$$\times \left(\frac{\boldsymbol{\xi}}{\xi} \cdot \frac{\boldsymbol{\xi}}{\xi}\right) \xi\, d\theta d\xi + 2\alpha T_{(k)},\tag{4.46a}$$

whose explicit evaluation leads to

$$\theta_{(k)} = \pi\, d\, h \delta^3 \zeta + 2\alpha T_{(k)}.\tag{4.46b}$$

Equating the expressions for dilatation from Eqs. 4.41a and 4.46b permits the determination of the peridynamic parameter $d$ as

$$d = \frac{2}{\pi\, h \delta^3}.\tag{4.47}$$

As illustrated in Fig. 4.5, a loading case of *simple shear* can be achieved by applying

$$\gamma_{xy(k)} = \zeta \quad \text{and} \quad \varepsilon_{xx(k)} = \varepsilon_{yy(k)} = T_{(k)} = 0,\tag{4.48}$$

for which the dilatation, $\theta_{(k)}$, and the strain energy density, $W_{(k)}$, within the realm of classical continuum mechanics become

$$\theta_{(k)} = 0 \quad \text{and} \quad W_{(k)} = \frac{1}{2}\mu\, \zeta^2.\tag{4.49a,b}$$

The relative position vector in the deformed state becomes

$$\left|\mathbf{y}_{(j)} - \mathbf{y}_{(k)}\right| = [1 + (\sin\theta\cos\theta)\zeta]\left|\mathbf{x}_{(j)} - \mathbf{x}_{(k)}\right|.\tag{4.50}$$

Therefore, the strain energy density, $W_{(k)}$, from Eq. 4.2 can be evaluated as

$$W_{(k)} = a\,(0) + bh \int_0^\delta \int_0^{2\pi} \frac{\delta}{\xi} ([1 + (\sin\theta\cos\theta)\zeta]\,\xi - \xi)^2 \xi d\theta d\xi,\tag{4.51a}$$

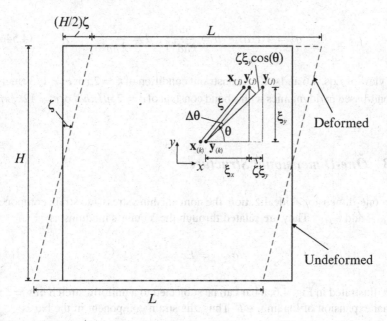

**Fig. 4.5** A two-dimensional plate subjected to simple shear

whose evaluation leads to

$$W_{(k)} = \frac{\pi h \delta^4 \zeta^2}{12} b. \tag{4.51b}$$

Equating the strain energy density expressions of Eqs. 4.49a,b and 4.51b obtained from classical continuum mechanics and the PD theory gives the relationship between the peridynamic parameter $b$ and shear modulus, $\mu$, as

$$b = \frac{6\mu}{\pi h \delta^4}. \tag{4.52}$$

Substituting from Eq. 4.52 into Eq. 4.45a results in the evaluation of the peridynamic parameter $a$ in terms of bulk modulus, $\kappa$, and shear modulus, $\mu$, as

$$a = \frac{1}{2}(\kappa - 2\mu). \tag{4.53}$$

In summary, the PD parameters for a two-dimensional analysis can be expressed as

$$a = \frac{1}{2}(\kappa - 2\mu), \quad a_2 = 4\alpha a, \tag{4.54a,b}$$

$$a_3 = 4\,\alpha^2\,a, \quad b = \frac{6\mu}{\pi h \delta^4}, \quad d = \frac{2}{\pi\,h\delta^3}. \tag{4.54c–e}$$

In view of Eqs. 4.6 and 4.12, a constraint condition of $\kappa = 2\mu$ or $\nu = 1/3$ emerges for bond-based peridynamics with a bond constant of $c = 24\mu/\pi h\delta^3$ or $c = 12\kappa/\pi h\delta^3$.

### 4.1.3   One-Dimensional Structures

Under one-dimensional idealization, the nonvanishing stress and strain components are $\sigma_{xx(k)}$ and $\varepsilon_{xx(k)}$. They are related through the Young's modulus as

$$\sigma_{xx(k)} = E\varepsilon_{xx(k)}. \tag{4.55}$$

As illustrated in Fig. 4.6, a bar can be subjected to a uniform stretch of $s = \zeta$ and thermal expansion of loading, $\alpha T$. Thus, the strain component in the bar is

$$\varepsilon_{xx(k)} = \zeta + \alpha T, \tag{4.56}$$

for which the dilatation, $\theta_{(k)}$, and strain energy density, $W_{(k)}$, within the realm of classical continuum mechanics become

$$\theta_{(k)} = \varepsilon_{xx(k)} = \zeta + \alpha T_{(k)} \tag{4.57a}$$

and

$$W_{(k)} = \frac{1}{2}E\,\zeta^2. \tag{4.57b}$$

As shown in Fig. 4.6, a one-dimensional structure is discretized with a single row of material points. The domain of integral $H$ in Eq. 2.22a becomes a line with a constant cross-sectional area, $A$.

The relative position vector between the material points at $\mathbf{x}_{(j)}$ and $\mathbf{x}_{(k)}$ in the deformed configuration becomes

$$\left|\mathbf{y}_{(j)} - \mathbf{y}_{(k)}\right| = \left(1 + \zeta + \alpha T_{(k)}\right)\left|\mathbf{x}_{(j)} - \mathbf{x}_{(k)}\right|, \tag{4.58}$$

in which $T_{(k)} = T$.

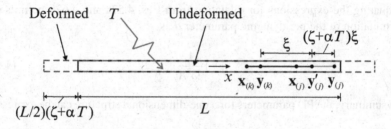

**Fig. 4.6** A one-dimensional bar subjected to isotropic expansion

The strain energy density, $W_{(k)}$, at material point $x_{(k)}$ that interacts with other material points within a line of length $\delta$ and area $A$ from Eq. 4.2 can be evaluated as

$$W_{(k)} = a\,\theta_{(k)}^2 - a_2\,\theta\,T_{(k)} + a_3\,T_{(k)}^2$$
$$+ 2bA\int_0^\delta \frac{\delta}{\xi}\left(\left[\left(1+\zeta+\alpha T_{(k)}\right)\xi - \xi\right] - \alpha T_{(k)}\,\xi\right)^2 d\xi, \tag{4.59}$$

in which $(\xi)$ serves as the coordinate. While invoking from Eq. 4.57a, its evaluation leads to

$$W_{(k)} = a\left(\zeta+\alpha T_{(k)}\right)^2 - a_2\left(\zeta+\alpha T_{(k)}\right)T_{(k)} + a_3\,T_{(k)}^2 + b\zeta^2\delta^3 A. \tag{4.60}$$

Assuming a = 0 due to the Poisson's ratio being zero, and equating the expressions for strain energy density from Eqs. 4.57b and 4.60 provides the relationships between the PD parameters and engineering material constants as

$$a_2 = a_3 = 0, \text{ and } b = \frac{E}{2A\delta^3}. \tag{4.61}$$

Similarly, the expression for $\theta_{(k)}$, from Eq. 4.3 can be recast as

$$\theta_{(k)} = 2dA\int_0^\delta \frac{\delta}{\xi}\left(\left[\left(1+\zeta+\alpha T_{(k)}\right)\xi - \xi\right] - \alpha T_{(k)}\,\xi\right)$$
$$\times \left(\frac{\xi}{\xi}\cdot\frac{\xi}{\xi}\right)d\xi + \alpha T_{(k)}, \tag{4.62a}$$

whose explicit evaluations leads to

$$\theta_{(k)} = 2d\delta^2\zeta A + \alpha T_{(k)}. \tag{4.62b}$$

Equating the expressions for dilatation from Eqs. 4.57a and 4.62b permits the determination of the peridynamic parameter $d$ as

$$d = \frac{1}{2\delta^2 A}. \tag{4.63}$$

In summary, the PD parameters for a one-dimensional structure can be expressed as

$$a = a_2 = a_3 = 0, \quad b = \frac{E}{2A\delta^3}, \quad d = \frac{1}{2\delta^2 A}. \tag{4.64a–c}$$

In view of Eq. 4.12, a bond constant for bond-based peridynamics becomes $c = 2E/A\delta^2$.

## 4.2  Surface Effects

The peridynamic material parameters $a$, $b$, and $d$ that appear in the peridynamic force-stretch relations are determined by computing both dilatation and strain energy density of a material point whose horizon is completely embedded in the material. The values of these parameters, except for $a$, depend on the domain of integration defined by the horizon. Therefore, the values of $b$ and $d$ require correction if the material point is close to free surfaces or material interfaces (Fig. 4.7). Since the presence of free surfaces is problem dependent, it is impractical to resolve this issue analytically. The correction of the material parameters is achieved by numerically integrating both dilatation and strain energy density at each material point inside the body for simple loading conditions and comparing them to their counterparts obtained from classical continuum mechanics.

For the first simple loading condition, the body is subjected to uniaxial stretch loadings in the $x$-, $y$-, and $z$-directions of the global coordinate system, i.e., $\varepsilon_{xx} \neq 0$, $\varepsilon_{\alpha\alpha} = \gamma_{\alpha\beta} = 0$ (shown in Fig. 4.8), $\varepsilon_{yy} \neq 0$, $\varepsilon_{\alpha\alpha} = \gamma_{\alpha\beta} = 0$, and $\varepsilon_{zz} \neq 0$, $\varepsilon_{\alpha\alpha} = \gamma_{\alpha\beta} = 0$, with $\alpha, \beta = x, y, z$.

The applied uniaxial stretch in the $x$-, $y$-, and $z$-directions is achieved through the constant displacement gradient, $\partial u_\alpha^*/\partial\alpha = \zeta$, with $\alpha = x, y, z$. The displacement field at material point $\mathbf{x}$ resulting from this loading can be expressed as

$$\mathbf{u}_1^T(\mathbf{x}) = \left\{ \frac{\partial u_x^*}{\partial x} x \quad 0 \quad 0 \right\}, \tag{4.65a}$$

$$\mathbf{u}_2^T(\mathbf{x}) = \left\{ 0 \quad \frac{\partial u_y^*}{\partial y} y \quad 0 \right\}, \tag{4.65b}$$

**Fig. 4.7** Surface effects in the domain of interest

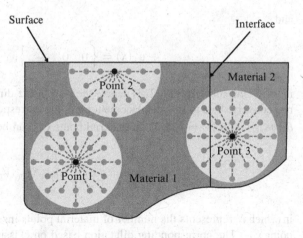

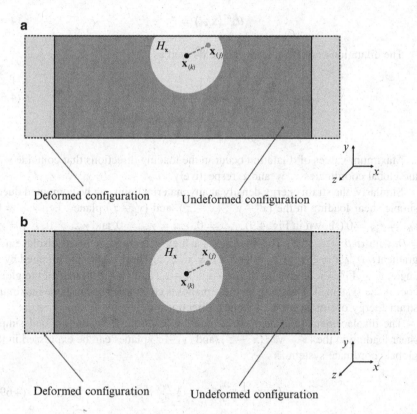

**Fig. 4.8** Material point **x** with (**a**) a truncated horizon and (**b**) far away from external surfaces of a material domain subjected to uniaxial stretch loading

and

$$\mathbf{u}_3^T(\mathbf{x}) = \left\{ 0 \quad 0 \quad \frac{\partial u_z^*}{\partial z} z \right\}, \tag{4.65c}$$

in which the subscripts $(1, 2, 3)$ denote the $x$-, $y$-, and $z$-directions of uniaxial stretch, respectively. Due to this displacement field, the corresponding PD dilatation term, $\theta_m^{PD}(\mathbf{x}_{(k)})$ with $(m = 1, 2, 3)$, at material point $\mathbf{x}_{(k)}$ can be obtained from Eq. 4.3 as

$$\theta_m^{PD}(\mathbf{x}_{(k)}) = d\,\delta \sum_{j=1}^{N} s_{(k)(j)} \Lambda_{(k)(j)} V_{(j)}, \tag{4.66}$$

in which $N$ represents the number of material points inside the horizon of material point $\mathbf{x}_{(k)}$. The corresponding dilatation based on classical continuum mechanics, $\theta_m^{CM}(\mathbf{x}_{(k)})$, is uniform throughout the domain and is determined as

$$\theta_m^{CM}(\mathbf{x}_{(k)}) = \zeta. \tag{4.67}$$

The dilatation correction term can be defined as

$$D_{m(k)} = \frac{\theta_m^{CM}(\mathbf{x}_{(k)})}{\theta_m^{PD}(\mathbf{x}_{(k)})} = \frac{\zeta}{d\,\delta \sum\limits_{j=1}^{N} s_{(k)(j)} \Lambda_{(k)(j)} V_{(j)}}. \tag{4.68}$$

Maximum values of dilatation occur in the loading directions that coincide with the global coordinates $x$, $y$, and $z$, respectively.

Similarly, the strain energy density at any material point can be computed due to simple shear loading in the $(x' - y')$, $(x' - z')$, and $(y' - z')$ planes, i.e., $\gamma_{x'y'} \neq 0$, $\varepsilon_{\alpha\alpha} = \gamma_{\alpha\beta} = 0$ (shown in Fig. 4.9), $\gamma_{x'z'} \neq 0$, $\varepsilon_{\alpha\alpha} = \gamma_{\alpha\beta} = 0$, and $\gamma_{y'z'} \neq 0$, $\varepsilon_{\alpha\alpha} = \gamma_{\alpha\beta} = 0$, with $\alpha, \beta = x', y', z'$. This loading is achieved through constant displacement gradient $\partial u_\alpha^* / \partial \beta = \zeta$, with $\alpha \neq \beta$ and $\alpha, \beta = x', y', z'$. These planes are oriented by an angle of $-45°$ in reference to the $(x - y)$, $(x - z)$, and $(y - z)$ planes of the global coordinate system. The loading on these planes is considered because the maximum strain energy occurs in the $x$-, $y$-, and $z$-directions.

The displacement field at material point $\mathbf{x}$ resulting from the applied simple shear loading in the $(x' - y')$, $(x' - z')$, and $(y' - z')$ planes can be expressed in the global coordinate system as

$$\mathbf{u}_1^T(\mathbf{x}) = \left\{ \frac{1}{2} \frac{\partial u_x^*}{\partial y'} x \quad -\frac{1}{2} \frac{\partial u_y^*}{\partial y'} y \quad 0 \right\}, \tag{4.69a}$$

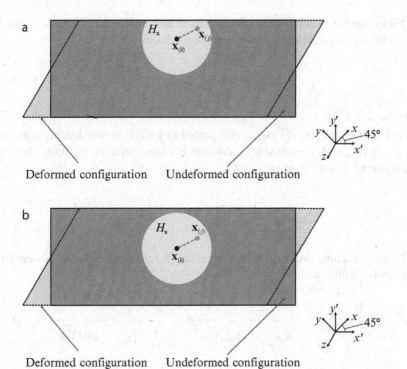

**Fig. 4.9** Material point $\mathbf{x}$ with (**a**) a truncated horizon and (**b**) far away from external surfaces of a material domain subjected to simple shear loading

$$\mathbf{u}_2^T(\mathbf{x}) = \left\{ 0 \quad \tfrac{1}{2}\tfrac{\partial u_y^*}{\partial z'}y \quad -\tfrac{1}{2}\tfrac{\partial u_y^*}{\partial z'}z \right\}, \tag{4.69b}$$

$$\mathbf{u}_3^T(\mathbf{x}) = \left\{ -\tfrac{1}{2}\tfrac{\partial u_z^*}{\partial x'}x \quad 0 \quad \tfrac{1}{2}\tfrac{\partial u_z^*}{\partial x'}z \right\}, \tag{4.69c}$$

in which the subscripts $(1,2,3)$ denote the applied simple shear loadings in the $(x'-y')$, $(y'-z')$, and $(x'-z')$ planes, respectively.

Due to these applied displacement fields, the PD strain energy density at material point $\mathbf{x}_{(k)}$ can be obtained from Eq. 4.2 as

$$W_m^{PD}(\mathbf{x}_{(k)}) = a\big(\theta_m^{PD}(\mathbf{x}_{(k)})\big)^2$$
$$+ b\delta \sum_{j=1}^{N} \frac{1}{\left|\mathbf{x}_{(j)} - \mathbf{x}_{(k)}\right|}\left(\left|\mathbf{y}_{(j)} - \mathbf{y}_{(k)}\right| - \left|\mathbf{x}_{(j)} - \mathbf{x}_{(k)}\right|\right)^2 V_{(j)}, \tag{4.70}$$

with $(m = 1,2,3)$.

Under simple shear loading, the dilatation and strain energy densities can be computed by using classical continuum mechanics as

$$\theta_m^{CM}(\mathbf{x}_{(k)}) = 0, \quad W_m^{CM}(\mathbf{x}_{(k)}) = \frac{1}{2}\mu\zeta^2, \qquad (4.71a,b)$$

with $(m = 1, 2, 3)$.

The dilatation term, $\theta_m^{PD}(\mathbf{x}_{(k)})$, is expected to vanish for this loading condition because it is already corrected with a dilatation correction term, Eq. 4.68. Thus, the strain energy density term reduces to

$$W_m^{PD}(\mathbf{x}_{(k)}) = b\delta \sum_{j=1}^{N} \frac{1}{|\mathbf{x}_{(j)} - \mathbf{x}_{(k)}|}\left(\left|\mathbf{y}_{(j)} - \mathbf{y}_{(k)}\right| - \left|\mathbf{x}_{(j)} - \mathbf{x}_{(k)}\right|\right)^2 V_{(j)}. \qquad (4.72)$$

Hence, the correction term is only necessary for the term including parameter $b$ and can be defined as

$$S_{m(k)} = \frac{W_{(m)}^{CM}(\mathbf{x}_{(k)})}{W_{(m)}^{PD}(\mathbf{x}_{(k)})} = \frac{\frac{1}{2}\mu\zeta^2}{b\delta \sum_{j=1}^{N} \frac{1}{|\mathbf{x}_{(j)} - \mathbf{x}_{(k)}|}\left(\left|\mathbf{y}_{(j)} - \mathbf{y}_{(k)}\right| - \left|\mathbf{x}_{(j)} - \mathbf{x}_{(k)}\right|\right)^2 V_{(j)}}. \qquad (4.73)$$

With these expressions, a vector of correction factors for the integral terms in dilatation and strain energy density at material point $\mathbf{x}_{(k)}$ can be written as

$$\mathbf{g}_{(d)}(\mathbf{x}_{(k)}) = \left\{g_{x(d)(k)}, \, g_{y(d)(k)}, g_{z(d)(k)}\right\}^T = \left\{D_{1(k)}, \, D_{2(k)}, D_{3(k)}\right\}^T, \qquad (4.74a)$$

$$\mathbf{g}_{(b)}(\mathbf{x}_{(k)}) = \left\{g_{x(b)(k)}, \, g_{y(b)(k)}, g_{z(b)(k)}\right\}^T = \left\{S_{1(k)}, \, S_{2(k)}, S_{3(k)}\right\}^T. \qquad (4.74b)$$

These correction factors are only valid in the $x$-, $y$-, and $z$-directions. However, they can be used as the principal values of an ellipsoid, as shown in Fig. 4.10, in order to approximate the surface correction factor in any direction. Arising from a general loading condition, the correction factor for interaction between material points $\mathbf{x}_{(k)}$ and $\mathbf{x}_{(j)}$, shown in Fig. 4.11a, can be obtained in the direction of their unit relative position vector, $\mathbf{n} = (\mathbf{x}_{(j)} - \mathbf{x}_{(k)})/|\mathbf{x}_{(j)} - \mathbf{x}_{(k)}| = \{n_x, \, n_y, \, n_z\}^T$.

A vector of correction factors for the integrals in the dilatation and strain energy density expressions at material point $\mathbf{x}_{(j)}$ can be similarly written as

$$\mathbf{g}_{(d)(j)}(\mathbf{x}_{(j)}) = \left\{g_{x(d)(j)}, \, g_{y(d)(j)}, g_{z(d)(j)}\right\}^T = \left\{D_{1(j)}, \, D_{2(j)}, D_{3(j)}\right\}^T, \qquad (4.75a)$$

**Fig. 4.10** Construction
of an ellipsoid for surface
correction factors

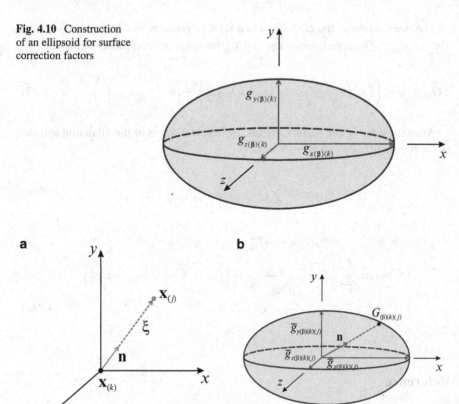

**Fig. 4.11** (**a**) PD interaction between material points at $\mathbf{x}_{(k)}$ and $\mathbf{x}_{(j)}$ and (**b**) the ellipsoid for the surface correction factors

$$\mathbf{g}_{(b)(j)}\left(\mathbf{x}_{(j)}\right) = \left\{g_{x(b)(j)}, \, g_{y(b)(j)}, \, g_{z(b)(j)}\right\}^{T} = \left\{S_{1(j)}, \, S_{2(j)}, \, S_{3(j)}\right\}^{T}. \qquad (4.75b)$$

These correction factors are, in general, different at material points $\mathbf{x}_{(k)}$ and $\mathbf{x}_{(j)}$. Therefore, the correction factor for an interaction between material points $\mathbf{x}_{(k)}$ and $\mathbf{x}_{(j)}$ can be obtained by their mean values as

$$\bar{\mathbf{g}}_{(\beta)(k)(j)} = \left\{\bar{g}_{x(\beta)(k)(j)}, \, \bar{g}_{y(\beta)(k)(j)}, \, \bar{g}_{z(\beta)(k)(j)}\right\}^{T} = \frac{\mathbf{g}_{(\beta)(k)} + \mathbf{g}_{(\beta)(j)}}{2},$$

$$(4.76)$$

with $\beta = d, b,$

which can be used as the principal values of an ellipsoid, as shown in Fig. 4.11b.

The intersection of the ellipsoid and a relative position vector, $\mathbf{n} = (\mathbf{x}_{(j)} - \mathbf{x}_{(k)})$ $/|\mathbf{x}_{(j)} - \mathbf{x}_{(k)}|$, of material points $\mathbf{x}_{(k)}$ and $\mathbf{x}_{(j)}$ provides the correction factors as

$$G_{(\beta)(k)(j)} = \left( \left[ n_x \big/ \bar{g}_{x(\beta)(k)(j)} \right]^2 + \left[ n_y \big/ \bar{g}_{y(\beta)(k)(j)} \right]^2 + \left[ n_z \big/ \bar{g}_{z(\beta)(k)(j)} \right]^2 \right)^{-1/2}. \quad (4.77)$$

After considering the surface effects, the discrete forms of the dilatation and the strain energy density can be corrected as

$$\theta_{(k)} = d\delta \sum_{j=1}^{N} G_{(d)(k)(j)} s_{(k)(j)} \Lambda_{(k)(j)} V_{(j)}, \quad (4.78a)$$

$$W_{(k)} = a\,\theta_{(k)}^2 - a_2 \theta_{(k)} T_{(k)} + a_3 T_{(k)}^2$$

$$+ b\delta \sum_{j=1}^{N} G_{(b)(k)(j)} \frac{1}{|\mathbf{x}_{(j)} - \mathbf{x}_{(k)}|} \left( |\mathbf{y}_{(j)} - \mathbf{y}_{(k)}| - |\mathbf{x}_{(j)} - \mathbf{x}_{(k)}| \right)^2 V_{(j)}.$$

$$(4.78b)$$

# Reference

Silling SA, Askari E (2005) A meshfree method based on the peridynamic model of solid mechanics. Comput Struct 83:1526–1535

# Chapter 5
# Peridynamics for Laminated Composite Materials

## 5.1 Basics

Fiber-reinforced laminated composites are generally constructed by bonding unidirectional laminae in a particular sequence. Each lamina has its own material properties and thickness. As shown in Fig. 5.1, the fiber orientation angle, $\theta$, is defined with respect to a reference axis, $x$. Fiber direction is commonly aligned with the $x_1 -$ axis, and transverse direction is aligned with the $x_2 -$ axis. A unidirectional lamina is specially orthotropic. Thus, a thin lamina has four independent material constants of elastic modulus in the fiber direction, $E_{11}$, elastic modulus in the transverse direction, $E_{22}$, in-plane shear modulus, $G_{12}$, and in-plane Poisson's ratio, $\nu_{12}$.

For a unidirectional lamina, the stiffness matrix, $\mathbf{Q}$, relates the stresses and strains at material point $\mathbf{x}_{(k)}$ in reference to the material (natural) coordinates, $(x_1, x_2)$ as

$$\left\{ \begin{array}{c} \sigma_{11} \\ \sigma_{22} \\ \sigma_{12} \end{array} \right\} = \left[ \begin{array}{ccc} Q_{11} & Q_{12} & 0 \\ Q_{12} & Q_{22} & 0 \\ 0 & 0 & Q_{66} \end{array} \right] \left\{ \begin{array}{c} \varepsilon_{11} \\ \varepsilon_{22} \\ \gamma_{12} \end{array} \right\}, \qquad (5.1a)$$

where

$$Q_{11} = \frac{E_{11}}{1 - \nu_{12}\nu_{21}}, \quad Q_{12} = \frac{\nu_{12}E_{22}}{1 - \nu_{12}\nu_{21}}, \quad Q_{22} = \frac{E_{22}}{1 - \nu_{12}\nu_{21}}, \quad Q_{66} = G_{12}, \qquad (5.1b)$$

with $\nu_{12}/E_{11} = \nu_{21}/E_{22}$.

The stress, $\sigma_{ij}$, and strain, $\varepsilon_{ij}$, components are referenced to the principal material (natural) coordinate system, $(x_1, x_2)$. The inverse of the lamina stiffness matrix, $\mathbf{Q}$, is referred to as the lamina compliance matrix, $\mathbf{S}$, whose coefficients are given as

E. Madenci and E. Oterkus, *Peridynamic Theory and Its Applications*,
DOI 10.1007/978-1-4614-8465-3_5, © Springer Science+Business Media New York 2014

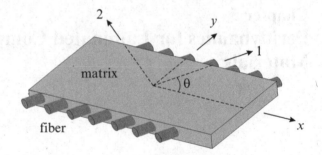

$$S_{11} = \frac{1}{E_{11}}, \; S_{12} = -\frac{\nu_{12}}{E_{11}} = -\frac{\nu_{21}}{E_{22}}, \; S_{22} = \frac{1}{E_{22}}, \; S_{66} = \frac{1}{G_{12}}. \tag{5.2}$$

Note that the coefficients of the stiffness and compliance matrices recover the relationship for an isotropic layer by specifying

$$Q_{11} = Q_{22} = \kappa + \mu, \quad Q_{12} = (\kappa - \mu), \quad Q_{66} = \mu \tag{5.3a}$$

and

$$S_{11} = S_{22} = \frac{\mu + \kappa}{4\kappa\mu}, \; S_{12} = \frac{\mu - \kappa}{4\kappa\mu}, \; S_{66} = \frac{1}{\mu}, \tag{5.3b}$$

where $\kappa$ and $\mu$ are bulk and shear modulus, respectively. The dilatation for a lamina based on classical continuum mechanics is

$$\theta = (\varepsilon_{11} + \varepsilon_{22}). \tag{5.4}$$

The strain energy density, $W$, based on classical continuum mechanics can be expressed as

$$W = \frac{1}{2}\sigma_{11}\varepsilon_{11} + \frac{1}{2}\sigma_{22}\varepsilon_{22} + \frac{1}{2}\sigma_{12}\gamma_{12} \tag{5.5a}$$

or

$$W = \frac{1}{2}\left(Q_{11}\varepsilon_{11}^2 + 2Q_{12}\varepsilon_{22}\varepsilon_{11} + Q_{66}\gamma_{12}^2 + Q_{22}\varepsilon_{22}^2\right). \tag{5.5b}$$

Under general loading conditions, the total deformation of a lamina cannot be decomposed as dilatational and distortional parts. Depending on the fiber orientation angle, the lamina may exhibit coupling of stretch and in-plane shear deformation.

## 5.2 Fiber-Reinforced Lamina

A lamina can be idealized as a two-dimensional structure, and is thus suitable for discretization with a single layer of material points in the thickness direction. In the case of an isotropic material, there is no directional dependence. However, the directional dependency of the interactions between the material points in a fiber-reinforced composite lamina must be included in the PD analysis.

As shown in Fig. 5.2, the material point $q$ represents material points that interact with material point $k$ only along the fiber direction with an orientation angle of $\theta$ in reference to the $x$-axis. Similarly, material point $r$ represents material points that interact with material point $k$ only along the transverse direction. However, the material point $p$ represents material points that interact with material point $k$ in any direction, including the fiber and transverse directions. The orientation of a PD interaction between the material point $k$ and the material point $p$ is defined by the angle $\phi$ with respect to the $x$-axis. The domain of integral $H$ in Eq. 2.22a is a disk with radius $\delta$ and thickness $h$.

The force density-stretch relations given by Eq. 2.48 must reflect the directional dependence of the PD material parameters for fiber-reinforced composite lamina. They can be defined in the form

$$\mathbf{t}_{(k)(j)}\left(\mathbf{u}_{(j)} - \mathbf{u}_{(k)}, \mathbf{x}_{(j)} - \mathbf{x}_{(k)}, t\right) = \frac{1}{2}A_{(k)(j)}\frac{\mathbf{y}_{(j)} - \mathbf{y}_{(k)}}{\left|\mathbf{y}_{(j)} - \mathbf{y}_{(k)}\right|} \tag{5.6a}$$

and

$$\mathbf{t}_{(j)(k)}\left(\mathbf{u}_{(k)} - \mathbf{u}_{(j)}, \mathbf{x}_{(k)} - \mathbf{x}_{(j)}, t\right) = -\frac{1}{2}B_{(j)(k)}\frac{\mathbf{y}_{(k)} - \mathbf{y}_{(j)}}{\left|\mathbf{y}_{(k)} - \mathbf{y}_{(j)}\right|}, \tag{5.6b}$$

where $A_{(k)(j)}$ and $B_{(j)(k)}$ are auxiliary parameters. As in the case of isotropic materials, these parameters can be determined by using Eq. 4.1, thus requiring an explicit form of the PD strain energy density at material point $\mathbf{x}_{(k)}$ for a unidirectional lamina.

In light of Eq. 4.2 and the directional dependency of a lamina, the PD strain energy density can be expressed as

**Fig. 5.2** PD horizon for a fiber-reinforced lamina and interaction of a family of material points

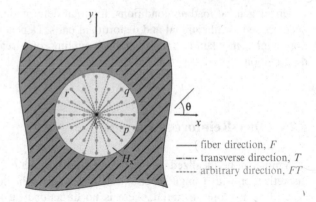

$$W_{(k)} = a\,\theta_{(k)}^2 + b_F \sum_{j=1}^{J} \frac{\delta}{\left|\mathbf{x}_{(j)} - \mathbf{x}_{(k)}\right|} \left(\left|\mathbf{y}_{(j)} - \mathbf{y}_{(k)}\right| - \left|\mathbf{x}_{(j)} - \mathbf{x}_{(k)}\right|\right)^2 V_{(j)}$$

$$+ b_{FT} \sum_{j=1}^{\infty} \frac{\delta}{\left|\mathbf{x}_{(j)} - \mathbf{x}_{(k)}\right|} \left(\left|\mathbf{y}_{(j)} - \mathbf{y}_{(k)}\right| - \left|\mathbf{x}_{(j)} - \mathbf{x}_{(k)}\right|\right)^2 V_{(j)} \qquad (5.7)$$

$$+ b_T \sum_{j=1}^{J} \frac{\delta}{\left|\mathbf{x}_{(j)} - \mathbf{x}_{(k)}\right|} \left(\left|\mathbf{y}_{(j)} - \mathbf{y}_{(k)}\right| - \left|\mathbf{x}_{(j)} - \mathbf{x}_{(k)}\right|\right)^2 V_{(j)},$$

in which the PD material parameter $a$ is associated with the deformation involving dilatation, $\theta_{(k)}$. The other material parameters, $b_F$, $b_T$, and $b_{FT}$, are associated with deformation of material points in the fiber direction, transverse direction, and arbitrary directions, respectively. The total number of material points within the family of material point $\mathbf{x}_{(k)}$ in either fiber or transverse directions is denoted by $J$. The PD dilatation, $\theta_{(k)}$, for a unidirectional lamina can be expressed as

$$\theta_{(k)} = d \sum_{j=1}^{\infty} \frac{\delta}{\left|\mathbf{x}_{(j)} - \mathbf{x}_{(k)}\right|} \left(\left|\mathbf{y}_{(j)} - \mathbf{y}_{(k)}\right| - \left|\mathbf{x}_{(j)} - \mathbf{x}_{(k)}\right|\right) \Lambda_{(k)(j)} V_{(j)}, \qquad (5.8)$$

in which $d$ is a PD parameter.

After substituting for $\theta_{(k)}$ from Eq. 5.8 in the expression for $W_{(k)}$ given by Eq. 5.7 and performing differentiation, the force density vector $\mathbf{t}_{(k)(j)}(\mathbf{u}_{(j)} - \mathbf{u}_{(k)}, \mathbf{x}_{(j)} - \mathbf{x}_{(k)}, t)$ from Eq. 4.1 can be rewritten in terms of PD material parameters as

$$\mathbf{t}_{(k)(j)}(\mathbf{u}_{(j)} - \mathbf{u}_{(k)}, \mathbf{x}_{(j)} - \mathbf{x}_{(k)}, t) = \frac{1}{2} A_{(k)(j)} \frac{\mathbf{y}_{(j)} - \mathbf{y}_{(k)}}{\left|\mathbf{y}_{(j)} - \mathbf{y}_{(k)}\right|}, \qquad (5.9a)$$

where

$$A_{(k)(j)} = 4ad \frac{\delta}{\left|\mathbf{x}_{(j)} - \mathbf{x}_{(k)}\right|} \Lambda_{(k)(j)}\theta_{(k)} + 4\delta(\mu_F b_F + b_{FT} + \mu_T b_T)s_{(k)(j)}, \qquad (5.9b)$$

with

$$\mu_F = \begin{cases} 1 & (\mathbf{x}_{(j)} - \mathbf{x}_{(k)})//\text{fiber direction} \\ 0 & \text{otherwise} \end{cases} \qquad (5.9c)$$

and

$$\mu_T = \begin{cases} 1 & (\mathbf{x}_{(j)} - \mathbf{x}_{(k)})\perp\text{fiber direction} \\ 0 & \text{otherwise}. \end{cases} \qquad (5.9d)$$

Similarly, the force density vector $\mathbf{t}_{(j)(k)}(\mathbf{u}_{(k)} - \mathbf{u}_{(j)}, \mathbf{x}_{(k)} - \mathbf{x}_{(j)}, t)$ can be expressed as

$$\mathbf{t}_{(j)(k)}\big(\mathbf{u}_{(k)} - \mathbf{u}_{(j)}, \mathbf{x}_{(k)} - \mathbf{x}_{(j)}, t\big) = -\frac{1}{2}B_{(j)(k)} \frac{\mathbf{y}_{(j)} - \mathbf{y}_{(k)}}{\left|\mathbf{y}_{(j)} - \mathbf{y}_{(k)}\right|}, \qquad (5.10a)$$

with

$$B_{(j)(k)} = 4ad \frac{\delta}{\left|\mathbf{x}_{(k)} - \mathbf{x}_{(j)}\right|} \Lambda_{(j)(k)}\theta_{(j)} + 4\delta(\mu_F b_F + b_{FT} + \mu_T b_T)s_{(j)(k)}. \qquad (5.10b)$$

Although Eqs. 5.9b and 5.10b appear to be similar, they are different because the dilatations $\theta_{(k)}$ and $\theta_{(j)}$ for the material points at $\mathbf{x}_{(k)}$ and $\mathbf{x}_{(j)}$, respectively, are different. This formulation can be extended to include the effect of thermal loading as described in Chap. 4. Oterkus and Madenci (2012) presented such an extension for the bond-based peridynamic formulation.

## 5.3   Laminated Composites

The laminae are perfectly bonded in the construction of a laminate; thus, there exists no slip among the laminae. Aside from the loading conditions, the deformation of a laminate is dependent on the lamina properties, thickness, and stacking sequence. There exists usually a resin-rich layer between the laminae; an inherent source for cracking and delamination. Therefore, transverse normal and shear deformations especially play a critical role in the initiation and growth of delamination. Also, in the presence of a nonsymmetric stacking sequence, the laminates exhibit coupling between in-plane and out-of-plane deformation, resulting in curvature.

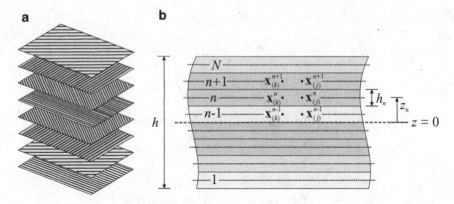

**Fig. 5.3** Elevation of each lamina in laminate

As shown in Fig. 5.3, the reference coordinate system $(x, y, z)$ is located on the midplane of the laminate. The laminate thickness, $h$, is given by

$$h = \sum_{n=1}^{N} h_n,  \tag{5.11}$$

where $N$ is the total number of lamina in the stacking sequence, and $h_n$ is the thickness of the $n^{th}$ lamina. With respect to the midplane, the position of each lamina, $z_n$, is defined as

$$z_n = -\frac{h}{2} + \sum_{m=1}^{n-1} h_m + \frac{1}{2} h_n.  \tag{5.12}$$

The presence of the transverse normal and transverse shear deformations in a laminate can be included in the derivation of the PD equation of motion under the assumption that material points in a particular lamina interact with the other material points of immediate neighboring laminae above and below it.

The total potential energy of a laminate with $N$ layers can be expressed in the form

$$U = \sum_{n=1}^{N} \sum_{i=1}^{\infty} W_{(i)}^n + \sum_{n=1}^{N-1} \sum_{i=1}^{\infty} \hat{W}_{(i)}^n + \sum_{n=1}^{N-1} \sum_{i=1}^{\infty} \tilde{W}_{(i)}^n - \sum_{n=1}^{N} \sum_{i=1}^{\infty} \mathbf{b}_{(i)}^n \cdot \mathbf{u}_{(i)}^n,  \tag{5.13}$$

where $W_{(i)}^n$, $\hat{W}_{(i)}^n$, and $\tilde{W}_{(i)}^n$ represent the contributions from the in-plane, transverse normal, and shear deformations, respectively, and $\mathbf{b}_{(i)}^n$ is the body load vector.

Using Eq. 5.7, the strain energy density, $W_{(k)}^n$, of material point $\mathbf{x}_{(k)}^n$ located on the $n^{th}$ layer, due to in-plane deformations, can be expressed as a summation of micropotentials, $w_{(k)(j)}$, arising from the interaction of material point $\mathbf{x}_{(k)}^n$ and the other material points $\mathbf{x}_{(j)}^n$ within its horizon in the form

$$W_{(k)}^n = \frac{1}{2} \sum_{j=1}^{\infty} \frac{1}{2} \left[ \begin{array}{l} w_{(k)(j)} \left( \mathbf{y}_{(1k)}^n - \mathbf{y}_{(k)}^n, \; \mathbf{y}_{(2k)}^n - \mathbf{y}_{(k)}^n, \cdots \right) + \\ w_{(j)(k)} \left( \mathbf{y}_{(1j)}^n - \mathbf{y}_{(j)}^n, \; \mathbf{y}_{(2j)}^n - \mathbf{y}_{(j)}^n, \cdots \right) \end{array} \right] V_{(j)}^n, \qquad (5.14)$$

in which $w_{(k)(j)} = 0$ for $k = j$. Due to transverse normal deformation, the strain energy density, $\hat{W}_{(k)}^n$, of material point $\mathbf{x}_{(k)}^n$ located on the $n^{th}$ layer can be expressed as a summation of micropotentials, $\hat{w}_{(k)}$, arising from the interaction of material point $\mathbf{x}_{(k)}^n$ and the adjacent material points, $\mathbf{x}_{(k)}^{(n+1)}$ and $\mathbf{x}_{(k)}^{(n-1)}$, located on $(n+1)^{th}$ and $(n-1)^{th}$ layers in the form

$$\hat{W}_{(k)}^n = \frac{1}{2} \sum_{m=n+1, n-1} \frac{1}{2} \left[ \hat{w}_{(k)} \left( \mathbf{y}_{(k)}^m - \mathbf{y}_{(k)}^n \right) V_{(k)}^m + \hat{w}_{(k)} \left( \mathbf{y}_{(k)}^n - \mathbf{y}_{(k)}^m \right) V_{(k)}^m \right]. \qquad (5.15)$$

Similarly, the strain energy density associated with transverse shear deformation, $\tilde{W}_{(k)}^n$, of material point $\mathbf{x}_{(k)}^n$ can be expressed as a summation of micropotentials, $\tilde{w}_{(k)(j)}$, arising from the interaction of material point $\mathbf{x}_{(k)}^n$ and the other material points (within its family), $\mathbf{x}_{(j)}^{(n+1)}$ and $\mathbf{x}_{(j)}^{(n-1)}$, and $\tilde{w}_{(j)(k)}$, arising from the interaction of material point $\mathbf{x}_{(j)}^n$ and the other material points (within its family), $\mathbf{x}_{(k)}^{(n+1)}$ and $\mathbf{x}_{(k)}^{(n-1)}$, in the form

$$\begin{aligned} \tilde{W}_{(k)}^n = \frac{1}{2} \Bigg\{ & \sum_{j=1}^{\infty} \frac{1}{2} \tilde{w}_{(k)(j)} \left( \mathbf{y}_{(j)}^{n+1} - \mathbf{y}_{(k)}^n, \mathbf{y}_{(k)}^{n+1} - \mathbf{y}_{(j)}^n \right) V_{(j)}^{n+1} \\ & + \sum_{j=1}^{\infty} \frac{1}{2} \tilde{w}_{(j)(k)} \left( \mathbf{y}_{(k)}^n - \mathbf{y}_{(j)}^{n+1}, \mathbf{y}_{(j)}^n - \mathbf{y}_{(k)}^{n+1} \right) V_{(j)}^{n+1} \\ & + \sum_{j=1}^{\infty} \frac{1}{2} \tilde{w}_{(j)(k)} \left( \mathbf{y}_{(k)}^n - \mathbf{y}_{(j)}^{n-1}, \mathbf{y}_{(j)}^n - \mathbf{y}_{(k)}^{n-1} \right) V_{(j)}^{n-1} \\ & + \sum_{j=1}^{\infty} \frac{1}{2} \tilde{w}_{(k)(j)} \left( \mathbf{y}_{(j)}^{n-1} - \mathbf{y}_{(k)}^n, \mathbf{y}_{(k)}^{n-1} - \mathbf{y}_{(j)}^n \right) V_{(j)}^{n-1} \\ & + \sum_{j=1}^{\infty} \frac{1}{2} \tilde{w}_{(j)(k)} \left( \mathbf{y}_{(k)}^{n+1} - \mathbf{y}_{(j)}^n, \mathbf{y}_{(j)}^{n+1} - \mathbf{y}_{(k)}^n \right) V_{(j)}^n \\ & + \sum_{j=1}^{\infty} \frac{1}{2} \tilde{w}_{(k)(j)} \left( \mathbf{y}_{(j)}^n - \mathbf{y}_{(k)}^{n+1}, \mathbf{y}_{(k)}^n - \mathbf{y}_{(j)}^{n+1} \right) V_{(j)}^n \\ & + \sum_{j=1}^{\infty} \frac{1}{2} \tilde{w}_{(k)(j)} \left( \mathbf{y}_{(j)}^n - \mathbf{y}_{(k)}^{n-1}, \mathbf{y}_{(k)}^n - \mathbf{y}_{(j)}^{n-1} \right) V_{(j)}^n \\ & + \sum_{j=1}^{\infty} \frac{1}{2} \tilde{w}_{(j)(k)} \left( \mathbf{y}_{(k)}^{n-1} - \mathbf{y}_{(j)}^n, \mathbf{y}_{(j)}^{n-1} - \mathbf{y}_{(k)}^n \right) V_{(j)}^n \Bigg\}. \end{aligned} \qquad (5.16)$$

Substituting for the strain energy densities, $W_{(i)}^n$, $\hat{W}_{(i)}^n$, and $\tilde{W}_{(i)}^n$, of material point $\mathbf{x}_{(i)}^n$ from Eqs. 5.14, 5.15 and 5.16, the potential energy of the laminate with $N$ layers can be rewritten as

$$
\begin{aligned}
U = & \sum_{n=1}^{N} \left\{ \frac{1}{2} \sum_{i=1}^{\infty} \sum_{j=1}^{\infty} \frac{1}{2} \left[ w_{(i)(j)} \left( \mathbf{y}_{(1^i)}^n - \mathbf{y}_{(i)}^n, \, \mathbf{y}_{(2^i)}^n - \mathbf{y}_{(i)}^n, \cdots \right) \right. \right. \\
& \left. \left. + w_{(j)(i)} \left( \mathbf{y}_{(1^j)}^n - \mathbf{y}_{(j)}^n, \, \mathbf{y}_{(2^j)}^n - \mathbf{y}_{(j)}^n, \cdots \right) \right] V_{(j)}^n V_{(i)}^n \right\} \\
& + \frac{1}{2} \sum_{n=1}^{N-1} \sum_{i=1}^{\infty} \sum_{m=n+1,n-1} \frac{1}{2} \left[ \hat{w}_{(i)} \left( \mathbf{y}_{(i)}^m - \mathbf{y}_{(i)}^n \right) + \hat{w}_{(i)} \left( \mathbf{y}_{(i)}^n - \mathbf{y}_{(i)}^m \right) \right] V_{(i)}^m V_{(i)}^n \\
& + \frac{1}{2} \sum_{n=1}^{N-1} \left\{ \sum_{i=1}^{\infty} \frac{1}{2} \left[ \sum_{j=1}^{\infty} \tilde{w}_{(i)(j)} \left( \mathbf{y}_{(j)}^{n+1} - \mathbf{y}_{(i)}^n, \mathbf{y}_{(i)}^{n+1} - \mathbf{y}_{(j)}^n \right) V_{(j)}^{n+1} V_{(i)}^n \right. \right. \\
& + \sum_{j=1}^{\infty} \tilde{w}_{(j)(i)} \left( \mathbf{y}_{(i)}^n - \mathbf{y}_{(j)}^{n+1}, \mathbf{y}_{(j)}^n - \mathbf{y}_{(i)}^{n+1} \right) V_{(j)}^{n+1} V_{(i)}^n \\
& + \sum_{j=1}^{\infty} \tilde{w}_{(j)(i)} \left( \mathbf{y}_{(i)}^n - \mathbf{y}_{(j)}^{n-1}, \mathbf{y}_{(j)}^n - \mathbf{y}_{(i)}^{n-1} \right) V_{(j)}^{n-1} V_{(i)}^n \\
& \left. \left. + \sum_{j=1}^{\infty} \tilde{w}_{(i)(j)} \left( \mathbf{y}_{(j)}^{n-1} - \mathbf{y}_{(i)}^n, \mathbf{y}_{(i)}^{n-1} - \mathbf{y}_{(j)}^n \right) V_{(j)}^{n-1} V_{(i)}^n \right] \right\} \\
& + \frac{1}{2} \sum_{n=1}^{N-1} \left\{ \sum_{i=1}^{\infty} \frac{1}{2} \left[ \sum_{j=1}^{\infty} \tilde{w}_{(j)(i)} \left( \mathbf{y}_{(i)}^{n+1} - \mathbf{y}_{(j)}^n, \mathbf{y}_{(j)}^{n+1} - \mathbf{y}_{(i)}^n \right) V_{(i)}^{n+1} V_{(j)}^n \right. \right. \\
& + \sum_{j=1}^{\infty} \tilde{w}_{(i)(j)} \left( \mathbf{y}_{(j)}^n - \mathbf{y}_{(i)}^{n+1}, \mathbf{y}_{(i)}^n - \mathbf{y}_{(j)}^{n+1} \right) V_{(i)}^{n+1} V_{(j)}^n \\
& + \sum_{j=1}^{\infty} \tilde{w}_{(i)(j)} \left( \mathbf{y}_{(j)}^n - \mathbf{y}_{(i)}^{n-1}, \mathbf{y}_{(i)}^n - \mathbf{y}_{(j)}^{n-1} \right) V_{(i)}^{n-1} V_{(j)}^n \\
& \left. \left. + \sum_{j=1}^{\infty} \tilde{w}_{(j)(i)} \left( \mathbf{y}_{(i)}^{n-1} - \mathbf{y}_{(j)}^n, \mathbf{y}_{(j)}^{n-1} - \mathbf{y}_{(i)}^n \right) V_{(i)}^{n-1} V_{(j)}^n \right] \right\} \\
& - \sum_{n=1}^{N} \left\{ \sum_{i=1}^{\infty} \left( \mathbf{b}_{(i)}^n \cdot \mathbf{u}_{(i)}^n \right) V_{(i)}^n \right\},
\end{aligned}
$$

(5.17a)

or exchanging the order of dummy indices $i$ and $j$ in the fourth summation of layers results in

$$
U = \sum_{n=1}^{N} \left\{ \frac{1}{2} \sum_{i=1}^{\infty} \sum_{j=1}^{\infty} \frac{1}{2} \left[ w_{(i)(j)} \left( \mathbf{y}_{(1^i)}^n - \mathbf{y}_{(i)}^n, \; \mathbf{y}_{(2^i)}^n - \mathbf{y}_{(i)}^n, \cdots \right) \right. \right.
$$

$$
\left. \left. + w_{(j)(i)} \left( \mathbf{y}_{(1^j)}^n - \mathbf{y}_{(j)}^n, \; \mathbf{y}_{(2^j)}^n - \mathbf{y}_{(j)}^n, \cdots \right) \right] V_{(j)}^n V_{(i)}^n \right\}
$$

$$
+ \frac{1}{2} \sum_{n=1}^{N-1} \sum_{i=1}^{\infty} \sum_{m=n+1,n-1} \frac{1}{2} \left[ \hat{w}_{(i)} \left( \mathbf{y}_{(i)}^m - \mathbf{y}_{(i)}^n \right) + \hat{w}_{(i)} \left( \mathbf{y}_{(i)}^n - \mathbf{y}_{(i)}^m \right) \right] V_{(i)}^m V_{(i)}^n
$$

$$
+ \frac{1}{2} \sum_{n=1}^{N-1} \sum_{i=1}^{\infty} \sum_{j=1}^{\infty} \sum_{m=n+1,n-1} \left[ \begin{array}{c} \tilde{w}_{(i)(j)} \left( \mathbf{y}_{(j)}^m - \mathbf{y}_{(i)}^n, \mathbf{y}_{(j)}^m - \mathbf{y}_{(i)}^n \right) \\ + \tilde{w}_{(j)(i)} \left( \mathbf{y}_{(i)}^n - \mathbf{y}_{(j)}^m, \mathbf{y}_{(j)}^m - \mathbf{y}_{(i)} \right) \end{array} \right] V_{(j)}^m V_{(i)}^n
$$

$$
- \sum_{n=1}^{N} \sum_{i=1}^{\infty} \left( \mathbf{b}_{(i)}^n \cdot \mathbf{u}_{(i)}^n \right) V_{(i)}^n \; .
$$

$$
(5.17b)
$$

As necessary for the derivation of equations of motion, the Lagrangian, Eq. 2.11, can be written in an expanded form by showing only the terms associated with the material point $\mathbf{x}_{(k)}^n$ located on the $n^{th}$ layer as

$$
L = \ldots + \frac{1}{2} \rho_{(k)}^n \, \dot{\mathbf{u}}_{(k)}^n \cdot \dot{\mathbf{u}}_{(k)}^n \, V_{(k)}^n + \ldots
$$

$$
\ldots - \frac{1}{2} \sum_{j=1}^{\infty} \left\{ w_{(k)(j)} \left( \mathbf{y}_{(1^k)}^n - \mathbf{y}_{(k)}^n, \; \mathbf{y}_{(2^k)}^n - \mathbf{y}_{(k)}^n, \cdots \right) V_{(j)}^n V_{(k)}^n \right\} \ldots
$$

$$
\ldots - \frac{1}{2} \sum_{j=1}^{\infty} \left\{ w_{(j)(k)} \left( \mathbf{y}_{(1^j)}^n - \mathbf{y}_{(j)}^n, \; \mathbf{y}_{(2^j)}^n - \mathbf{y}_{(j)}^n, \cdots \right) V_{(j)}^n V_{(k)}^n \right\} \ldots
$$

$$
\ldots - \frac{1}{2} \hat{w}_{(k)} \left( \mathbf{y}_{(k)}^{n+1} - \mathbf{y}_{(k)}^n \right) V_{(k)}^{n+1} V_{(k)}^n \ldots - \frac{1}{2} \hat{w}_{(k)} \left( \mathbf{y}_{(k)}^n - \mathbf{y}_{(k)}^{n+1} \right) V_{(k)}^n V_{(k)}^{n+1}
$$

$$
\ldots - \frac{1}{2} \hat{w}_{(k)} \left( \mathbf{y}_{(k)}^n - \mathbf{y}_{(k)}^{n-1} \right) V_{(k)}^n V_{(k)}^{n-1} \ldots - \frac{1}{2} \hat{w}_{(k)} \left( \mathbf{y}_{(k)}^{n-1} - \mathbf{y}_{(k)}^n \right) V_{(k)}^{n-1} V_{(k)}^n
$$

$$
\ldots - \sum_{j=1}^{\infty} \tilde{w}_{(k)(j)} \left( \mathbf{y}_{(j)}^{n+1} - \mathbf{y}_{(k)}^n, \mathbf{y}_{(k)}^{n+1} - \mathbf{y}_{(j)}^n \right) V_{(j)}^{n+1} V_{(k)}^n \ldots
$$

$$
(5.18)
$$

$$
\ldots - \sum_{j=1}^{\infty} \tilde{w}_{(j)(k)} \left( \mathbf{y}_{(k)}^n - \mathbf{y}_{(j)}^{n+1}, \mathbf{y}_{(j)}^n - \mathbf{y}_{(k)}^{n+1} \right) V_{(k)}^n V_{(j)}^{n+1}
$$

$$
\ldots - \sum_{j=1}^{\infty} \tilde{w}_{(j)(k)} \left( \mathbf{y}_{(k)}^n - \mathbf{y}_{(j)}^{n-1}, \mathbf{y}_{(j)}^n - \mathbf{y}_{(k)}^{n-1} \right) V_{(k)}^n V_{(j)}^{n-1} \ldots
$$

$$
\ldots - \sum_{j=1}^{\infty} \tilde{w}_{(k)(j)} \left( \mathbf{y}_{(j)}^{n-1} - \mathbf{y}_{(k)}^n, \mathbf{y}_{(k)}^{n-1} - \mathbf{y}_{(j)}^n \right) V_{(j)}^{n-1} V_{(k)}^n
$$

$$
\ldots \ldots + \mathbf{b}_{(k)}^n \cdot \mathbf{u}_{(k)}^n V_{(k)}^n \ldots \ldots
$$

Substituting from Eq. 5.18 into Eq. 2.10 results in the Lagrange's equation of the material point $\mathbf{x}_{(k)}^n$ located on the $n^{th}$ layer as

$$
\begin{aligned}
\Bigg\{ \rho_{(k)}^n \ddot{\mathbf{u}}_{(k)}^n &+ \sum_{j=1}^{\infty} \frac{1}{2} \left( \sum_{i=1}^{\infty} \frac{\partial w_{(k)(i)}}{\partial \left( \mathbf{y}_{(j)}^n - \mathbf{y}_{(k)}^n \right)} V_{(i)}^n \right) \frac{\partial \left( \mathbf{y}_{(j)}^n - \mathbf{y}_{(k)}^n \right)}{\partial \mathbf{u}_{(k)}^n} \\
&+ \sum_{j=1}^{\infty} \frac{1}{2} \left( \sum_{i=1}^{\infty} \frac{\partial w_{(i)(k)}}{\partial \left( \mathbf{y}_{(k)}^n - \mathbf{y}_{(j)}^n \right)} V_{(i)}^n \right) \frac{\partial \left( \mathbf{y}_{(k)}^n - \mathbf{y}_{(j)}^n \right)}{\partial \mathbf{u}_{(k)}^n} \\
&+ \sum_{m=n+1,n-1} \frac{1}{2} \frac{\partial \hat{w}_{(k)}}{\partial \left( \mathbf{y}_{(k)}^m - \mathbf{y}_{(k)}^n \right)} \frac{\partial \left( \mathbf{y}_{(k)}^m - \mathbf{y}_{(k)}^n \right)}{\partial \mathbf{u}_{(k)}^n} V_{(k)}^m \\
&+ \sum_{m=n+1,n-1} \frac{1}{2} \frac{\partial \hat{w}_{(k)}}{\partial \left( \mathbf{y}_{(k)}^n - \mathbf{y}_{(k)}^m \right)} \frac{\partial \left( \mathbf{y}_{(k)}^n - \mathbf{y}_{(k)}^m \right)}{\partial \mathbf{u}_{(k)}^n} V_{(k)}^m \\
&+2 \sum_{m=n+1,n-1} \sum_{j=1}^{\infty} \frac{1}{2} \frac{\partial \tilde{w}_{(k)(j)}}{\partial \left( \mathbf{y}_{(j)}^m - \mathbf{y}_{(k)}^n \right)} \frac{\partial \left( \mathbf{y}_{(j)}^m - \mathbf{y}_{(k)}^n \right)}{\partial \mathbf{u}_{(k)}^n} V_{(j)}^m \\
&+2 \sum_{m=n+1,n-1} \sum_{j=1}^{\infty} \frac{1}{2} \frac{\partial \tilde{w}_{(j)(k)}}{\partial \left( \mathbf{y}_{(k)}^n - \mathbf{y}_{(j)}^m \right)} \frac{\partial \left( \mathbf{y}_{(k)}^n - \mathbf{y}_{(j)}^m \right)}{\partial \mathbf{u}_{(k)}^n} V_{(j)}^m \\
&- b_{(k)}^n \Bigg\} V_{(k)}^n = 0,
\end{aligned}
\tag{5.19}
$$

in which it is assumed that the interactions not involving material point $\mathbf{x}_{(k)}^n$ do not have any effect on material point $\mathbf{x}_{(k)}^n$. With the interpretation that the derivatives of the micropotentials represent the force densities that material points exert upon each other, this equation can be rewritten as

$$
\begin{aligned}
\rho_{(k)}^n \ddot{\mathbf{u}}_{(k)}^n = &\sum_{j=1}^{\infty} \left[ \mathbf{t}_{(k)(j)}^n \left( \mathbf{u}_{(j)}^n - \mathbf{u}_{(k)}^n, \mathbf{x}_{(j)}^n - \mathbf{x}_{(k)}^n, t \right) \right. \\
&\left. - \mathbf{t}_{(j)(k)}^n \left( \mathbf{u}_{(k)}^n - \mathbf{u}_{(j)}^n, \mathbf{x}_{(k)}^n - \mathbf{x}_{(j)}^n, t \right) \right] V_{(j)}^n \\
&+ \sum_{m=n+1,n-1} \left[ \mathbf{r}_{(k)}^{(n)(m)} \left( \mathbf{u}_{(k)}^m - \mathbf{u}_{(k)}^n, \mathbf{x}_{(k)}^m - \mathbf{x}_{(k)}^n, t \right) \right. \\
&\left. - \mathbf{r}_{(k)}^{(m)(n)} \left( \mathbf{u}_{(k)}^n - \mathbf{u}_{(k)}^m, \mathbf{x}_{(k)}^n - \mathbf{x}_{(k)}^m, t \right) \right] V_{(k)}^m \\
&+2 \sum_{m=n+1,n-1} \sum_{j=1}^{\infty} \left[ \mathbf{s}_{(k)(j)}^{(n)(m)} \left( \mathbf{u}_{(j)}^m - \mathbf{u}_{(k)}^n, \mathbf{u}_{(k)}^m - \mathbf{u}_{(j)}^n, \mathbf{x}_{(j)}^m - \mathbf{x}_{(k)}^n, \mathbf{x}_{(k)}^m - \mathbf{x}_{(j)}^n, t \right) \right. \\
&\left. - \mathbf{s}_{(j)(k)}^{(m)(n)} \left( \mathbf{u}_{(k)}^n - \mathbf{u}_{(j)}^m, \mathbf{u}_{(j)}^n - \mathbf{u}_{(k)}^m, \mathbf{x}_{(k)}^n - \mathbf{x}_{(j)}^m, \mathbf{x}_{(j)}^n - \mathbf{x}_{(k)}^m, t \right) \right] V_{(j)}^m \\
&+ \mathbf{b}_{(k)}^n .
\end{aligned}
\tag{5.20}
$$

Arising from in-plane deformation, $t^n_{(k)(j)}$ represents the force density that material point $x^n_{(j)}$ exerts upon material point $x^n_{(k)}$. Similarly, $t^n_{(j)(k)}$ represents the force density that material point $x^n_{(k)}$ exerts upon material point $x^n_{(j)}$. The force density vectors, $r^{(n)(m)}_{(k)}$ and $r^{(m)(n)}_{(k)}$ with $m = (n+1), (n-1)$, develop due to the transverse normal deformation between the material points $x^n_{(k)}$ and $x^m_{(k)}$. The force density vector $r^{(n)(m)}_{(k)}$ represents the force exerted by material point $x^m_{(k)}$ upon the material point $x^n_{(k)}$, and $r^{(m)(n)}_{(k)}$ represents the opposite. The force density vectors $s^{(n)(m)}_{(k)(j)}$ and $s^{(m)(n)}_{(j)(k)}$, with $m = (n+1), (n-1)$, are associated with transverse shear deformation between the material points $x^m_{(j)}$ and $x^n_{(k)}$. The force density vector $s^{(n)(m)}_{(k)(j)}$ represents the force exerted by material point $x^m_{(j)}$ on the material point $x^n_{(k)}$, and $s^{(m)(n)}_{(j)(k)}$ represents the other way around. These force density vectors are defined as

$$
\left.
\begin{aligned}
t^n_{(k)(j)}\left(u^n_{(j)} - u^n_{(k)}, x^n_{(j)} - x^n_{(k)}, t\right) &= \frac{1}{2}\frac{1}{V^n_{(j)}}\left(\sum_{i=1}^{\infty}\frac{\partial w_{(k)(i)}}{\partial\left(y^n_{(j)} - y^n_{(k)}\right)}V^n_{(i)}\right) \\
t^n_{(j)(k)}\left(u^n_{(k)} - u^n_{(j)}, x^n_{(k)} - x^n_{(j)}, t\right) &= \frac{1}{2}\frac{1}{V^n_{(j)}}\left(\sum_{i=1}^{\infty}\frac{\partial w_{(i)(k)}}{\partial\left(y^n_{(k)} - y^n_{(j)}\right)}V^n_{(i)}\right)
\end{aligned}
\right\}, \quad (5.21a)
$$

$$
\left.
\begin{aligned}
r^{(n)(m)}_{(k)}\left(u^m_{(k)} - u^n_{(k)}, x^m_{(k)} - x^n_{(k)}, t\right) &= \frac{1}{2}\frac{\partial \hat{w}_{(k)}}{\partial\left(y^m_{(k)} - y^n_{(k)}\right)} \\
r^{(m)(n)}_{(k)}\left(u^n_{(k)} - u^m_{(k)}, x^n_{(k)} - x^m_{(k)}, t\right) &= \frac{1}{2}\frac{\partial \hat{w}_{(k)}}{\partial\left(y^n_{(k)} - y^m_{(k)}\right)}
\end{aligned}
\right\}, \quad (5.21b)
$$

and

$$
\left.
\begin{aligned}
&s^{(n)(m)}_{(k)(j)}\left(u^m_{(j)} - u^n_{(k)}, u^m_{(k)} - u^n_{(j)}, x^m_{(j)} - x^n_{(k)}, x^m_{(k)} - x^n_{(j)}, t\right) \\
&= \frac{1}{2}\frac{\partial \tilde{w}_{(k)(j)}}{\partial\left(y^m_{(j)} - y^n_{(k)}\right)} \\
&s^{(m)(n)}_{(j)(k)}\left(u^n_{(k)} - u^m_{(j)}, u^n_{(j)} - u^m_{(k)}, x^n_{(k)} - x^m_{(j)}, x^n_{(j)} - x^m_{(k)}, t\right) \\
&= \frac{1}{2}\frac{\partial \tilde{w}_{(j)(k)}}{\partial\left(y^n_{(k)} - y^m_{(j)}\right)}
\end{aligned}
\right\}, \quad (5.21c)
$$

with $m = (n+1), (n-1)$. As derived in Sect. 2.8, in order to satisfy the balance of angular momentum, the equation of motion, Eq. 5.20 must satisfy the requirements of

$$\int_H \left( \left( \mathbf{y}^n_{(j)} - \mathbf{y}^n_{(k)} \right) \times \mathbf{t}^n_{(k)(j)} \left( \mathbf{u}^n_{(j)} - \mathbf{u}^n_{(k)}, \mathbf{x}^n_{(j)} - \mathbf{x}^n_{(k)}, t \right) \right) dH = 0, \qquad (5.22a)$$

$$\int_H \left( \left( \mathbf{y}^m_{(k)} - \mathbf{y}^n_{(k)} \right) \times \mathbf{r}^{(n)(m)}_{(k)} \left( \mathbf{u}^m_{(k)} - \mathbf{u}^n_{(k)}, \mathbf{x}^m_{(k)} - \mathbf{x}^n_{(k)}, t \right) \right) dH = 0, \qquad (5.22b)$$

$$\int_H \left( \left( \mathbf{y}^m_{(j)} - \mathbf{y}^n_{(k)} \right) \times \mathbf{s}^{(n)(m)}_{(k)(j)} \left( \mathbf{u}^m_{(j)} - \mathbf{u}^n_{(k)}, \mathbf{u}^m_{(k)} - \mathbf{u}^n_{(j)}, \mathbf{x}^m_{(j)} \right. \right.$$

$$\left. \left. -\mathbf{x}^n_{(k)}, \mathbf{x}^m_{(k)} - \mathbf{x}^n_{(j)}, t \right) \right) dH = 0. \qquad (5.22c)$$

It is apparent that these requirements are automatically satisfied if the force vectors, $\mathbf{t}^n_{(k)(j)}$, $\mathbf{r}^{(n)(m)}_{(k)}$, and $\mathbf{s}^{(n)(m)}_{(k)(j)}$, are aligned with the relative position vector of the material points in the deformed state, $(\mathbf{y}^n_{(j)} - \mathbf{y}^n_{(k)})$, $(\mathbf{y}^m_{(k)} - \mathbf{y}^n_{(k)})$, and $(\mathbf{y}^m_{(j)} - \mathbf{y}^n_{(k)})$, respectively. Therefore, they can be expressed in the form

$$\mathbf{t}^n_{(k)(j)} = \frac{1}{2} A^n_{(k)(j)} \frac{\mathbf{y}^n_{(j)} - \mathbf{y}^n_{(k)}}{\left| \mathbf{y}^n_{(j)} - \mathbf{y}^n_{(k)} \right|}, \qquad (5.23a)$$

$$\mathbf{t}^n_{(j)(k)} = -\frac{1}{2} B^n_{(j)(k)} \frac{\mathbf{y}^n_{(j)} - \mathbf{y}^n_{(k)}}{\left| \mathbf{y}^n_{(j)} - \mathbf{y}^n_{(k)} \right|}, \qquad (5.23b)$$

and

$$\mathbf{r}^{(n)(m)}_{(k)} = \frac{1}{2} C^{(n)(m)}_{(k)} \frac{\mathbf{y}^m_{(k)} - \mathbf{y}^n_{(k)}}{\left| \mathbf{y}^m_{(k)} - \mathbf{y}^n_{(k)} \right|} = \frac{1}{2} \mathbf{P}^{(n)(m)}_{(k)}, \qquad (5.24a)$$

$$\mathbf{r}^{(m)(n)}_{(k)} = -\frac{1}{2} C^{(n)(m)}_{(k)} \frac{\mathbf{y}^m_{(k)} - \mathbf{y}^n_{(k)}}{\left| \mathbf{y}^m_{(k)} - \mathbf{y}^n_{(k)} \right|} = -\frac{1}{2} \mathbf{P}^{(n)(m)}_{(k)}, \qquad (5.24b)$$

and

$$\mathbf{s}^{(n)(m)}_{(k)(j)} = \frac{1}{2} D^{(n)(m)}_{(k)(j)} \frac{\mathbf{y}^m_{(j)} - \mathbf{y}^n_{(k)}}{\left| \mathbf{y}^m_{(j)} - \mathbf{y}^n_{(k)} \right|} = \frac{1}{2} \mathbf{q}^{(n)(m)}_{(k)(j)}, \qquad (5.25a)$$

$$\mathbf{s}^{(m)(n)}_{(j)(k)} = -\frac{1}{2} D^{(n)(m)}_{(k)(j)} \frac{\mathbf{y}^m_{(j)} - \mathbf{y}^n_{(k)}}{\left| \mathbf{y}^m_{(j)} - \mathbf{y}^n_{(k)} \right|} = -\frac{1}{2} \mathbf{q}^{(n)(m)}_{(k)(j)}, \qquad (5.25b)$$

where $A_{(k)(j)}^n$, $B_{(j)(k)}^n$, $C_{(k)}^{(n)(m)}$, and $D_{(k)(j)}^{(n)(m)}$ are auxiliary parameters. With these representations of the force density vectors, the equation of motion for material point $\mathbf{x}_{(k)}^n$ located on the $n^{th}$ layer can be further simplified as

$$
\begin{aligned}
\rho_{(k)}^n \ddot{\mathbf{u}}_{(k)}^n =\ & \sum_{j=1}^{\infty} \left[ \mathbf{t}_{(k)(j)}^n \left( \mathbf{u}_{(j)}^n - \mathbf{u}_{(k)}^n, \mathbf{x}_{(j)}^n - \mathbf{x}_{(k)}^n, t \right) \right. \\
& \left. - \mathbf{t}_{(j)(k)}^n \left( \mathbf{u}_{(k)}^n - \mathbf{u}_{(j)}^n, \mathbf{x}_{(k)}^n - \mathbf{x}_{(j)}^n, t \right) \right] V_{(j)}^n \\
& + \sum_{m=n+1,n-1} \mathbf{p}_{(k)}^{(n)(m)} V_{(k)}^m + 2 \sum_{m=n+1,n-1} \sum_{j=1}^{\infty} \mathbf{q}_{(k)(j)}^{(n)(m)} V_{(j)}^m + \mathbf{b}_{(k)}^n .
\end{aligned}
\tag{5.26}
$$

The auxiliary parameters, $A_{(k)(j)}^n$, $B_{(j)(k)}^n$ $C_{(k)}^{(n)(m)}$, and $D_{(k)(j)}^{(n)(m)}$, can be determined by using the relationship between the force density vector and the strain energy density, $W_{(k)}$. The explicit expressions for the auxiliary parameters $A_{(k)(j)}^n$ and $B_{(j)(k)}^n$ are already given by Eqs. 5.9b and 5.10b. The remaining auxiliary parameters, $C_{(k)}^{(n)(m)}$ and $D_{(k)(j)}^{(n)(m)}$, can be determined by using the relationships

$$
\mathbf{r}_{(k)}^{(n)(m)} = \frac{1}{V_{(k)}^m} \frac{\partial \hat{W}_{(k)}^n}{\partial \left( \left| \mathbf{y}_{(k)}^m - \mathbf{y}_{(k)}^n \right| \right)} \frac{\mathbf{y}_{(k)}^m - \mathbf{y}_{(k)}^n}{\left| \mathbf{y}_{(k)}^m - \mathbf{y}_{(k)}^n \right|}
\tag{5.27a}
$$

and

$$
\mathbf{s}_{(k)(j)}^{(n)(m)} = \frac{1}{V_{(j)}^m} \frac{\partial \tilde{W}_{(k)}^n}{\partial \left( \left| \mathbf{y}_{(j)}^m - \mathbf{y}_{(k)}^n \right| \right)} \frac{\mathbf{y}_{(j)}^m - \mathbf{y}_{(k)}^n}{\left| \mathbf{y}_{(j)}^m - \mathbf{y}_{(k)}^n \right|},
\tag{5.27b}
$$

in which $V_{(k)}^m$ and $V_{(j)}^m$ represent the volume of material points $\mathbf{x}_{(k)}^m$ and $\mathbf{x}_{(j)}^m$ respectively, and the direction of the force density vector is aligned with the relative position vector in the deformed configuration. However, determination of the auxiliary parameters requires an explicit form of the strain energy density function. For transverse normal and shear deformations of an isotropic and elastic material (resin-rich layer), the explicit form of the strain energy density functions, $\hat{W}_{(k)}^n$ and $\tilde{W}_{(k)}^n$, can be written as

$$
\hat{W}_{(k)}^n = b_N \sum_{m=n+1,n-1} \frac{\hat{\delta}}{\left| \mathbf{x}_{(k)}^m - \mathbf{x}_{(k)}^n \right|} \left( \left| \mathbf{y}_{(k)}^m - \mathbf{y}_{(k)}^n \right| - \left| \mathbf{x}_{(k)}^m - \mathbf{x}_{(k)}^n \right| \right)^2 V_{(k)}^m
\tag{5.28a}
$$

and

$$\tilde{W}_{(k)}^n = b_S \sum_{m=n+1,n-1} \sum_{j=1}^{\infty} \frac{\tilde{\delta}}{\left|\mathbf{x}_{(j)}^m - \mathbf{x}_{(k)}^n\right|} \left[ \left( \left|\mathbf{y}_{(j)}^m - \mathbf{y}_{(k)}^n\right| - \left|\mathbf{x}_{(j)}^m - \mathbf{x}_{(k)}^n\right| \right) \right.$$

$$\left. - \left( \left|\mathbf{y}_{(k)}^m - \mathbf{y}_{(j)}^n\right| - \left|\mathbf{x}_{(k)}^m - \mathbf{x}_{(j)}^n\right| \right) \right]^2 V_{(j)}^m, \tag{5.28b}$$

in which the PD material parameters $b_N$ and $b_S$ are associated with the transverse normal and shear deformations of the matrix material, but are yet to be determined in terms of Young's modulus and shear modulus. The horizon size in the thickness direction is $\hat{\delta}$, and $\tilde{\delta}$ is defined as $\tilde{\delta} = \sqrt{\delta^2 + \hat{\delta}^2}$. Note that $\left|\mathbf{x}_{(j)}^m - \mathbf{x}_{(k)}^n\right|$ and $\left|\mathbf{x}_{(k)}^m - \mathbf{x}_{(j)}^n\right|$ are equivalent quantities. Substituting for strain energy density from Eqs. 5.28a, b in Eqs. 5.27a, b and performing differentiation result in

$$\mathbf{p}_{(k)}^{(n)(m)} = 4b_N \hat{\delta} \left( \frac{\left|\mathbf{y}_{(k)}^m - \mathbf{y}_{(k)}^n\right| - \left|\mathbf{x}_{(k)}^m - \mathbf{x}_{(k)}^n\right|}{\left|\mathbf{x}_{(k)}^m - \mathbf{x}_{(k)}^n\right|} \right) \frac{\mathbf{y}_{(k)}^m - \mathbf{y}_{(k)}^n}{\left|\mathbf{y}_{(k)}^m - \mathbf{y}_{(k)}^n\right|} \tag{5.29a}$$

and

$$\mathbf{q}_{(k)(j)}^{(n)(m)} = 4b_S \tilde{\delta} \left[ \left( \frac{\left|\mathbf{y}_{(j)}^m - \mathbf{y}_{(k)}^n\right| - \left|\mathbf{x}_{(j)}^m - \mathbf{x}_{(k)}^n\right|}{\left|\mathbf{x}_{(j)}^m - \mathbf{x}_{(k)}^n\right|} \right) \right.$$

$$\left. - \left( \frac{\left|\mathbf{y}_{(k)}^m - \mathbf{y}_{(j)}^n\right| - \left|\mathbf{x}_{(k)}^m - \mathbf{x}_{(j)}^n\right|}{\left|\mathbf{x}_{(k)}^m - \mathbf{x}_{(j)}^n\right|} \right) \right] \frac{\mathbf{y}_{(j)}^m - \mathbf{y}_{(k)}^n}{\left|\mathbf{y}_{(j)}^m - \mathbf{y}_{(k)}^n\right|}. \tag{5.29b}$$

Comparisons of Eqs. 5.24a and 5.29a and 5.25a and 5.29b lead to the determination of $C_{(k)}^{(n)(m)}$ and $D_{(k)(j)}^{(n)(m)}$ as

$$C_{(k)}^{(n)(m)} = 4b_N \hat{\delta} \left( \frac{\left|\mathbf{y}_{(k)}^m - \mathbf{y}_{(k)}^n\right| - \left|\mathbf{x}_{(k)}^m - \mathbf{x}_{(k)}^n\right|}{\left|\mathbf{x}_{(k)}^m - \mathbf{x}_{(k)}^n\right|} \right) \tag{5.30a}$$

$$D_{(k)(j)}^{(n)(m)} = 4b_S \tilde{\delta} \left[ \left( \frac{\left|\mathbf{y}_{(j)}^m - \mathbf{y}_{(k)}^n\right| - \left|\mathbf{x}_{(j)}^m - \mathbf{x}_{(k)}^n\right|}{\left|\mathbf{x}_{(j)}^m - \mathbf{x}_{(k)}^n\right|} \right) - \left( \frac{\left|\mathbf{y}_{(k)}^m - \mathbf{y}_{(j)}^n\right| - \left|\mathbf{x}_{(k)}^m - \mathbf{x}_{(j)}^n\right|}{\left|\mathbf{x}_{(k)}^m - \mathbf{x}_{(j)}^n\right|} \right) \right]. \tag{5.30b}$$

## 5.4   Peridynamic Material Parameters

The peridynamic material parameters that appear in force density vector-stretch relations for in-plane and transverse normal and shear deformations can be determined in terms of engineering material constants of classical laminate theory by considering simple loading conditions.

### 5.4.1   Material Parameters for a Lamina

The PD material parameters, $a$, $d$, $b_F$, $b_T$, and $b_{FT}$, that appear in the force density vector-stretch relations for in-plane deformation of a lamina, Eqs. 5.9b and 5.10b, are related to the engineering constants by considering four different simple loading conditions as

1. Simple shear: $\gamma_{12} = \zeta$
2. Uniaxial stretch in fiber direction: $\varepsilon_{11} = \zeta$, $\varepsilon_{22} = 0$
3. Uniaxial stretch in transverse direction: $\varepsilon_{11} = 0$, $\varepsilon_{22} = \zeta$
4. Biaxial stretch: $\varepsilon_{11} = \zeta$, $\varepsilon_{22} = \zeta$

#### 5.4.1.1   Simple Shear: $\gamma_{12} = \zeta$

Using Eq. 5.1a, the stresses in the lamina due to this loading are obtained as

$$\begin{Bmatrix} \sigma_{11} \\ \sigma_{22} \\ \sigma_{12} \end{Bmatrix} = \begin{bmatrix} Q_{11} & Q_{12} & 0 \\ Q_{12} & Q_{22} & 0 \\ 0 & 0 & Q_{66} \end{bmatrix} \begin{Bmatrix} 0 \\ 0 \\ \zeta \end{Bmatrix} \quad \text{or} \quad \begin{Bmatrix} \sigma_{11} \\ \sigma_{22} \\ \sigma_{12} \end{Bmatrix} = \begin{Bmatrix} 0 \\ 0 \\ Q_{66}\zeta \end{Bmatrix}. \tag{5.31}$$

Based on Eqs. 5.4 and 5.5b, the corresponding dilatation and strain energy density from the classical continuum mechanics at material point $\mathbf{x}_{(k)}^n$ are

$$\theta_{(k)} = 0 \tag{5.32a}$$

and

$$W_{(k)} = \frac{1}{2} Q_{66} \zeta^2. \tag{5.32b}$$

As illustrated in Fig. 5.4, the length of the relative position of material points $\mathbf{y}_{(j)}$ and $\mathbf{y}_{(k)}$ in the deformed state becomes

$$|\mathbf{y}' - \mathbf{y}| = [1 + (\sin\phi\cos\phi)\zeta] \, |\mathbf{x}' - \mathbf{x}| \tag{5.33a}$$

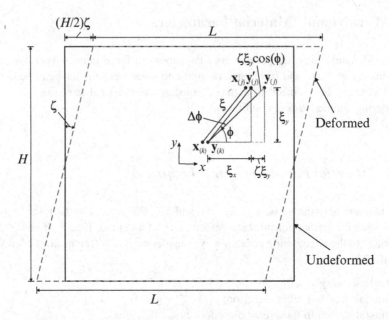

**Fig. 5.4** Simple shear

or

$$\left| \mathbf{y}_{(j)}^n - \mathbf{y}_{(k)}^n \right| = \left[ 1 + \left( \sin \phi_{(j)(k)} \cos \phi_{(j)(k)} \right) \zeta \right] \left| \mathbf{x}_{(j)}^n - \mathbf{x}_{(k)}^n \right|. \qquad (5.33b)$$

Note that if the material points $\mathbf{y}_{(j)}^n$ and $\mathbf{y}_{(k)}^n$ are aligned with the fiber and transverse directions, the angles become $\phi_{(j)(k)} = 0°$ and $\phi_{(j)(k)} = 90°$, respectively. For this deformation, the dilatation, Eq. 5.8, is evaluated as

$$\theta_{(k)} = d \int_H \frac{\delta}{\xi} \{ [1 + (\sin \phi \cos \phi)\zeta]\xi - \xi \} dH, \qquad (5.34)$$

in which $\xi = \left| \mathbf{x}_{(j)}^n - \mathbf{x}_{(k)}^n \right|$.

As expected, this loading condition results in no dilatation. The strain energy density, Eq. 5.7, is evaluated as

$$W_{(k)} = a(0) + b_F(0) + b_{FT} \int_H \frac{\delta}{\xi} ([1 + (\sin \phi \cos \phi)\zeta]\xi - \xi)^2 dH + b_T(0) \qquad (5.35a)$$

or

$$W_{(k)} = b_{FT} h \int_0^\delta \int_0^{2\pi} \frac{\delta}{\xi} ([1 + (\sin \phi \cos \phi)\zeta]\xi - \xi)^2 \xi d\xi d\phi = \frac{\pi h \delta^4 \zeta^2}{12} b_{FT}. \qquad (5.35b)$$

Equating the expressions for strain energy density from the classical and PD formulations, Eqs. 5.32b and 5.35b, results in

$$b_{FT} = \frac{6Q_{66}}{\pi h \delta^4}.$$ (5.36)

### 5.4.1.2   Uniaxial Stretch in the Fiber Direction: $\varepsilon_{11} = \zeta$, $\varepsilon_{22} = 0$

Using Eq. 5.1a, the stresses in the lamina due to this loading becomes

$$\left\{ \begin{array}{c} \sigma_{11} \\ \sigma_{22} \\ \sigma_{12} \end{array} \right\} = \left\{ \begin{array}{c} Q_{11}\zeta \\ Q_{12}\zeta \\ 0 \end{array} \right\}.$$ (5.37)

Based on Eqs. 5.4 and 5.5b, the corresponding dilatation and strain energy density from the classical continuum mechanics at material point $\mathbf{x}^n_{(k)}$ are

$$\theta_{(k)} = \zeta$$ (5.38a)

and

$$W_{(k)} = \frac{1}{2}Q_{11}\zeta^2.$$ (5.38b)

As illustrated in Fig. 5.5, the length of the relative position of material points $\mathbf{y}^n_{(j)}$ and $\mathbf{y}^n_{(k)}$ in the deformed state becomes

$$|\mathbf{y}' - \mathbf{y}| = \left[1 + (\cos^2 \phi)\zeta\right] |\mathbf{x}' - \mathbf{x}|$$ (5.39a)

or

$$\left|\mathbf{y}^n_{(j)} - \mathbf{y}^n_{(k)}\right| = \left[1 + \left(\cos^2 \phi_{(j)(k)}\right)\zeta\right] \left|\mathbf{x}^n_{(j)} - \mathbf{x}^n_{(k)}\right|.$$ (5.39b)

Due to this deformation, the dilatation is evaluated as

$$\theta_{(k)} = d \int_H \frac{\delta}{\xi} \left\{\left[1 + (\cos^2 \phi)\zeta\right]\xi - \xi\right\} dH$$ (5.40a)

or

$$\theta_{(k)} = \frac{\pi d h \delta^3 \zeta}{2}.$$ (5.40b)

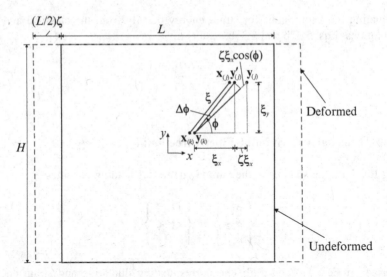

**Fig. 5.5** Uniaxial stretch in fiber direction

Equating the expressions for dilatation from the classical and PD formulations, Eqs. 5.38a and 5.40b, results in

$$d = \frac{2}{\pi h \delta^3}. \tag{5.41}$$

The strain energy density for this deformation is evaluated as

$$W_{(k)} = a\zeta^2 + b_F \sum_{j=1}^{J} \frac{\delta}{\left|\mathbf{x}_{(j)}^n - \mathbf{x}_{(k)}^n\right|} \left(\left(\cos^2 \phi_{(j)(k)}\right)\zeta \left|\mathbf{x}_{(j)}^n - \mathbf{x}_{(k)}^n\right|\right)^2 V_{(j)}^n$$

$$+ b_{FT} \int_H \frac{\delta}{\xi} \left(\left[1 + (\cos^2 \phi)\zeta\right] \xi - \xi\right)^2 dH + b_T(0) \tag{5.42a}$$

or

$$W_{(k)} = a\zeta^2 + b_F\, \delta\zeta^2 \sum_{j=1}^{J} \left(\left|\mathbf{x}_{(j)}^n - \mathbf{x}_{(k)}^n\right|\right) V_{(j)} + \frac{\pi h \delta^4 \zeta^2}{4} b_{FT}. \tag{5.42b}$$

After substituting for $b_{FT}$ from Eq. 5.36, it takes the final form

$$W_{(k)} = a\zeta^2 + b_F\, \delta\zeta^2 \left(\sum_{j=1}^{J} \left|\mathbf{x}_{(j)}^n - \mathbf{x}_{(k)}^n\right| V_{(j)}\right) + \frac{3Q_{66}\zeta^2}{2}. \tag{5.43}$$

Equating the expressions for strain energy density from the classical and PD formulations, Eqs. 5.38b and 5.43, results in

$$a + \delta \left( \sum_{j=1}^{J} \left| \mathbf{x}_{(j)}^n - \mathbf{x}_{(k)}^n \right| V_{(j)} \right) b_F = \frac{1}{2} (Q_{11} - 3Q_{66}). \qquad (5.44)$$

### 5.4.1.3   Uniaxial Stretch in the Transverse Direction: $\varepsilon_{11} = 0, \ \varepsilon_{22} = \zeta$

Using Eq. 5.1, the stresses in the lamina due to this loading become

$$\begin{Bmatrix} \sigma_{11} \\ \sigma_{22} \\ \sigma_{12} \end{Bmatrix} = \begin{Bmatrix} Q_{12}\zeta \\ Q_{22}\zeta \\ 0 \end{Bmatrix}. \qquad (5.45)$$

Based on Eqs. 5.4 and 5.5b, the corresponding dilatation and strain energy density from classical continuum mechanics at material point $\mathbf{x}_{(k)}$ are

$$\theta_{(k)} = \zeta, \qquad (5.46a)$$

$$W_{(k)} = \frac{1}{2} Q_{22} \zeta^2. \qquad (5.46b)$$

As illustrated in Fig. 5.6, the length of the relative position of material points $\mathbf{y}_{(j)}$ and $\mathbf{y}_{(k)}$ in the deformed state becomes

$$|\mathbf{y}' - \mathbf{y}| = \left[ 1 + (\sin^2 \phi) \zeta \right] |\mathbf{x}' - \mathbf{x}| \qquad (5.47a)$$

or

$$\left| \mathbf{y}_{(j)}^n - \mathbf{y}_{(k)}^n \right| = \left[ 1 + \left( \sin^2 \phi_{(j)(k)} \right) \zeta \right] \left| \mathbf{x}_{(j)}^n - \mathbf{x}_{(k)}^n \right|. \qquad (5.47b)$$

For this deformation, the dilatation is evaluated as

$$\theta_{(k)} = d \int_{H} \frac{\delta}{\xi} \left( \left[ 1 + (\sin^2 \phi) \zeta \right] \xi - \xi \right) dH \qquad (5.48a)$$

or

$$\theta_{(k)} = \frac{\pi d h \delta^3 \zeta}{2}. \qquad (5.48b)$$

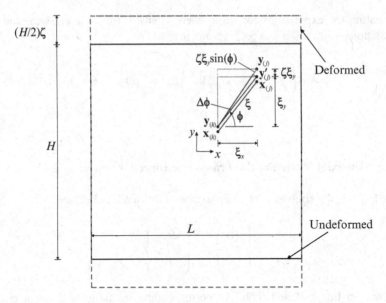

**Fig. 5.6** Uniaxial stretch in transverse direction

Equating the expressions for dilatation from the classical and PD formulations, Eqs. 5.46a and 5.48b, results in

$$d = \frac{2}{\pi h \delta^3}.$$ (5.49)

As expected, the PD parameter $d$ obtained from the uniform stretch in the fiber direction, Eq. 5.41, and that in the transverse direction, Eq, 5.49, are equal to each other and are independent of material properties.

The strain energy density for this deformation is evaluated as

$$W_{(k)} = a\,\zeta^2 + b_F(0) + b_{FT} \int_H \frac{\delta}{\xi} \left( \left[ 1 + (\sin^2 \phi)\zeta \right] \xi - \xi \right)^2 dH$$

$$+ b_T\,\delta\zeta^2 \left( \sum_{j=1}^{J} \left| \mathbf{x}_{(j)}^n - \mathbf{x}_{(k)}^n \right| V_{(j)}^n \right)$$ (5.50a)

or

$$W_{(k)} = a\,\zeta^2 + b_{FT} \frac{\pi h \delta^4 \zeta^2}{4} + b_T\,\delta\zeta^2 \left( \sum_{j=1}^{J} \left| \mathbf{x}_{(j)}^n - \mathbf{x}_{(k)}^n \right| V_{(j)}^n \right).$$ (5.50b)

After substituting for $b_{FT}$ from Eq. 5.36, it takes the final form

$$W_{(k)} = a\,\zeta^2 + \frac{3Q_{66}\zeta^2}{2} + b_T\,\delta\zeta^2\left(\sum_{j=1}^{J}\left|\mathbf{x}_{(j)}^n - \mathbf{x}_{(k)}^n\right|V_{(j)}^n\right). \tag{5.51}$$

Equating the expressions for strain energy density from the classical and PD formulations, Eqs. 5.46b and 5.51, results in

$$\frac{1}{2}(Q_{22} - 3Q_{66}) = a + \delta\left(\sum_{j=1}^{J}\left|\mathbf{x}_{(j)}^n - \mathbf{x}_{(k)}^n\right|V_{(j)}\right)b_T. \tag{5.52}$$

### 5.4.1.4   Biaxial Stretch: $\varepsilon_{11} = \zeta$, $\varepsilon_{22} = \zeta$

Using Eq. 5.1a, the stresses in the lamina due to this loading become

$$\left\{\begin{array}{c}\sigma_{11}\\\sigma_{22}\\\sigma_{12}\end{array}\right\} = \begin{bmatrix}Q_{11}&Q_{12}&0\\Q_{12}&Q_{22}&0\\0&0&Q_{66}\end{bmatrix}\left\{\begin{array}{c}\zeta\\\zeta\\0\end{array}\right\} \text{ or } \left\{\begin{array}{c}\sigma_{11}\\\sigma_{22}\\\sigma_{12}\end{array}\right\} = \left\{\begin{array}{c}(Q_{11}+Q_{12})\zeta\\(Q_{12}+Q_{22})\zeta\\0\end{array}\right\}. \tag{5.53}$$

Based on Eqs. 5.4 and 5.5b, the corresponding dilatation and strain energy density from classical continuum mechanics at material point $\mathbf{x}_{(k)}$ are

$$\theta_{(k)} = 2\zeta \tag{5.54a}$$

and

$$W_{(k)} = \frac{1}{2}(Q_{11} + 2Q_{12} + Q_{22})\zeta^2. \tag{5.54b}$$

As illustrated in Fig. 5.7, the length of the relative position of material points $\mathbf{y}_{(j)}$ and $\mathbf{y}_{(k)}$ in the deformed state becomes

$$|\mathbf{y}' - \mathbf{y}| = \left[1 + (\cos^2\phi + \sin^2\phi)\zeta\right]|\mathbf{x}' - \mathbf{x}| \tag{5.55a}$$

or

$$\left|\mathbf{y}_{(j)}^n - \mathbf{y}_{(k)}^n\right| = \left[1 + \left(\cos^2\phi_{(j)(k)} + \sin^2\phi_{(j)(k)}\right)\zeta\right]\left|\mathbf{x}_{(j)}^n - \mathbf{x}_{(k)}^n\right|. \tag{5.55b}$$

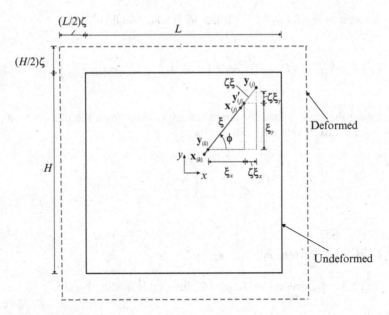

**Fig. 5.7** Biaxial stretch

For this deformation, the dilatation is evaluated as

$$\theta_{(k)} = d \int_H \frac{\delta}{\xi} ([1+\zeta]\,\xi - \xi) dH \tag{5.56a}$$

or

$$\theta_{(k)} = \pi d h \delta^3 \zeta. \tag{5.56b}$$

Equating the dilatation contributions from the classical and PD formulations, Eqs. 5.54a and 5.56b, also results in the same value of the PD parameter

$$d = \frac{2}{\pi h \delta^3}. \tag{5.57}$$

For this deformation given by Eq. 5.55b, the strain energy density is evaluated as

$$\begin{aligned} W_{(k)} = {} & 4a\,\zeta^2 + b_F\,\zeta^2\delta \left( \sum_{j=1}^{J} \left( \left| \mathbf{x}_{(j)}^n - \mathbf{x}_{(k)}^n \right| \right) V_{(j)}^n \right) \\ & + b_{FT}\frac{2\pi h \delta^4 \zeta^2}{3} + b_T \zeta^2 \delta \left( \sum_{j=1}^{J} \left( \left| \mathbf{x}_{(j)}^n - \mathbf{x}_{(k)}^n \right| \right) V_{(j)}^n \right). \end{aligned} \tag{5.58}$$

After substituting for $b_{FT}$ from Eq. 5.36, it takes the final form

$$W_{(k)} = 4a\,\zeta^2 + b_F\,\zeta^2\delta\left(\sum_{j=1}^{J}\left(\left|\mathbf{x}_{(j)}^n - \mathbf{x}_{(k)}^n\right|\right)V_{(j)}^n\right)$$
$$+ 4Q_{66}\zeta^2 + b_T\zeta^2\delta\left(\sum_{j=1}^{J}\left(\left|\mathbf{x}_{(j)}^n - \mathbf{x}_{(k)}^n\right|\right)V_{(j)}\right). \tag{5.59}$$

Equating the expressions for strain energy density from the classical and PD formulations, Eqs. 5.54b and 5.59, results in

$$\frac{1}{2}(Q_{11} + 2Q_{12} + Q_{22} - 8Q_{66}) = 4a + b_F\,\delta\left(\sum_{j=1}^{J}\left(\left|\mathbf{x}_{(j)}^n - \mathbf{x}_{(k)}^n\right|\right)V_{(j)}^n\right)$$
$$+ b_T\delta\left(\sum_{j=1}^{J}\left(\left|\mathbf{x}_{(j)}^n - \mathbf{x}_{(k)}^n\right|\right)V_{(j)}^n\right). \tag{5.60}$$

The remaining peridynamic parameters in the strain energy density expression can now be evaluated by using the previous two relations obtained from the uniform stretch in the fiber and transverse directions, Eqs. 5.44 and 5.52, in conjunction with Eq. 5.60, as

$$a = \frac{1}{2}(Q_{12} - Q_{66}), \tag{5.61a}$$

$$b_F = \frac{(Q_{11} - Q_{12} - 2Q_{66})}{2\delta\left(\sum_{j=1}^{N}\left|\mathbf{x}_{(j)}^n - \mathbf{x}_{(k)}^n\right|V_{(j)}^n\right)}, \tag{5.61b}$$

$$b_T = \frac{(Q_{22} - Q_{12} - 2Q_{66})}{2\delta\left(\sum_{j=1}^{N}\left|\mathbf{x}_{(j)}^n - \mathbf{x}_{(k)}^n\right|V_{(j)}^n\right)}, \tag{5.61c}$$

$$b_{FT} = \frac{6Q_{66}}{\pi h\delta^4}. \tag{5.61d}$$

For bond-based peridynamics, the parameter $a$ associated with dilatation and the parameter $b_T$ associated with the transverse direction should both vanish, thus leading to constraint equations, previously derived by Oterkus and Madenci (2012), as

$$Q_{12} = Q_{66} \quad \text{and} \quad Q_{22} = 3Q_{12}. \tag{5.62}$$

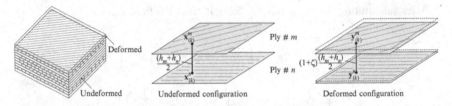

**Fig. 5.8**  A composite laminate subjected to transverse normal stretch

The nonvanishing peridynamic parameters, $b_F$ and $b_{FT}$ in the fiber and remaining directions, respectively, also recover the expressions derived by Oterkus and Madenci (2012) as

$$b_F = \frac{(Q_{11} - Q_{22})}{2\delta\left(\sum\limits_{j=1}^{N}\left|\mathbf{x}_{(j)}^n - \mathbf{x}_{(k)}^n\right|V_{(j)}\right)} \quad \text{and} \quad b_{FT} = \frac{6Q_{66}}{\pi h \delta^4}. \tag{5.63}$$

For isotropic materials with $Q_{11} = Q_{22} = \kappa + \mu$, $Q_{12} = (\kappa - \mu)$, and $Q_{66} = \mu$, these peridynamic parameters recover Eqs. 4.52 and 4.53 as

$$a = \frac{1}{2}(\kappa - 2\mu), \ b_F = 0, \quad b_T = 0 \text{ and } \quad b_{FT} = b = \frac{6\mu}{\pi h \delta^4}, \tag{5.64}$$

and the parameter $d$ is also equal to that of isotropic material given by Eq. 4.47.

### 5.4.2  Material Parameters for Transverse Deformation

The peridynamic material parameters $b_N$ and $b_S$ in the force density vector-stretch relations, Eqs. 5.29a, b associated with transverse deformation in a laminate are determined by considering two simple loading conditions as

1. Transverse normal stretch: $\varepsilon_{33} = \zeta$
2. Simple transverse shear: $\gamma_{13} = \zeta$

#### 5.4.2.1  Transverse Normal Stretch: $\varepsilon_{33} = \zeta$

In order to obtain the peridynamic material parameter $b_N$, the laminate is subjected to a uniform transverse normal strain of $\zeta$, as shown in Fig. 5.8. The corresponding strain energy density from the classical continuum mechanics at material point $\mathbf{x}_{(k)}$ is

$$\hat{W}_{(k)} = \frac{1}{2}E_m \zeta^2, \tag{5.65}$$

with $E_m$ representing the Young's modulus of matrix material.

The relative distance between the material points at $\mathbf{x}_{(k)}^m$ and $\mathbf{x}_{(k)}^n$, before and after deformation, can be expressed as

$$\left|\mathbf{x}_{(k)}^m - \mathbf{x}_{(k)}^n\right| = \frac{1}{2}(h_m + h_n) \tag{5.66a}$$

and

$$\left|\mathbf{y}_{(k)}^m - \mathbf{y}_{(k)}^n\right| = (1 + \zeta)\left|\mathbf{x}_{(k)}^m - \mathbf{x}_{(k)}^n\right|. \tag{5.66b}$$

Defining $\boldsymbol{\xi} = \mathbf{x}_{(k)}^m - \mathbf{x}_{(k)}^n$ and noting that its length is equal to half of the sum of the two neighboring ply thicknesses, i.e., $\xi = |\boldsymbol{\xi}| = (h_m + h_n)/2$, with $m = (n+1)$, $(n-1)$, and substituting for the relative position vector, from Eq. 5.66a, in the expression for the strain energy density, $\hat{W}_{(k)}$, Eq. 5.28a, at material point $\mathbf{x}_{(k)}^n$ result in

$$\hat{W}_{(k)}^n = \frac{1}{2}\zeta^2 b_N \hat{\delta}\left[(h_{n+1} + h_n)V_{(k)}^{n+1} + (h_{n-1} + h_n)V_{(k)}^{n-1}\right]. \tag{5.67}$$

Equating the expressions for strain energy density from Eqs. 5.65 and 5.67 provides the relationship between the PD parameters, $b_N$, and the Young's modulus of the matrix material as

$$b_N = \frac{E_m}{\hat{\delta}\left[(h_{n+1} + h_n)V_{(k)}^{n+1} + (h_{n-1} + h_n)V_{(k)}^{n-1}\right]}. \tag{5.68}$$

#### 5.4.2.2 Simple Transverse Shear: $\gamma_{13} = \zeta$

Similarly, the peridynamic material parameter $b_S$ is evaluated by subjecting the laminate to a simple transverse shear loading of $\zeta$, as shown in Fig. 5.9. The corresponding strain energy density from classical continuum mechanics at material point $\mathbf{x}_{(k)}$ is

$$\tilde{W}_{(k)} = \frac{1}{2}G_m\zeta^2, \tag{5.69}$$

with $G_m$ representing the shear modulus of matrix material.

As shown in Fig. 5.10, the relative distance between the material points at $\mathbf{x}_{(j)}^m$ and $\mathbf{x}_{(k)}^n$, before and after deformation, can be expressed as

$$\left|\mathbf{x}_{(j)}^m - \mathbf{x}_{(k)}^n\right| = \sqrt{\ell^2 + \frac{(h_m + h_n)^2}{4}} \tag{5.70a}$$

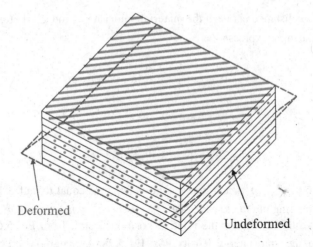

**Fig. 5.9**  A composite laminate subjected to simple transverse shear

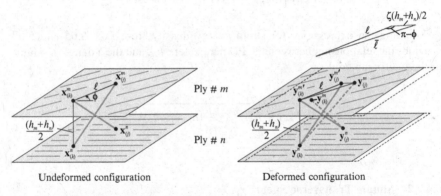

Undeformed configuration                        Deformed configuration

**Fig. 5.10**  Position of material points before and after deformation due to simple transverse shear

$$\left| \mathbf{y}_{(j)}^m - \mathbf{y}_{(k)}^n \right| = \sqrt{\bar{\ell}^2 + \frac{(h_m + h_n)^2}{4}}, \tag{5.70b}$$

in which $\bar{\ell}$ can be obtained from the law of cosines as

$$\bar{\ell}^2 = \ell^2 + \zeta^2 \frac{(h_m + h_n)^2}{4} - \ell\zeta(h_m + h_n)\cos(\pi - \phi). \tag{5.71}$$

Thus, the distance between $\mathbf{x}_{(j)}^m$ and $\mathbf{x}_{(k)}^n$ in the deformed state can be rewritten as

$$\left| \mathbf{y}_{(j)}^m - \mathbf{y}_{(k)}^n \right| = \sqrt{\left( \ell^2 + \frac{(h_m + h_n)^2}{4} \right) + \ell\zeta(h_m + h_n)\cos(\phi)}. \tag{5.72}$$

In deriving this expression, the $\zeta^2(h_m + h_n)^2/4$ term is disregarded with respect to $(h_m + h_n)^2/4$ because $\zeta$ is much less than unity. Also, this expression can be further simplified by using the square root approximation because $\ell\zeta(h_m + h_n)\cos(\phi) \ll (\ell^2 + (h_m + h_n)^2/4)$, leading to

$$\left| \mathbf{y}_{(j)}^m - \mathbf{y}_{(k)}^n \right| = \sqrt{\ell^2 + \frac{(h_m + h_n)^2}{4}} + \frac{\ell\zeta(h_m + h_n)\cos(\phi)}{2\sqrt{\ell^2 + \frac{(h_m+h_n)^2}{4}}}. \tag{5.73}$$

Thus, the extension between these material points is obtained as

$$\left| \mathbf{y}_{(j)}^m - \mathbf{y}_{(k)}^n \right| - \left| \mathbf{x}_{(j)}^m - \mathbf{x}_{(k)}^n \right| = \frac{\ell\zeta(h_m + h_n)\cos(\phi)}{2\sqrt{\ell^2 + \frac{(h_m+h_n)^2}{4}}}. \tag{5.74}$$

Similarly, the distance between the material points $\mathbf{x}_{(k)}^m$ and $\mathbf{x}_{(j)}^n$ before and after deformation can be obtained as

$$\left| \mathbf{x}_{(k)}^m - \mathbf{x}_{(j)}^n \right| = \sqrt{\ell^2 + \frac{(h_m + h_n)^2}{4}} \tag{5.75a}$$

and

$$\left| \mathbf{y}_{(k)}^m - \mathbf{y}_{(j)}^n \right| = \sqrt{\ell^2 + \frac{(h_m + h_n)^2}{4}} - \frac{\ell\zeta(h_m + h_n)\cos(\phi)}{2\sqrt{\ell^2 + \frac{(h_m+h_n)^2}{4}}}, \tag{5.75b}$$

in which the minus sign emerges due to the contraction between material points $\mathbf{x}_{(k)}^m$ and $\mathbf{x}_{(j)}^n$ in the deformed state, whereas extension occurs between material points $\mathbf{x}_{(j)}^m$ and $\mathbf{x}_{(k)}^n$. Thus, the contraction between these material points is obtained as

$$\left| \mathbf{y}_{(k)}^m - \mathbf{y}_{(j)}^n \right| - \left| \mathbf{x}_{(k)}^m - \mathbf{x}_{(j)}^n \right| = -\frac{\ell\zeta(h_m + h_n)\cos(\phi)}{2\sqrt{\ell^2 + \frac{(h_m+h_n)^2}{4}}}. \tag{5.76}$$

Prior to substituting for the stretch between the material points $\mathbf{x}_{(j)}^m$ and $\mathbf{x}_{(k)}^n$ and $\mathbf{x}_{(k)}^m$ and $\mathbf{x}_{(j)}^n$, the strain energy expression can be rewritten in a slightly different form as

$$\tilde{W}_{(k)}^n = b_S \sum_{m=n+1,n-1} \left( \frac{h_m + h_n}{2} \right)^2$$

$$\times \sum_{j=1}^{\infty} \frac{\tilde{\delta}}{\left| \mathbf{x}_{(j)}^m - \mathbf{x}_{(k)}^n \right|} \left[ \frac{\left| \mathbf{y}_{(j)}^m - \mathbf{y}_{(k)}^n \right| - \left| \mathbf{x}_{(j)}^m - \mathbf{x}_{(k)}^n \right|}{\left(\frac{h_m+h_n}{2}\right)} - \frac{\left| \mathbf{y}_{(k)}^m - \mathbf{y}_{(j)}^n \right| - \left| \mathbf{x}_{(k)}^m - \mathbf{x}_{(j)}^n \right|}{\left(\frac{h_m+h_n}{2}\right)} \right]^2 V_{(j)}^m,$$

$$\tag{5.77}$$

Fig. 5.11 Change in angle
after deformation

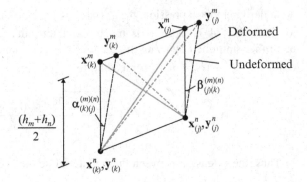

in which the ratios in the summation can be interpreted as the change in angle from $\pi/2$ provided that $\left|\mathbf{x}_{(j)}^m - \mathbf{x}_{(k)}^n\right| \gg h$ and $\left|\mathbf{x}_{(k)}^m - \mathbf{x}_{(j)}^n\right| \gg h$, as depicted in Fig. 5.11. With this interpretation, this expression can be rewritten as

$$\tilde{W}_{(k)}^n = b_S \sum_{m=n+1,n-1} \left(\frac{h_m + h_n}{2}\right)^2 \sum_{j=1}^{\infty} \frac{\tilde{\delta}}{\left|\mathbf{x}_{(j)}^m - \mathbf{x}_{(k)}^n\right|} \left[\alpha_{(k)(j)}^{(m)(n)} + \beta_{(j)(k)}^{(m)(n)}\right]^2 V_{(j)}^m, \quad (5.78)$$

with

$$\alpha_{(k)(j)}^{(m)(n)} = \frac{\left|\mathbf{y}_{(j)}^m - \mathbf{y}_{(k)}^n\right| - \left|\mathbf{x}_{(j)}^m - \mathbf{x}_{(k)}^n\right|}{\left(\frac{h_m + h_n}{2}\right)} \qquad (5.79a)$$

$$\beta_{(j)(k)}^{(m)(n)} = -\frac{\left|\mathbf{y}_{(k)}^m - \mathbf{y}_{(j)}^n\right| - \left|\mathbf{x}_{(k)}^m - \mathbf{x}_{(j)}^n\right|}{\left(\frac{h_m + h_n}{2}\right)}. \qquad (5.79b)$$

The average change in angle, $\varphi_{(j)(k)}^{(m)(n)}$, corresponding to the shear strain in classical continuum mechanics becomes

$$\begin{aligned}
\varphi_{(k)(j)}^{(m)(n)} &= \frac{\alpha_{(k)(j)}^{(m)(n)} + \beta_{(j)(k)}^{(m)(n)}}{2} \\
&= \frac{\left(\left|\mathbf{y}_{(j)}^m - \mathbf{y}_{(k)}^n\right| - \left|\mathbf{x}_{(j)}^m - \mathbf{x}_{(k)}^n\right|\right) - \left(\left|\mathbf{y}_{(k)}^m - \mathbf{y}_{(j)}^n\right| - \left|\mathbf{x}_{(k)}^m - \mathbf{x}_{(j)}^n\right|\right)}{(h_m + h_n)}.
\end{aligned} \qquad (5.80)$$

Substituting for the stretch between the material points, $\mathbf{x}_{(j)}^m$ and $\mathbf{x}_{(k)}^n$ and $\mathbf{x}_{(k)}^m$ and $\mathbf{x}_{(j)}^n$, the average change in angle, $\varphi_{(k)(j)}^{(m)(n)}$, for the applied simple shear loading can be determined as

$$\varphi^{(m)(n)}_{(k)(j)} = \frac{\ell\zeta\,\cos(\phi)}{\sqrt{\ell^2 + \frac{(h_m+h_n)^2}{4}}}.$$

(5.81)

Therefore, the strain energy density function can be rewritten in terms of the average change in angle as

$$\tilde{W}^n_{(k)} = 4b_S \sum_{m=n+1,n-1} \left(\frac{h_m + h_n}{2}\right)^2 \sum_{j=1}^{\infty} \frac{\tilde{\delta}}{\left|\mathbf{x}^m_{(j)} - \mathbf{x}^n_{(k)}\right|} \left[\varphi^{(m)(n)}_{(k)(j)}\right]^2 V^m_{(j)}$$

(5.82a)

or

$$\tilde{W}^n_{(k)} = 4b_S \left( \left(\frac{h_{n+1} + h_n}{2}\right)^2 \sum_{j=1}^{\infty} \frac{\tilde{\delta}}{\left|\mathbf{x}^{n+1}_{(j)} - \mathbf{x}^n_{(k)}\right|} \left[\varphi^{(n+1)(n)}_{(k)(j)}\right]^2 V^{n+1}_{(j)} \right.$$

$$\left. + \left(\frac{h_{n-1} + h_n}{2}\right)^2 \sum_{j=1}^{\infty} \frac{\tilde{\delta}}{\left|\mathbf{x}^{n-1}_{(j)} - \mathbf{x}^n_{(k)}\right|} \left[\varphi^{(n-1)(n)}_{(k)(j)}\right]^2 V^{n-1}_{(j)} \right)$$

(5.82b)

or

$$\tilde{W}^n_{(k)} = 4\zeta^2 b_S \tilde{\delta} \left( \left(\frac{h_{n+1} + h_n}{2}\right)^2 \sum_{j=1}^{\infty} \frac{\ell^2 \cos^2(\phi)}{\left[\ell^2 + \left(\frac{h_{n+1}+h_n}{2}\right)^2\right]^{3/2}} V^{n+1}_{(j)} \right.$$

$$\left. + \left(\frac{h_{n-1} + h_n}{2}\right)^2 \sum_{j=1}^{\infty} \frac{\ell^2 \cos^2(\phi)}{\left[\ell^2 + \left(\frac{h_{n-1}+h_n}{2}\right)^2\right]^{3/2}} V^{n-1}_{(j)} \right).$$

(5.82c)

Converting summation to integration leads to

$$\tilde{W}^n_{(k)} = 4\zeta^2 b_S \tilde{\delta} \left( \left(\frac{h_{n+1} + h_n}{2}\right)^3 \int_0^\delta \int_0^{2\pi} \frac{\ell^2 \cos^2(\phi)}{\left[\ell^2 + \left(\frac{h_{n+1}+h_n}{2}\right)^2\right]^{3/2}} \ell d\ell d\phi \right.$$

$$\left. + \left(\frac{h_{n-1} + h_n}{2}\right)^3 \int_0^\delta \int_0^{2\pi} \frac{\ell^2 \cos^2(\phi)}{\left[\ell^2 + \left(\frac{h_{n-1}+h_n}{2}\right)^2\right]^{3/2}} \ell d\ell d\phi \right).$$

(5.83)

Performing the integration results in

$$
\begin{aligned}
\tilde{W}^n_{(k)} = 4\zeta^2 b_S \pi \tilde{\delta} \Bigg( &\left(\frac{h_{n+1}+h_n}{2}\right)^3 \left(\frac{\delta^2 + 2\left(\frac{h_{n+1}+h_n}{2}\right)^2}{\sqrt{\delta^2 + \left(\frac{h_{n+1}+h_n}{2}\right)^2}} - (h_{n+1}+h_n)\right) \\
&+ \left(\frac{h_{n-1}+h_n}{2}\right)^3 \left(\frac{\delta^2 + 2\left(\frac{h_{n-1}+h_n}{2}\right)^2}{\sqrt{\delta^2 + \left(\frac{h_{n-1}+h_n}{2}\right)^2}} - (h_{n-1}+h_n)\right) \Bigg) .
\end{aligned}
\tag{5.84}
$$

Equating the expressions for strain energy density from Eqs. 5.69 and 5.84 provides the relationship between the PD parameter $b_S$ and the shear modulus of the matrix material as

$$
b_S = \frac{G_m}{8\pi\tilde{\delta}\left(\left(\frac{h_{n+1}+h_n}{2}\right)^3 \left(\frac{\delta^2 + 2\left(\frac{h_{n+1}+h_n}{2}\right)^2}{\sqrt{\delta^2 + \left(\frac{h_{n+1}+h_n}{2}\right)^2}} - (h_{n+1}+h_n)\right) + \left(\frac{h_{n-1}+h_n}{2}\right)^3 \left(\frac{\delta^2 + 2\left(\frac{h_{n-1}+h_n}{2}\right)^2}{\sqrt{\delta^2 + \left(\frac{h_{n-1}+h_n}{2}\right)^2}} - (h_{n-1}+h_n)\right)\right)} .
\tag{5.85}
$$

## 5.5　Surface Effects

The peridynamic material parameters $a$, $d$, $b_F$, $b_T$, $b_{FT}$, $b_N$, and $b_S$ that appear in the peridynamic force-stretch relations are determined by computing both dilatation and strain energy density of a material point whose horizon is completely embedded in the material. The values of these parameters, except for $a$, depend on the accuracy of integration and domain of integration defined by the horizon. Therefore, the values of these parameters will be different for a material point located near a boundary, Fig. 5.12. Thus, these parameters need to be corrected near the free surfaces.

Since the presence of free surfaces is problem dependent, it is impractical to resolve this issue analytically. The correction of the material parameters is achieved by numerically integrating both dilatation and strain energy density at each material point inside the body for simple loading conditions and comparing them to their counterparts obtained from classical continuum mechanics. After determining the correction factor for each parameter, the force density vector is modified in the PD equations of motion.

**Fig. 5.12** Surface effects in the domain of interest

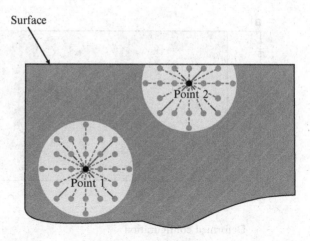

In order to determine the surface correction factors for the peridynamic parameters $d$ and $b_\ell$ $(\ell = F, T, FT)$, two simple loading conditions are achieved by applying uniaxial stretch first in the fiber direction, and then in the transverse direction, i.e., $\varepsilon_{11} \neq 0$, $\varepsilon_{22} = \gamma_{12} = 0$ (shown in Fig. 5.13) and $\varepsilon_{22} \neq 0$, $\varepsilon_{11} = \gamma_{12} = 0$. The fiber and transverse directions coincide with the axes of the natural (material) coordinate system, $(1, 2)$.

The applied uniaxial stretch in the fiber and transverse directions is achieved through a constant displacement gradient, $\partial u_\alpha^* / \partial x_\alpha = \zeta$ with $(\alpha = 1, 2)$. The displacement field at material point $\mathbf{x}$ arising from these two loading conditions can be expressed as

$$\mathbf{u}_1^T(\mathbf{x}) = \left\{ \frac{\partial u_1^*}{\partial x_1} x_1 \quad 0 \right\} \quad \text{and} \quad \mathbf{u}_2^T(\mathbf{x}) = \left\{ 0 \quad \frac{\partial u_2^*}{\partial x_2} x_2 \right\}. \tag{5.86a,b}$$

Due to these displacement fields, the peridynamic dilatation term, $\theta_\alpha^{PD}(\mathbf{x}_{(i)}^n)$, at material point $\mathbf{x}_{(i)}^n$ can be obtained from Eq. 5.8 as

$$\theta_\alpha^{PD}(\mathbf{x}_{(i)}^n) = d \sum_{j=1}^{N} \frac{\delta}{\left| \mathbf{x}_{(j)}^n - \mathbf{x}_{(i)}^n \right|} \left( \left| \mathbf{y}_{(j)}^n - \mathbf{y}_{(i)}^n \right| - \left| \mathbf{x}_{(j)}^n - \mathbf{x}_{(i)}^n \right| \right) \Lambda_{(i)(j)}^n V_{(j)}, \tag{5.87}$$

in which $N$ represents the number of material points inside the horizon of material point $\mathbf{x}_{(i)}^n$. The corresponding dilatation based on classical continuum mechanics, $\theta_\alpha^{CM}(\mathbf{x}_{(i)}^n)$, is uniform throughout the domain, and is determined as

$$\theta_\alpha^{CM}(\mathbf{x}_{(i)}^n) = \varepsilon_{\alpha\alpha} = \zeta, \quad \text{with } (\alpha = 1, 2), \tag{5.88}$$

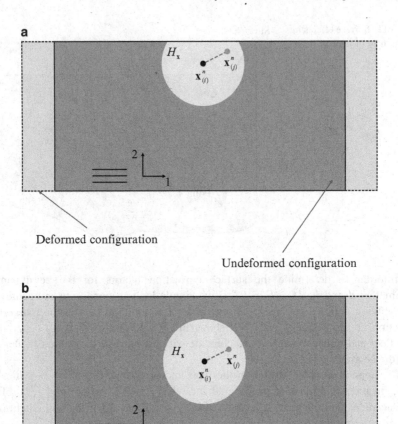

**Fig. 5.13** Material point **x** in lamina subjected to uniaxial stretch: (**a**) a truncated horizon, and (**b**) far away from external surfaces

The dilatation correction term can be defined as

$$D_{\alpha(i)} = \frac{\theta_\alpha^{CM}(\mathbf{x}_{(i)}^n)}{\theta_\alpha^{PD}(\mathbf{x}_{(i)}^n)} = \frac{\zeta}{d\,\delta\sum\limits_{j=1}^{N} s_{(i)(j)}^n \Lambda_{(i)(j)}^n V_{(j)}^n}. \qquad (5.89)$$

Maximum values of dilatation occur in the loading directions that coincide with the natural coordinates 1 and 2, respectively.

The peridynamic strain energy density at material point $\mathbf{x}_{(i)}^n$ can be obtained from Eq. 5.7 as

$$W_\alpha^{PD}(\mathbf{x}_{(i)}^n) = W_{\alpha\theta}^{PD}(\mathbf{x}_{(i)}^n) + W_{\alpha F}^{PD}(\mathbf{x}_{(i)}^n) + W_{\alpha FT}^{PD}(\mathbf{x}_{(i)}^n) + W_{\alpha T}^{PD}(\mathbf{x}_{(i)}^n), \tag{5.90}$$

where $(\alpha = 1, 2)$, $W_{\alpha\theta}^{PD}$ is associated with the dilatation term, and $W_{\alpha F}^{PD}$, $W_{\alpha T}^{PD}$, and $W_{\alpha FT}^{PD}$ represent contributions from the deformation in the fiber direction, transverse direction, and arbitrary directions, respectively. Based on Eq. 5.7, each of these terms is expressed as

$$W_{\alpha\theta}^{PD}(\mathbf{x}_{(i)}^n) = a\left(\theta_\alpha^{PD}(\mathbf{x}_{(i)}^n)\right)^2, \tag{5.91a}$$

$$W_{\alpha F}^{PD}(\mathbf{x}_{(i)}^n) = b_F \, \delta \sum_{j=1}^{M} \frac{1}{\left|\mathbf{x}_{(j)}^n - \mathbf{x}_{(i)}^n\right|}\left(\left|\mathbf{y}_{(j)}^n - \mathbf{y}_{(i)}^n\right| - \left|\mathbf{x}_{(j)}^n - \mathbf{x}_{(i)}^n\right|\right)^2 V_{(j)}^n, \tag{5.91b}$$

$$W_{\alpha T}^{PD}(\mathbf{x}_{(i)}^n) = b_T \, \delta \sum_{j=1}^{N} \frac{1}{\left|\mathbf{x}_{(j)}^n - \mathbf{x}_{(i)}^n\right|}\left(\left|\mathbf{y}_{(j)}^n - \mathbf{y}_{(i)}^n\right| - \left|\mathbf{x}_{(j)}^n - \mathbf{x}_{(i)}^n\right|\right)^2 V_{(j)}^n, \tag{5.91c}$$

$$W_{\alpha FT}^{PD}(\mathbf{x}_{(i)}^n) = b_{FT} \, \delta \sum_{j=1}^{P} \frac{1}{\left|\mathbf{x}_{(j)}^n - \mathbf{x}_{(i)}^n\right|}\left(\left|\mathbf{y}_{(j)}^n - \mathbf{y}_{(i)}^n\right| - \left|\mathbf{x}_{(j)}^n - \mathbf{x}_{(i)}^n\right|\right)^2 V_{(j)}^n. \tag{5.91d}$$

Based on classical continuum mechanics, the strain energy density corresponding to uniaxial stretch in the fiber, $W_1^{CM}(\mathbf{x}_{(i)}^n)$, and transverse directions, $W_2^{CM}(\mathbf{x}_{(i)}^n)$, is uniform, and can be determined from

$$W_\alpha^{CM}(\mathbf{x}_{(i)}^n) = \frac{1}{2}Q_{\alpha\alpha}\zeta^2 \ (\alpha = 1, 2), \tag{5.92}$$

which can be decomposed as

$$W_\alpha^{CM}(\mathbf{x}_{(i)}^n) = W_{\alpha\theta}^{CM}(\mathbf{x}_{(i)}^n) + W_{\alpha F}^{CM}(\mathbf{x}_{(i)}^n) + W_{\alpha T}^{CM}(\mathbf{x}_{(i)}^n) + W_{\alpha FT}^{CM}(\mathbf{x}_{(i)}^n), \tag{5.93}$$

where $W_{\alpha\theta}^{CM}$ is associated with the dilatation terms, and $W_{\alpha F}^{CM}$, $W_{\alpha T}^{CM}$, and $W_{\alpha FT}^{CM}$ represent strain energy densities arising from the deformation in the fiber direction, transverse direction, and arbitrary directions, respectively. From Eq. 5.42b in conjunction with Eqs. 5.61a, b, d for uniaxial stretch in the fiber direction, i.e., $(\alpha = 1)$, each strain energy density component can be expressed as

$$W_{1\theta}^{CM}(\mathbf{x}_{(i)}^n) = \frac{1}{2}(Q_{12} - Q_{66})\zeta^2, \tag{5.94a}$$

$$W_{1F}^{CM}(\mathbf{x}_{(i)}^n) = \frac{1}{2}(Q_{11} - Q_{12} - 2Q_{66})\zeta^2, \tag{5.94b}$$

$$W_{1T}^{CM}(\mathbf{x}_{(i)}^n) = 0, \tag{5.94c}$$

$$W_{1FT}^{CM}(\mathbf{x}_{(i)}^n) = \frac{3}{2}Q_{66}\zeta^2. \tag{5.94d}$$

From Eq. 5.51 in conjunction with Eqs. 5.61a, c, for uniaxial stretch in the transverse direction, i.e., $(\alpha = 2)$, each strain energy component can be expressed as

$$W_{2\theta}^{CM}(\mathbf{x}_{(i)}^n) = \frac{1}{2}(Q_{12} - Q_{66})\zeta^2, \tag{5.95a}$$

$$W_{2F}^{CM}(\mathbf{x}_{(i)}^n) = 0, \tag{5.95b}$$

$$W_{2T}^{CM}(\mathbf{x}_{(i)}^n) = \frac{1}{2}(Q_{22} - Q_{12} - 2Q_{66})\zeta^2, \tag{5.95c}$$

$$W_{2FT}^{CM}(\mathbf{x}_{(i)}^n) = \frac{3}{2}Q_{66}\zeta^2. \tag{5.95d}$$

Because the dilatation term, $\theta_\alpha^{PD}(\mathbf{x}_{(i)}^n)$, is corrected with a dilatation correction term in the peridynamic computation, it is expected that Eq. 5.91a is automatically corrected for this loading condition. Hence, the correction is only necessary for the terms including parameter $b_\ell$, with $\ell = F, FT, T$. For the uniaxial stretch in the fiber direction, the correction terms for these parameters can be defined as

$$S_{1F(i)} = \frac{W_{1F}^{CM}(\mathbf{x}_{(i)}^n)}{W_{1F}^{PD}(\mathbf{x}_{(i)}^n)}$$
$$= \frac{\frac{1}{2}(Q_{11} - Q_{12} - 2Q_{66})\zeta^2}{b_F \delta \sum_{j=1}^{M} \frac{1}{|\mathbf{x}_{(j)}^n - \mathbf{x}_{(i)}^n|}\left(\left|\mathbf{y}_{(j)}^n - \mathbf{y}_{(i)}^n\right| - \left|\mathbf{x}_{(j)}^n - \mathbf{x}_{(i)}^n\right|\right)^2 V_{(j)}^n}, \tag{5.96a}$$

$$S_{1T(i)} = 1, \tag{5.96b}$$

$$S_{1FT(i)} = \frac{W_{1FT}^{CM}(\mathbf{x}_{(i)}^n)}{W_{1FT}^{PD}(\mathbf{x}_{(i)}^n)}$$
$$= \frac{\frac{3}{2}Q_{66}\zeta^2}{b_{FT} \delta \sum_{j=1}^{P} \frac{1}{|\mathbf{x}_{(j)}^n - \mathbf{x}_{(i)}^n|}\left(\left|\mathbf{y}_{(j)}^n - \mathbf{y}_{(i)}^n\right| - \left|\mathbf{x}_{(j)}^n - \mathbf{x}_{(i)}^n\right|\right)^2 V_{(j)}^n}. \tag{5.96c}$$

For the uniaxial stretch in the transverse direction, the correction terms for these parameters can be defined as

$$S_{2F(i)} = 1, \tag{5.97a}$$

$$S_{2T(i)} = \frac{W_{2T}^{CM}(\mathbf{x}_{(i)}^n)}{W_{2T}^{PD}(\mathbf{x}_{(i)}^n)}$$

$$= \frac{\frac{1}{2}(Q_{22} - Q_{12} - 2Q_{66})\zeta^2}{b_T \delta \sum\limits_{j=1}^{N} \frac{1}{\left|\mathbf{x}_{(j)}^n - \mathbf{x}_{(i)}^n\right|}\left(\left|\mathbf{y}_{(j)}^n - \mathbf{y}_{(i)}^n\right| - \left|\mathbf{x}_{(j)}^n - \mathbf{x}_{(i)}^n\right|\right)^2 V_{(j)}^n}, \tag{5.97b}$$

$$S_{2FT(i)} = \frac{W_{2FT}^{CM}(\mathbf{x}_{(i)}^n)}{W_{2FT}^{PD}(\mathbf{x}_{(i)}^n)}$$

$$= \frac{\frac{3}{2}Q_{66}\zeta^2}{b_{FT} \delta \sum\limits_{j=1}^{P} \frac{1}{\left|\mathbf{x}_{(j)}^n - \mathbf{x}_{(i)}^n\right|}\left(\left|\mathbf{y}_{(j)}^n - \mathbf{y}_{(i)}^n\right| - \left|\mathbf{x}_{(j)}^n - \mathbf{x}_{(i)}^n\right|\right)^2 V_{(j)}^n}. \tag{5.97c}$$

With these correction factors, a vector of correction factors for the integral and summation terms that appear in dilatation and the strain energy density at material point $\mathbf{x}_{(i)}^n$ can be written as

$$\mathbf{g}_{(d)(i)}(\mathbf{x}_{(i)}^n) = \left\{g_{1(d)}(\mathbf{x}_{(i)}^n),\ g_{2(d)}(\mathbf{x}_{(i)}^n)\right\}^T = \left\{D_{1(i)},\ D_{2(i)}\right\}^T, \tag{5.98a}$$

$$\mathbf{g}_{(b)\ell(i)}(\mathbf{x}_{(i)}^n) = \left\{g_{1(b)\ell}(\mathbf{x}_{(i)}^n),\ g_{2(b)\ell}(\mathbf{x}_{(i)}^n)\right\}^T = \left\{S_{1\ell(i)},\ S_{2\ell(i)}\right\}^T, \tag{5.98b}$$

with $\ell = F, FT, T$.

These correction factors are only based on loading in the fiber and transverse directions. However, they can be used as the principal values of an ellipse as shown in Fig. 5.14 in order to approximate the surface correction factor in any direction. Arising from a general loading condition, the correction factor for interaction between material points $\mathbf{x}_{(i)}^n$ and $\mathbf{x}_{(j)}^n$, shown in Fig. 5.15a, can be obtained in the direction of their unit relative position vector, $\mathbf{n} = (\mathbf{x}_{(j)}^n - \mathbf{x}_{(i)}^n)/|\mathbf{x}_{(j)}^n - \mathbf{x}_{(i)}^n| = \{n_1,\ n_2\}^T$.

A vector of correction factors for the integrals in the dilatation and strain energy density expressions at material point $\mathbf{x}_{(j)}^n$ can be similarly written as

$$\mathbf{g}_{(d)(j)}(\mathbf{x}_{(j)}^n) = \left\{g_{1(d)}(\mathbf{x}_{(j)}^n),\ g_{2(d)}(\mathbf{x}_{(j)}^n)\right\}^T = \left\{D_{1(j)},\ D_{2(j)}\right\}^T, \tag{5.99a}$$

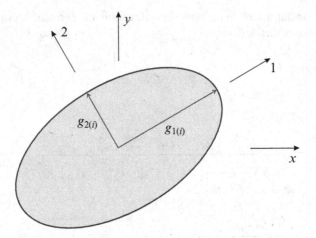

**Fig. 5.14** Construction of an ellipse for surface correction factors

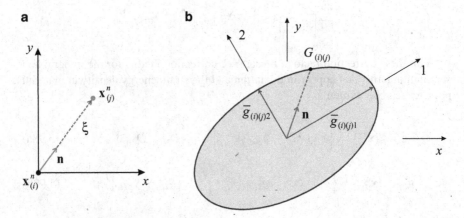

**Fig. 5.15** (a) PD interaction between material points at $\mathbf{x}_{(i)}^n$ and $\mathbf{x}_{(j)}^n$, and (b) the ellipse for the surface correction factors

$$\mathbf{g}_{(b)\ell(j)}(\mathbf{x}_{(j)}^n) = \left\{ g_{1(b)\ell}(\mathbf{x}_{(j)}^n),\ g_{2(b)\ell}(\mathbf{x}_{(j)}^n) \right\}^T = \left\{ S_{1\ell(j)},\ S_{2\ell(j)} \right\}^T. \qquad (5.99b)$$

These correction factors are, in general, different at material points $\mathbf{x}_{(i)}^n$ and $\mathbf{x}_{(j)}^n$. Therefore, the correction factor for an interaction between material points $\mathbf{x}_{(i)}^n$ and $\mathbf{x}_{(j)}^n$ can be obtained by their mean values as

$$\bar{\mathbf{g}}_{(d)(i)(j)} = \left\{ \bar{g}_{(d)(i)(j)1}, \bar{g}_{(d)(i)(j)2} \right\}^T = \frac{\mathbf{g}_{(d)(i)} + \mathbf{g}_{(d)(j)}}{2} \qquad (5.100a)$$

and

$$\bar{\mathbf{g}}_{(b)\ell(i)(j)} = \left\{ \bar{g}_{(b)\ell(i)(j)1}, \bar{g}_{(b)\ell(i)(j)2} \right\}^{T} = \frac{\mathbf{g}_{(b)\ell(i)} + \mathbf{g}_{(b)\ell(j)}}{2}, \quad (5.100b)$$

which can be used as the principal values of an ellipse for the interactions other than in the fiber and transverse directions, as shown in Fig. 5.15b. The intersection of the ellipse and a relative position vector, $\mathbf{n}$, of material points $\mathbf{x}^n_{(i)}$ and $\mathbf{x}^n_{(j)}$, provides the correction factors as

$$G_{(d)(i)(j)} = \left( \left[ n_1 \big/ \bar{g}_{(d)(i)(j)1} \right]^2 + \left[ n_2 \big/ \bar{g}_{(d)(i)(j)2} \right]^2 \right)^{-1/2} \quad (5.101a)$$

and

$$G_{(b)\ell(i)(j)} = \left( \left[ n_1 \big/ \bar{g}_{(b)\ell(i)(j)1} \right]^2 + \left[ n_2 \big/ \bar{g}_{(b)\ell(i)(j)2} \right]^2 \right)^{-1/2}. \quad (5.101b)$$

After considering the surface effects, the discrete forms of the dilatation and the strain energy density are corrected as

$$\theta^n_{(i)} = d \sum_{j=1}^{P} G_{(d)(i)(j)} \frac{\delta}{\left| \mathbf{x}^n_{(j)} - \mathbf{x}^n_{(i)} \right|} \left( \left| \mathbf{y}^n_{(j)} - \mathbf{y}^n_{(i)} \right| - \left| \mathbf{x}^n_{(j)} - \mathbf{x}^n_{(i)} \right| \right)$$

$$\times \left( \frac{\mathbf{y}^n_{(j)} - \mathbf{y}^n_{(i)}}{\left| \mathbf{y}^n_{(j)} - \mathbf{y}^n_{(i)} \right|} \cdot \frac{\mathbf{x}^n_{(j)} - \mathbf{x}^n_{(i)}}{\left| \mathbf{x}^n_{(j)} - \mathbf{x}^n_{(i)} \right|} \right) V^n_{(j)},$$

(5.102a)

$$W^n_{(i)} = a\,\theta^2_{(i)} + b_F \delta \sum_{j=1}^{M} G_{(b)F(i)(j)} \frac{1}{\left| \mathbf{x}^n_{(j)} - \mathbf{x}^n_{(i)} \right|}$$

$$\times \left( \left| \mathbf{y}^n_{(j)} - \mathbf{y}^n_{(i)} \right| - \left| \mathbf{x}^n_{(j)} - \mathbf{x}^n_{(i)} \right| \right)^2 V^n_{(j)}$$

$$+ b_T \delta \sum_{j=1}^{N} G_{(b)T(i)(j)} \frac{1}{\left| \mathbf{x}^n_{(j)} - \mathbf{x}^n_{(i)} \right|}$$

(5.102b)

$$\times \left( \left| \mathbf{y}^n_{(j)} - \mathbf{y}^n_{(i)} \right| - \left| \mathbf{x}^n_{(j)} - \mathbf{x}^n_{(i)} \right| \right)^2 V^n_{(j)}$$

$$+ b_{FT} \delta \sum_{j=1}^{P} G_{(b)FT(i)(j)} \frac{1}{\left| \mathbf{x}^n_{(j)} - \mathbf{x}^n_{(i)} \right|}$$

$$\times \left( \left| \mathbf{y}^n_{(j)} - \mathbf{y}^n_{(i)} \right| - \left| \mathbf{x}^n_{(j)} - \mathbf{x}^n_{(i)} \right| \right)^2 V^n_{(j)}.$$

The peridynamic material parameters $b_N$ and $b_S$ for a material point located on the bounding laminae, such as $n = 1$ or $n = N$, also require correction. However, the correction factors for $b_N$ and $b_S$ are not necessary for material points $x_{(i)}^n$ for $n \neq 1, N$ because they are imbedded in the laminate, as shown in Fig. 5.3.

Simple loading conditions of uniform transverse stretch, $\partial u_3^* / \partial x_3 = \zeta$, and simple transverse shear, $\partial u_1^* / \partial x_3 = \zeta$, are applied to the laminate separately to determine the correction factors.

The corresponding displacement fields at material point $\mathbf{x}$ as a result of these loading conditions can be expressed as

$$\mathbf{u}_3^T = \left\{ 0 \quad 0 \quad \frac{\partial u_3^*}{\partial x_3} x_3 \right\} \tag{5.103a}$$

and

$$\mathbf{u}_S^T = \left\{ \frac{\partial u_1^*}{\partial x_3} x_3 \quad 0 \quad 0 \right\}. \tag{5.103b}$$

The PD strain energy density of material point $\mathbf{x}_{(i)}^n$ with $n = 1, N$ due to these loading conditions, respectively, can be expressed as

$$\left. \begin{aligned} W_3^{PD}\left(\mathbf{x}_{(i)}^1\right) &= \frac{1}{4} \zeta^2 b_N (h_{n+1} + h_n)^2 V_{(i)}^{n+1} \\ W_3^{PD}\left(\mathbf{x}_{(i)}^N\right) &= \frac{1}{4} \zeta^2 b_N (h_{n-1} + h_n)^2 V_{(i)}^{n-1} \end{aligned} \right\} \tag{5.104a}$$

and

$$\left. \begin{aligned} W_S^{PD}\left(\mathbf{x}_{(i)}^1\right) &= 4\zeta^2 b_S \left(\frac{h_{n+1} + h_n}{2}\right)^2 \sum_{j=1}^{N} \frac{\ell^2 \cos^2(\phi)}{\ell^2 + \left(\frac{h_{n+1}+h_n}{2}\right)^2} V_{(j)}^{n+1} \\ W_S^{PD}\left(\mathbf{x}_{(i)}^N\right) &= 4\zeta^2 b_S \left(\frac{h_{n-1} + h_n}{2}\right)^2 \sum_{j=1}^{N} \frac{\ell^2 \cos^2(\phi)}{\ell^2 + \left(\frac{h_{n-1}+h_n}{2}\right)^2} V_{(j)}^{n-1} \end{aligned} \right\}. \tag{5.104b}$$

The corresponding strain energy density expressions based on classical continuum mechanics can be expressed as

$$W_3^{CM}\left(\mathbf{x}_{(i)}^n\right) = \frac{1}{2} E_m \zeta^2 \qquad n = 1, N \tag{5.105a}$$

and

$$W_S^{CM}\left(\mathbf{x}_{(i)}^n\right) = \frac{1}{2} G_m \zeta^2 \qquad n = 1, N \tag{5.105b}$$

Therefore, the correction factors associated with the material parameters, $b_N$ and $b_S$, at material point $\mathbf{x}_{(i)}^n$ for $n = 1, N$ can be defined as

$$S_{3(i)}^n = \frac{W_3^{CM}\left(\mathbf{x}_{(i)}^n\right)}{W_3^{PD}\left(\mathbf{x}_{(i)}^n\right)} \tag{5.106a}$$

and

$$S_{S(i)}^n = \frac{W_S^{CM}\left(\mathbf{x}_{(i)}^n\right)}{W_S^{PD}\left(\mathbf{x}_{(i)}^n\right)}. \tag{5.106b}$$

Correction factors for $b_N$ and $b_S$ are not necessary for material points $\mathbf{x}_{(i)}^n$ for $n \neq 1, N$. Therefore, the correction factor for an interaction between material points $\mathbf{x}_{(i)}^n$ for $n = 1, N$ and $\mathbf{x}_{(j)}^m$ for $m \neq 1, N$ can be obtained by their mean values as

$$\left.\begin{aligned} \bar{S}_{3(i)}^{(n)(m)} &= \left(S_{3(i)}^n + 1\right)/2 \quad \text{for } n = 1, N \text{ and } m \neq 1, N \\ \bar{S}_{3(i)}^{(n)(m)} &= 1 \quad \text{for } n, m \neq 1, N \end{aligned}\right\}, \tag{5.107a}$$

$$\left.\begin{aligned} \bar{S}_{S(i)(j)}^{(n)(m)} &= \left(S_{S(i)}^n + 1\right)/2 \quad \text{for } n = 1, N \text{ and } m \neq 1, N \\ \bar{S}_{S(i)(j)}^{(n)(m)} &= 1 \quad \text{for } n, m \neq 1, N \end{aligned}\right\}. \tag{5.107b}$$

After considering the surface effects, the discrete form of the strain energy density functions $\hat{W}_{(i)}^n$ and $\tilde{W}_{(i)}^n$ are corrected as

$$\hat{W}_{(i)}^n = b_N \sum_{m=n+1,n-1} \bar{S}_{3(i)}^{(n)(m)} \left(\left|\mathbf{y}_{(i)}^m - \mathbf{y}_{(i)}^n\right| - \left|\mathbf{x}_{(i)}^m - \mathbf{x}_{(i)}^n\right|\right)^2 V_{(i)}^m \tag{5.108a}$$

and

$$\tilde{W}_{(i)}^n = b_S \sum_{m=n+1,n-1} \sum_{j=1}^{\infty} \bar{S}_{S(i)(j)}^{(n)(m)} \left[\left(\left|\mathbf{y}_{(j)}^m - \mathbf{y}_{(i)}^n\right| - \left|\mathbf{x}_{(j)}^m - \mathbf{x}_{(i)}^n\right|\right)\right.$$
$$\left. - \left(\left|\mathbf{y}_{(i)}^m - \mathbf{y}_{(i)}^n\right| - \left|\mathbf{x}_{(i)}^m - \mathbf{x}_{(j)}^n\right|\right)\right]^2 V_{(j)}^m. \tag{5.108b}$$

# Reference

Oterkus E, Madenci E (2012) Peridynamic analysis of fiber reinforced composite materials. J Mech Mater Struct 7(1):45–84

# Chapter 6
# Damage Prediction

Material damage in peridynamics (PD) is introduced through elimination of interactions (micropotentials) among the material points. It is assumed that when the stretch, $s_{(k)(j)}$, between two material points, $k$ and $j$, exceeds its critical value, $s_c$, the onset of damage occurs. Damage is reflected in the equations of motion by removing the force density vectors between the material points in an irreversible manner. As a result, the load is redistributed among the material points in the body, leading to progressive damage growth in an autonomous fashion.

## 6.1 Critical Stretch

In order to create a new crack surface, $A$, all of the micropotentials (interactions) between the material points $\mathbf{x}_{(k^+)}$ and $\mathbf{x}_{(j^-)}$ whose line of action crosses this new surface must be terminated, as sketched in Fig. 6.1. The material points $\mathbf{x}_{(k^+)}$ and $\mathbf{x}_{(j^-)}$ are located above and below the new crack surface, respectively.

The micropotentials for linear elastic deformation can be obtained from Eq. 2.17 as

$$w_{(k^+)(j^-)} = 2\mathbf{t}_{(k^+)(j^-)} \bullet \left( \mathbf{u}_{(j^-)} - \mathbf{u}_{(k^+)} \right), \tag{6.1a}$$

$$w_{(j^-)(k^+)} = 2\mathbf{t}_{(j^-)(k^+)} \bullet \left( \mathbf{u}_{(k^+)} - \mathbf{u}_{(j^-)} \right), \tag{6.1b}$$

or

$$w_{(k^+)(j^-)} = A\left( \left| \mathbf{y}_{(j^-)} - \mathbf{y}_{(k^+)} \right| - \Lambda_{(k^+)(j^-)} \left| \mathbf{x}_{(j^-)} - \mathbf{x}_{(k^+)} \right| \right), \tag{6.2a}$$

$$w_{(j^-)(k^+)} = B\left( \left| \mathbf{y}_{(k^+)} - \mathbf{y}_{(j^-)} \right| - \Lambda_{(j^-)(k^+)} \left| \mathbf{x}_{(k^+)} - \mathbf{x}_{(j^-)} \right| \right), \tag{6.2b}$$

E. Madenci and E. Oterkus, *Peridynamic Theory and Its Applications*, DOI 10.1007/978-1-4614-8465-3_6, © Springer Science+Business Media New York 2014

**Fig. 6.1** Interaction
between material points $\mathbf{x}_{(k)}^{+}$
and $\mathbf{x}_{(j)}^{-}$, whose line of action
crosses the crack surface

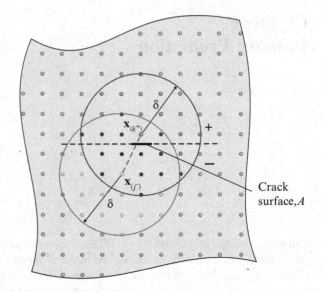

Crack
surface,$A$

with

$$A = \frac{4ad\delta}{\left|\mathbf{x}_{(j-)} - \mathbf{x}_{(k+)}\right|}\Lambda_{(k+)(j-)}\theta_{(k+)} + 4\delta\,bs_{(k+)(j-)}, \tag{6.3a}$$

$$B = \frac{4ad\delta}{\left|\mathbf{x}_{(k+)} - \mathbf{x}_{(j-)}\right|}\Lambda_{(j-)(k+)}\theta_{(j-)} + 4\delta bs_{(j-)(k+)}, \tag{6.3b}$$

in which

$$\theta_{(k+)} = d\delta \sum_{i=1}^{N} \Lambda_{(k+)(i)}s_{(k+)(i)} V_{(i)}, \tag{6.4a}$$

$$\theta_{(j-)} = d\delta \sum_{i=1}^{N} \Lambda_{(j-)(i)}s_{(j-)(i)} V_{(i)}, \tag{6.4b}$$

and

$$\Lambda_{(k+)(j-)} = \frac{\mathbf{y}_{(j-)} - \mathbf{y}_{(k+)}}{\left|\mathbf{y}_{(j-)} - \mathbf{y}_{(k+)}\right|} \bullet \frac{\mathbf{x}_{(j-)} - \mathbf{x}_{(k+)}}{\left|\mathbf{x}_{(j-)} - \mathbf{x}_{(k+)}\right|}, \tag{6.5a}$$

$$\Lambda_{(j-)(k+)} = \frac{\mathbf{y}_{(k+)} - \mathbf{y}_{(j-)}}{\left|\mathbf{y}_{(k+)} - \mathbf{y}_{(j-)}\right|} \bullet \frac{\mathbf{x}_{(k+)} - \mathbf{x}_{(j-)}}{\left|\mathbf{x}_{(k+)} - \mathbf{x}_{(j-)}\right|}. \tag{6.5b}$$

Under the assumption of linear elastic deformation, i.e., $\Lambda_{(j^-)(k^+)} \approx 1$ and $\Lambda_{(k^+)(j^-)} \approx 1$, the expressions for the micropotentials can be rewritten as

$$w_{(k^+)(j^-)} = 4ad^2\delta^2 \left( \sum_{i=1}^{N-K^-} s_{(k^+)(i)}s_{(k^+)(j^-)}V_{(i)} + \sum_{i=1}^{K^-} s_{(k^+)(i)}s_{(k^+)(j^-)}V_{(i)} \right)$$
$$+ 4\delta\, bs^2_{(k^+)(j^-)} |\mathbf{x}_{(j^-)} - \mathbf{x}_{(k^+)}|, \tag{6.6a}$$

$$w_{(j^-)(k^+)} = 4ad^2\delta^2 \left( \sum_{i=1}^{N-J^+} s_{(j^-)(i)}s_{(j^-)(k^+)}V_{(i)} + \sum_{i=1}^{J^+} s_{(j^-)(i)}s_{(j^-)(k^+)}V_{(i)} \right)$$
$$+ 4\delta bs^2_{(j^-)(k^+)} |\mathbf{x}_{(k^+)} - \mathbf{x}_{(j^-)}|, \tag{6.6b}$$

in which $N$ represents the total number of material points within the family of $\mathbf{x}_{(k^+)}$ and $\mathbf{x}_{(j^-)}$.

The number of material points within the family of $\mathbf{x}_{(k^+)}$ below the crack surface and intersecting with the crack is denoted by $K^-$. Similarly, $J^+$ represents the number of material points above the crack surface within the family of $\mathbf{x}_{(j^-)}$ and intersecting with the crack. Even at the critical stretch, these micropotentials do not completely vanish because of the contribution of the material points to the micropotential through the first term arising from dilatation. Retaining only the interactions crossing the crack surface, the critical values of these micropotentials can be obtained by substituting the critical value, $s_c$, of the stretch $s_{(k^+)(j^-)}$ and $s_{(j^-)(k^+)}$ as

$$w^c_{(k^+)(j^-)} = \left( 4ad^2\delta^2 \left( \sum_{i=1}^{K^-} s_c^2 V_{(i)} \right) + 4\delta\, bs_c^2 |\mathbf{x}_{(j^-)} - \mathbf{x}_{(k^+)}| \right) \tag{6.7a}$$

and

$$w^c_{(j^-)(k^+)} = \left( 4ad^2\delta^2 \left( \sum_{i=1}^{J^+} s_c^2 V_{(i)} \right) + 4\delta bs_c^2 |\mathbf{x}_{(k^+)} - \mathbf{x}_{(j^-)}| \right). \tag{6.7b}$$

Hence, the strain energy required to remove the interaction between two material points, $\mathbf{x}_{(k^+)}$ and $\mathbf{x}_{(j^-)}$, can be expressed as

$$W^c_{(k^+)(j^-)} = \frac{1}{2} \frac{w^c_{(k^+)(j^-)} + w^c_{(j^-)(k^+)}}{2} V_{(k^+)}V_{(j^-)}. \tag{6.8}$$

Furthermore, the total strain energy required to remove all of the interactions across the newly created crack surface $A$ can be obtained as

$$W^c = \frac{1}{2}\sum_{k=1}^{K^+}\frac{1}{2}\sum_{j=1}^{J^-} w^c_{(k^+)(j^-)} V_{(k^+)} V_{(j^-)} + \frac{1}{2}\sum_{k=1}^{K^+}\frac{1}{2}\sum_{j=1}^{J^-} w^c_{(j^-)(k^+)} V_{(j^-)} V_{(k^+)}, \qquad (6.9)$$

for which the line of interaction defined by $|\mathbf{x}_{(k^+)} - \mathbf{x}_{(j^-)}|$ and the crack surface intersect, and $K^+$ and $J^-$ indicate the number of material points, above and below the crack surface, within the families of $\mathbf{x}_{(k^+)}$ and $\mathbf{x}_{(j^-)}$, respectively. If this line of interaction and crack surface intersect at the crack tip, only half of the critical micropotential is considered in the summation. Substituting for micropotentials given by Eqs. 6.7a, b in Eq. 6.9 results in the critical strain energy required to eliminate all of the interactions across the newly created crack surface $A$ as

$$W^c = s_c^2 \sum_{k=1}^{K^+}\sum_{j=1}^{J^-}\left(2\delta b\left|\mathbf{x}_{(j^-)} - \mathbf{x}_{(k^+)}\right| + ad^2\delta^2\left(\sum_{i=1}^{K^-} V_{(i)} + \sum_{i=1}^{J^+} V_{(i)}\right)\right) V_{(k^+)} V_{(j^-)}.$$

$$(6.10)$$

The total work, $W^c$, required to eliminate all interactions across this new surface can be equated to the critical energy release rate, $G_c$, in order to establish the value of critical stretch, $s_c$, as

$$G_c = \frac{s_c^2 \sum_{k=1}^{K^+}\sum_{j=1}^{J^-}\left(2\delta b\left|\mathbf{x}_{(j^-)} - \mathbf{x}_{(k^+)}\right| + ad^2\delta^2\left(\sum_{i=1}^{K^-} V_{(i)} + \sum_{i=1}^{J^+} V_{(i)}\right)\right) V_{(k^+)} V_{(j^-)}}{A},$$

$$(6.11)$$

which yields the critical stretch, $s_c$, expression of

$$s_c = \sqrt{\frac{G_c A}{\sum_{k=1}^{K^+}\sum_{j=1}^{J^-}\left(2\delta b\left|\mathbf{x}_{(j^-)} - \mathbf{x}_{(k^+)}\right| + ad^2\delta^2\left(\sum_{i=1}^{K^-} V_{(i)} + \sum_{i=1}^{J^+} V_{(i)}\right)\right) V_{(k^+)} V_{(j^-)}}}.$$

$$(6.12)$$

Setting $a = 0$ and $4\delta b = c$ reduces this expression to bond-based peridynamics

$$G_c = \frac{1}{2} c s_c^2 \frac{\left\{\sum_{k=1}^{K^+}\sum_{j=1}^{J^-}\left|\mathbf{x}_{(j^-)} - \mathbf{x}_{(k^+)}\right| V_{(k^+)} V_{(j^-)}\right\}}{A}. \qquad (6.13)$$

For three-dimensional analysis, the critical energy release rate for bond-based peridynamics was derived by Silling and Askari (2005) in integral form as

**Fig. 6.2** -Integration domain of the micropotentials crossing a fracture surface

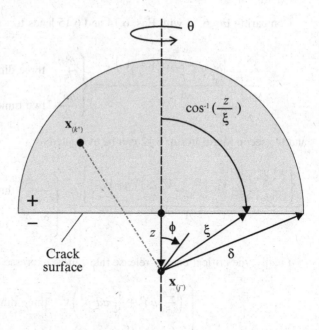

$$G_c = \int_0^{\delta} \left\{ \int_0^{2\pi} \int_z^{\delta} \int_0^{\cos^{-1}z/\xi} \left( \frac{1}{2}c\xi s_c^2 \xi^2 \right) \sin\phi \, d\phi \, d\xi \, d\theta \right\} dz = \frac{1}{2}cs_c^2 \left( \frac{\delta^5 \pi}{5} \right). \quad (6.14)$$

This integral represents the summation of the work required to terminate all interactions (micropotentials) between point $\mathbf{x}_{(j^-)}$ (below the fracture surface) and all of the points $\mathbf{x}_{(k^+)}$ (above the fracture surface) within its horizon, as shown in Fig. 6.2. The integration in spherical coordinates, $(\xi, \theta, \phi)$, results in the volume of all the points $\mathbf{x}_{(k^+)}$ that are above the fracture surface and within the horizon of point $\mathbf{x}_{(j^-)}$. The line integral includes the contribution of all the points $\mathbf{x}_{(j^-)}$ from 0 to the horizon, $\delta$.

In the case of two-dimensional analysis, the expression for the critical energy release rate for bond-based peridynamics becomes

$$G_c = 2h \int_0^{\delta} \left\{ \int_z^{\delta} \int_0^{\cos^{-1}z/\xi} \left( \frac{1}{2}c\xi s_c^2 \xi \right) d\phi \, d\xi \right\} dz = \frac{1}{2}cs_c^2 \left( \frac{h\delta^4}{2} \right), \quad (6.15)$$

in which $h$ represents the thickness of the material. The integration is performed in polar coordinates, $(\xi, \phi)$.

Comparing Eq. 6.13 with Eqs. 6.14 and 6.15 leads to

$$
\frac{\sum_{k=1}^{K^+} \sum_{j=1}^{J^-} \left| \mathbf{x}_{(j-)} - \mathbf{x}_{(k+)} \right| V_{(k+)} V_{(j-)}}{A} = \begin{cases} \dfrac{\delta^5 \pi}{5} & \text{three dimensions} \\[2ex] \dfrac{h\delta^4}{2} & \text{two dimensions} \end{cases} \tag{6.16}
$$

and the second term in Eq. 6.12 can be evaluated as

$$
\frac{\left\{ \sum_{k=1}^{K^+} \sum_{j=1}^{J^-} \left( \sum_{i=1}^{K^-} V_{(i-)} + \sum_{i=1}^{J^+} V_{(i+)} \right) V_{(k+)} V_{(j-)} \right\}}{A} = \begin{cases} \dfrac{\delta^7 \pi^2}{8} & \text{three dimensions} \\[2ex] \dfrac{8h^2\delta^5}{9} & \text{two dimensions .} \end{cases} \tag{6.17}
$$

Finally, the critical energy release rate can be expressed as

$$
G_c = \begin{cases} \left( \dfrac{2\pi}{5} b\delta^6 + \dfrac{\pi^2}{8} ad^2\delta^9 \right) s_c^2 & \text{three dimensions} \\[2ex] \left( bh\delta^5 + \dfrac{8}{9} ad^2 h^2 \delta^7 \right) s_c^2 & \text{two dimensions .} \end{cases} \tag{6.18}
$$

After substituting for the peridynamic parameters, $a$, $b$, and $d$, the critical stretch can be expressed as

$$
s_c = \begin{cases} \sqrt{\dfrac{G_c}{\left( 3\mu + \left( \frac{3}{4} \right)^4 \left( \kappa - \frac{5\mu}{3} \right) \right)\delta}} & \text{three dimensions} \\[3ex] \sqrt{\dfrac{G_c}{\left( \frac{6}{\pi}\mu + \frac{16}{9\pi^2} (\kappa - 2\mu) \right)\delta}} & \text{two dimensions .} \end{cases} \tag{6.19}
$$

It is worth noting that the critical stretch is a function of the horizon. The value of the horizon brings in the effect of the physical material characteristics, nature of loading, length scale, and the computational cut-off radius. This simple relationship provides the value of critical stretch for a linear elastic brittle material with a known critical energy release rate. If the material exhibits time-dependent nonlinear behavior such as viscoplasticity, a single critical stretch value is not a viable failure criterion. Foster et al. (2011) proposed the use of the critical energy density as a failure criterion in rate-dependent situations. For complex material behavior, there is no simple approach for determining the critical stretch value or critical energy. An inverse approach can be adopted to extract their critical values by performing PD simulations of the fracture experiments with measured failure loads. After each PD simulation with a trial critical value, the PD failure load prediction is compared with that of the measured value, and PD simulations continue with updated critical values until the PD prediction and measured values are within an acceptable range.

## 6.2   Damage Initiation

In order to include damage initiation in the material response, the force density vector can be modified through a history-dependent scalar-valued function $\mu$ (Silling and Bobaru 2005) as

$$\mathbf{t}_{(k)(j)} = 2\delta \left\{ ad \frac{\Lambda_{(k)(j)}}{\left|\mathbf{x}_{(j)} - \mathbf{x}_{(k)}\right|} \theta_{(k)} + b\mu\left(\mathbf{x}_{(j)} - \mathbf{x}_{(k)}, t\right) s_{(k)(j)} \right\} \frac{\mathbf{y}_{(j)} - \mathbf{y}_{(k)}}{\left|\mathbf{y}_{(j)} - \mathbf{y}_{(k)}\right|}, \qquad (6.20)$$

with the dilatation term

$$\theta_{(k)} = d\delta \sum_{\ell=1}^{N} \Lambda_{(k)(\ell)} \mu\left(\mathbf{x}_{(\ell)} - \mathbf{x}_{(k)}, t\right) s_{(k)(\ell)} V_{(\ell)}, \qquad (6.21)$$

where $\mu$ can be written as

$$\mu\left(\mathbf{x}_{(j)} - \mathbf{x}_{(k)}, t\right) = \begin{cases} 1 & \text{if } s_{(k)(j)}\left(\mathbf{x}_{(j)} - \mathbf{x}_{(k)}, t'\right) < s_c & \text{for all } 0 < t' \\ 0 & \text{otherwise .} \end{cases} \qquad (6.22)$$

During the solution process, the displacements of each material point, as well as the stretch, $s_{(k)(j)}$, between pairs of material points, $\mathbf{x}_{(k)}$ and $\mathbf{x}_{(j)}$, are computed and monitored. When the stretch between these material points exceeds its critical stretch, failure occurs; thus, the history-dependent scalar-valued function $\mu$ is zero, rendering the associated part of the force density vector to be zero.

## 6.3   Local Damage

Local damage at a point is defined as the weighted ratio of the number of eliminated interactions to the total number of initial interactions of a material point with its family members. The local damage at a point can be quantified as (Silling and Askari 2005)

$$\varphi(\mathbf{x}, t) = 1 - \frac{\int_{H} \mu(\mathbf{x}' - \mathbf{x}, t) dV'}{\int_{H} dV'}. \qquad (6.23)$$

The local damage ranges from 0 to 1. When the local damage is one, all the interactions initially associated with the point have been eliminated, while a local damage of zero means that all interactions are intact. The measure of local damage

**Fig. 6.3** (a) All
interactions are intact
(no damage); (**b**) half
of the terminated
interactions create a crack

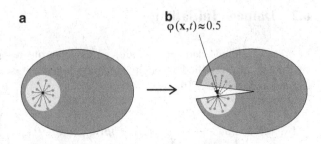

is an indicator of possible crack formation within a body. For example, initially a
material point interacts with all materials in its horizon, as shown in Fig. 6.3a; thus,
the local damage has a value of zero. However, the creation of a crack terminates
half of the interactions within its horizon, resulting in a local damage value of
one-half, as shown in Fig. 6.3b.

## 6.4  Failure Load and Crack Path Prediction

The applicability of the critical stretch as a failure parameter is demonstrated for a
linear elastic material by considering the experimental study conducted by
Ayatollahi and Aliha (2009). They considered diagonally loaded square plate
specimens, shown in Fig. 6.4, to investigate the effect of mode mixity ranging
from pure mode I to pure mode II. They provided the failure loads, crack propagation paths for each of the specimens, and fracture toughness of the material, $K_{IC}$.
The edge length of the diagonal square is $2W = 0.15$ m and its thickness is $h =
0.005$ m. The length of the crack is $2a = 0.045$ m, with an orientation angle of $\alpha$.
The material has an elastic modulus of $E = 2940$ MPa, Poisson's ratio of $\nu = 0.38$,
and fracture toughness of $K_{IC} = 1.33$ MPa$\sqrt{\text{m}}$. This corresponds to a critical stretch
value of 0.089. They also reported the failure loads for varying crack orientation
angles of $\alpha = 0°$ (Mode I), $15°$, $30°$, $45°$, and $62.5°$ (Mode II). The center of the
crack coincides with the origin of the Cartesian coordinate system.

The applied load is introduced through a velocity constraint of $10^{-9}$ m/s along
the circular regions in opposite directions. The initial crack is inserted in the PD
model by removing the interactions across the crack surface. The force is monitored
by summing the forces between the interactions crossing the dotted black line.

As demonstrated in Fig. 6.5, the crack propagation paths obtained from the
peridynamic simulations and those of the experimental results agree well with each
other for all crack orientation angles. Crack growth initiation angles are also
compared between the predictions and measurements. Again a good comparison
is obtained, as shown in Fig. 6.6. Finally, the failure loads are compared and it is
observed that the failure loads obtained from the peridynamic simulations are
within 15 % of the experimental values for all crack inclination angles, as depicted
in Fig. 6.7. While the peridynamic simulations closely match the experimental

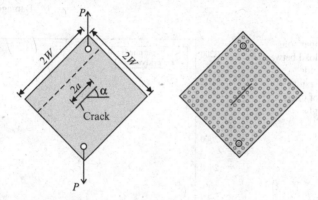

**Fig. 6.4** Peridynamic model of the specimen

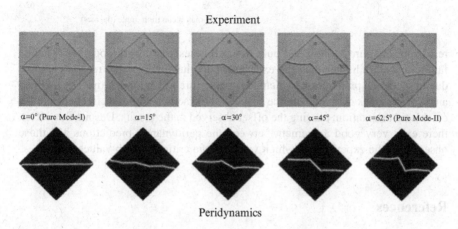

Experiment

$\alpha = 0°$ (Pure Mode-I)    $\alpha = 15°$    $\alpha = 30°$    $\alpha = 45°$    $\alpha = 62.5°$ (Pure Mode-II)

Peridynamics

**Fig. 6.5** Comparison of experimental and peridynamic crack propagation paths

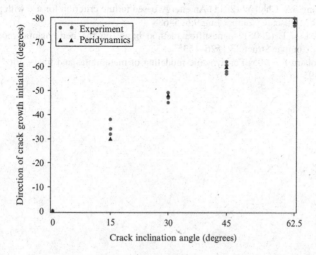

**Fig. 6.6** Comparison of crack growth initiation angle between peridynamic and experimental results as a function of crack inclination angle

**Fig. 6.7** Comparison
of the failure load between
peridynamic and
experimental results
as a function of crack
inclination angle

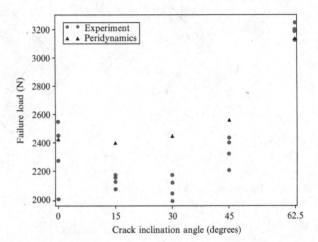

results for the pure Mode I and pure Mode II cases, the mixed mode peridynamic
failure loads are higher than the experimental values. A possible reason could be
due to specimen preparation, which does not ensure a sharp crack tip. The inclined
angle of the crack coupled with the shape of the crack tip could in effect change the
crack's tip orientation, causing the offset observed in the results. Despite this offset,
there exist very good agreement between the peridynamic predictions and those
observed in the experiments, which validates the critical stretch values.

# References

Ayatollahi MR, Aliha MRM (2009) Analysis of a new specimen for mixed mode fracture tests on
    brittle materials. Eng Fract Mech 76:1563–1573
Foster JT, Silling SA, Chen W (2011) An energy based failure criterion for use with peridynamic
    states. Int J Multiscale Comput Eng 9:675–688
Silling SA, Askari E (2005) A meshfree method based on the peridynamic model of solid
    mechanics. Comput Struct 83:1526–1535
Silling SA, Bobaru F (2005) Peridynamic modeling of membranes and fibers. Int J Non-Linear
    Mech 40:395–409

# Chapter 7
# Numerical Solution Method

The peridynamic (PD) equation of motion is an integro-differential equation, which is not usually amenable for analytical solutions. Therefore, its solution is constructed by using numerical techniques for spatial and time integrations. The spatial integration can be performed by using the collocation method of a meshless scheme due to its simplicity. Hence, the domain can be divided into a finite number of subdomains, with integration or collocation (material) points associated with specific volumes (Sect. 7.1). Associated with a particular material point, numerical implementation of spatial integration involves the summation of the volumes of material points within its horizon. However, the volume of each material point may not be embedded in the horizon in its entirety, i.e., the material points located near the surface of the horizon may have truncated volumes. As a result, the volume integration over the horizon may be incorrect if the entire volume of each material point is included in the numerical implementation. Therefore, a volume correction factor is necessary to correct for the extra volume. A volume correction procedure required for such a case is described in Sect. 7.2.

Numerical time integration can be performed by using backward and forward difference explicit integration schemes, although other techniques are also applicable, such as the Adams-Bashforth method, Adams-Moulton method, and Runge–Kutta method. If an explicit integration scheme is adopted, a stability criterion on the value of the incremental time step is necessary to ensure convergence. Details of the time integration scheme and stability criterion are given in Sects. 7.3 and 7.4, respectively.

The PD equation of motion includes the inertial terms; it is not directly applicable to static and quasi-static problems. Hence, a special treatment is required so that the system will converge to a static condition in a short amount of computational time. Although there are different techniques available for this purpose, adaptive dynamic relaxation (ADR) can be utilized (Kilic and Madenci 2010), and it is described in detail in Sect. 7.5.

Another important concern when using a numerical technique is the convergence of the results. It is important to use optimum values of parameters to achieve sufficient accuracy within a suitable amount of computational time. The determination of such PD parameters is described in Sect. 7.6.

As described in Sect. 4.2, the interactions associated with material points close to the free surfaces are truncated, and this causes a reduction of the stiffness of these material points. In other words, these material points do not represent the accurate bulk behavior and require a correction. The correction can be imposed by introducing surface correction factors that can be directly inserted in the equation of motion, as described in Sect. 7.7.

Solution to the PD equation of motion requires initial conditions on displacement and velocity, as well as boundary conditions, as described in Sect. 2.7. Numerical implementations of the initial and boundary conditions are given in Sect. 7.8. If necessary, the introduction of a pre-existing crack is rather straightforward, as explained in Sect. 7.9. Moreover, as a result of extreme loading conditions, such as high velocity boundary conditions, large displacement boundary constraints, impact problems, etc., unexpected damage patterns may occur, especially close to the boundary region. This problem can be overcome by defining "no fail zones" and is also explained in Sect. 7.9. The measure of local damage for crack growth is explained in Sect. 7.10.

Each material point has its own particular family members defined by its horizon. For domains including a large number of material points, it is important to utilize an efficient process to search and establish the family members, and store their information, as presented in Sect. 7.11. Utilization of parallel computing is a crucial process to achieve significant computational efficiency. A brief discussion on parallel computing is given in Sect. 7.12.

The development of a solution algorithm for the PD equation of motion may involve the following steps:

- Specify the input parameters and initialize the matrices.
- Determine a stable time step size for the time integration. If the analysis involves the adaptive dynamic relaxation technique, the time step size is equal to 1.
- Generate the material points.
- Determine the material points inside the horizon of each material point and store them.
- In the case of a pre-existing crack problem, remove the PD interactions that are passing through the crack surfaces.
- Compute the surface correction factor for each material point.
- Apply initial conditions.
- If the analysis involves the adaptive dynamic relaxation technique, construct the stable mass matrix.
- Start time integration.
- Apply boundary conditions.
- Compute the total PD interaction forces acting on each material (collocation) point.
- Terminate the PD interaction if its stretch exceeds the critical stretch.

- If the analysis involves the adaptive dynamic relaxation technique, compute the adaptive dynamic relaxation technique parameters.
- Perform time integration to obtain displacements and velocities.

## 7.1   Spatial Discretization

In order to solve Eq. 2.22, a collocation method is adopted and the numerical treatment involves the discretization of the domain of interest into subdomains, as shown in Fig. 7.1. The domain can be discretized into subdomains by employing line subdomains for one-dimensional geometries, triangular and quadrilateral subdomains for two-dimensional regions, and hexahedron, tetrahedron, and wedge subdomains for three-dimensional regions, as shown in Fig. 7.2.

After discretizing the domain, the collocation points are placed in the subdomains, as shown in Fig. 7.1. With this meshless discretization scheme, the volume integration in Eq. 2.22 can be approximated as

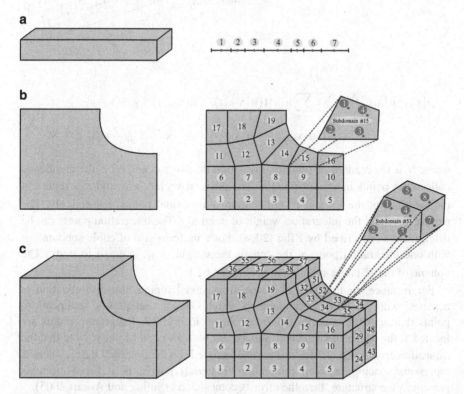

**Fig. 7.1** Discretization of the domain of interest for (**a**) one-dimensional, (**b**) two-dimensional, and (**c**) three-dimensional regions

128      7 Numerical Solution Method

**Fig. 7.2** Subdomain shapes
for one-, two-, and three-
dimensional geometries

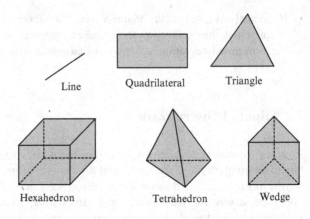

Line     Quadrilateral     Triangle

Hexahedron     Tetrahedron     Wedge

**Fig. 7.3** Discretization
and material points in a
one-dimensional region

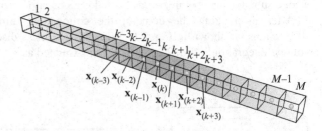

$$\rho\big(\mathbf{x}_{(k)}\big)\,\ddot{\mathbf{u}}\big(\mathbf{x}_{(k)},t\big)=\sum_{e=1}^{N}\sum_{j=1}^{N_e}w_{(j)}\big[\mathbf{t}\big(\mathbf{u}\big(\mathbf{x}_{(j)},t\big)-\mathbf{u}\big(\mathbf{x}_{(k)},t\big),\mathbf{x}_{(j)}-\mathbf{x}_{(k)}\big)$$
$$-\mathbf{t}\big(\mathbf{u}\big(\mathbf{x}_{(k)},t\big)-\mathbf{u}\big(\mathbf{x}_{(j)},t\big),\mathbf{x}_{(k)}-\mathbf{x}_{(j)}\big)\big]V_{(j)}+\mathbf{b}\big(\mathbf{x}_{(k)},t\big),$$

(7.1)

where $N$ is the number of subdomains within the horizon and $N_e$ is the number of collocation points in $e^{th}$ subdomain. The position vectors $\mathbf{x}_{(k)}$ and $\mathbf{x}_{(j)}$ represent the locations of the $k^{th}$ and $j^{th}$ collocation (integration) points, respectively. The parameter $w_{(j)}$ is the integration weight of point $\mathbf{x}_{(j)}$. The integration points can be determined as described by Kilic (2008). For a uniform grid of cubic subdomains with one integration point at the center, the weight, $w_{(j)}$, is equal to unity. The volume of the $j^{th}$ cubic subdomain is denoted by $V_{(j)}$

For instance, in the case of a one-dimensional region, the discretization is achieved with $M$ cubic subdomains in which Gaussian integration (collocation) points represent the material points, as shown in Fig. 7.3. Integration points are located at the center of each cubic subdomain with a weight of unity. Note that the truncation error in Eq. 7.1 for this particular case is on the order of $O(\Delta^2)$, where $\Delta$ represents spacing between integration (material) points. If a discontinuity is present in the structure, then the error becomes $O(\Delta)$ (Silling and Askari 2005).

## 7.2   Volume Correction Procedure

Associated with a material point, $\mathbf{x}_{(k)}$, the numerical integration over its horizon is approximated by considering the entire volume of each material point, $\mathbf{x}_{(j)}$, within its horizon. As illustrated in Fig. 7.4, in the case of a uniform spacing of $\Delta$ between the material points leading to cubic subdomains $(w_{(j)} = 1)$, and for a horizon of $\delta = 3\Delta$, this numerical approximation leads to summation of the material point volumes within the range of $\xi_{(k)(j)} = |\mathbf{x}_{(j)} - \mathbf{x}_{(k)}| < \delta$. As implemented in the EMU code (Silling 2004), this approximation can be improved by considering the entire volume of the material points within the range of $\xi_{(k)(j)} = |\mathbf{x}_{(j)} - \mathbf{x}_{(k)}| < \delta - r$, in which $r = \Delta/2$, the distance from the surface of the horizon. For the material points that are within the range of $\delta - r < \xi_{(k)(j)} < \delta$, a volume correction factor of $v_{c(j)} = (\delta + r - \xi_{(k)(j)})/2r$ is introduced by using a linear variation between a factor of 1 and ½ depending on the family member's location with respect to the horizon boundary. For the material points that are located outside of this region, the volume correction factor is $v_{c(j)} = 1$.

Thus, the discretized equation of motion, Eq. 7.1, for material point $\mathbf{x}_{(k)}$ including the volume correction can be rewritten as

$$\rho(\mathbf{x}_{(k)})\,\ddot{\mathbf{u}}(\mathbf{x}_{(k)}, t) = \sum_{e=1}^{N} \big[\mathbf{t}\big(\mathbf{u}(\mathbf{x}_{(j)}, t) - \mathbf{u}(\mathbf{x}_{(k)}, t), \mathbf{x}_{(j)} - \mathbf{x}_{(k)}\big)$$
$$- \mathbf{t}\big(\mathbf{u}(\mathbf{x}_{(k)}, t) - \mathbf{u}(\mathbf{x}_{(j)}, t), \mathbf{x}_{(k)} - \mathbf{x}_{(j)}\big)\big]\big(v_{c(j)}V_{(j)}\big) + \mathbf{b}(\mathbf{x}_{(k)}, t).$$

$$(7.2)$$

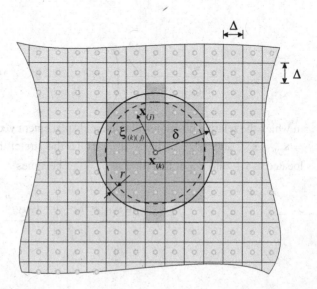

**Fig. 7.4** Volume correction for the collocation points inside the horizon

## 7.3  Time Integration

The time integration of the PD equation of motion in Eq. 7.2 can be performed by using explicit forward and backward difference techniques (Silling 2004). If the solution to Eq. 7.2 at the $n^{th}$ time step of $\Delta t$ (i.e., $t = n\,\Delta t$) is represented as $\mathbf{u}_{(k)}^{n}$ $= \mathbf{u}_{(k)}(t = n\Delta t)$, Eq. 7.2 can be rewritten for this time step in the form

$$\rho_{(k)}\ddot{\mathbf{u}}_{(k)}^{n} = \sum_{j=1}^{N} \left( \mathbf{t}_{(k)(j)}^{n} - \mathbf{t}_{(j)(k)}^{n} \right) \left( v_{c(j)} V_{(j)} \right) + \mathbf{b}_{(k)}^{n}, \tag{7.3}$$

where

$$\mathbf{t}_{(k)(j)}^{n} = \mathbf{t}_{(k)(j)}^{n} \left( \mathbf{u}_{(j)}^{n} - \mathbf{u}_{(k)}^{n}, \mathbf{x}_{(j)} - \mathbf{x}_{(k)} \right)$$

and

$$\mathbf{t}_{(j)(k)}^{n} = \mathbf{t}_{(j)(k)}^{n} \left( \mathbf{u}_{(k)}^{n} - \mathbf{u}_{(j)}^{n}, \mathbf{x}_{(k)} - \mathbf{x}_{(j)} \right)$$

represent the force density vectors between the material points located at $\mathbf{x}_{(k)}$ and $\mathbf{x}_{(j)}$. Using Eqs. 4.4 and 4.5, the force density vectors can be explicitly written as

$$\mathbf{t}_{(k)(j)}^{n} = \frac{\boldsymbol{\xi}_{(k)(j)} + \boldsymbol{\eta}_{(k)(j)}^{n}}{\left| \boldsymbol{\xi}_{(k)(j)} + \boldsymbol{\eta}_{(k)(j)}^{n} \right|} \left( 2ad\,\delta \frac{\Lambda_{(k)(j)}^{n}}{\left| \boldsymbol{\xi}_{(k)(j)} \right|} \theta_{(k)}^{n} + 2b\delta s_{(k)(j)}^{n} \right) \tag{7.4a}$$

and

$$\mathbf{t}_{(j)(k)}^{n} = -\frac{\boldsymbol{\xi}_{(k)(j)} + \boldsymbol{\eta}_{(k)(j)}^{n}}{\left| \boldsymbol{\xi}_{(k)(j)} + \boldsymbol{\eta}_{(k)(j)}^{n} \right|} \left( 2ad\,\delta \frac{\Lambda_{(k)(j)}^{n}}{\left| \boldsymbol{\xi}_{(k)(j)} \right|} \theta_{(j)}^{n} + 2b\delta s_{(k)(j)}^{n} \right), \tag{7.4b}$$

in which the relative position and relative displacement vectors are defined as $\boldsymbol{\xi}_{(k)(j)}$ $= \mathbf{x}_{(j)} - \mathbf{x}_{(k)}$ and $\boldsymbol{\eta}_{(k)(j)}^{n} = \mathbf{u}_{(j)}^{n} - \mathbf{u}_{(k)}^{n}$. Thus, the stretch between material points located at $\mathbf{x}_{(k)}$ and $\mathbf{x}_{(j)}$ at this time step, $s_{(k)(j)}^{n}$, becomes

$$s_{(k)(j)}^{n} = \frac{\left| \boldsymbol{\xi}_{(k)(j)} + \boldsymbol{\eta}_{(k)(j)}^{n} \right| - \left| \boldsymbol{\xi}_{(k)(j)} \right|}{\left| \boldsymbol{\xi}_{(k)(j)} \right|}. \tag{7.5}$$

**Fig. 7.5** Interaction of material points within the horizon

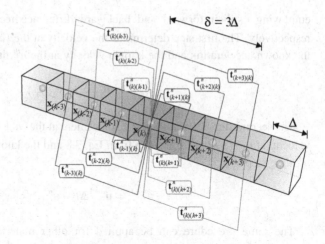

Furthermore, the dilatation at material points located at $\mathbf{x}_{(k)}$ and $\mathbf{x}_{(j)}$ can be computed from

$$\theta^n_{(k)} = d\delta \sum_{\ell=1}^{N} s^n_{(k)(\ell)} \Lambda^n_{(k)(\ell)} \left( v_{c(\ell)} V_{(\ell)} \right) \tag{7.6a}$$

and

$$\theta^n_{(j)} = d\delta \sum_{\ell=1}^{N} s^n_{(j)(\ell)} \Lambda^n_{(j)(\ell)} \left( v_{c(\ell)} V_{(\ell)} \right). \tag{7.6b}$$

As shown in Fig. 7.5, if the material point $k$ interacts with other material points within a horizon of $\delta = 3\Delta$, the peridynamic equation becomes

$$
\begin{aligned}
\rho_{(k)} \ddot{\mathbf{u}}^n_{(k)} = {} & \left( \mathbf{t}^n_{(k)(k+1)} - \mathbf{t}^n_{(k+1)(k)} \right) \left( v_{c(k+1)} V_{(k+1)} \right) \\
& + \left( \mathbf{t}^n_{(k)(k+2)} - \mathbf{t}^n_{(k+2)(k)} \right) \left( v_{c(k+2)} V_{(k+2)} \right) \\
& + \left( \mathbf{t}^n_{(k)(k+3)} - \mathbf{t}^n_{(k+3)(k)} \right) \left( v_{c(k+3)} V_{(k+3)} \right) \\
& + \left( \mathbf{t}^n_{(k)(k-1)} - \mathbf{t}^n_{(k-1)(k)} \right) \left( v_{c(k-1)} V_{(k-1)} \right) \\
& + \left( \mathbf{t}^n_{(k)(k-2)} - \mathbf{t}^n_{(k-2)(k)} \right) \left( v_{c(k-2)} V_{(k-2)} \right) \\
& + \left( \mathbf{t}^n_{(k)(k-3)} - \mathbf{t}^n_{(k-3)(k)} \right) \left( v_{c(k-3)} V_{(k-3)} \right) + \mathbf{b}^n_{(k)}.
\end{aligned}
\tag{7.7}
$$

After determining the acceleration of a material point at the $n^{th}$ time step from Eq. 7.3, the velocity and displacement at the next time step can be obtained by employing explicit forward and backward difference techniques in two steps, respectively. The first step determines the velocity at the $(n + 1)^{th}$ time step using the known acceleration and the known velocity at the $n^{th}$ time step as

$$\dot{\mathbf{u}}_{(k)}^{n+1} = \ddot{\mathbf{u}}_{(k)}^{n}\Delta t + \dot{\mathbf{u}}_{(k)}^{n}. \tag{7.8}$$

The second step determines the displacement at the $(n + 1)^{th}$ time step using the velocity at the $(n + 1)^{th}$ time step from Eq. 7.8 and the known displacement at the $n^{th}$ time step as

$$\mathbf{u}_{(k)}^{n+1} = \dot{\mathbf{u}}_{(k)}^{n+1}\Delta t + \mathbf{u}_{(k)}^{n}. \tag{7.9}$$

The same procedure can be applied for other material points as well. For instance, the displacement and velocity of the $(k + 1)^{th}$ material point can be obtained as

$$\mathbf{u}_{(k+1)}^{n+1} = \dot{\mathbf{u}}_{(k+1)}^{n+1}\Delta t + \mathbf{u}_{(k+1)}^{n} \tag{7.10a}$$

and

$$\dot{\mathbf{u}}_{(k+1)}^{n+1} = \ddot{\mathbf{u}}_{(k+1)}^{n}\Delta t + \dot{\mathbf{u}}_{(k+1)}^{n}. \tag{7.10b}$$

Note that the numerical error to obtain the displacement value by integrating the computed acceleration value from Eq. 7.3 is on the order of $O(\Delta t^2)$. Hence, the overall numerical error becomes $O(\Delta^2) + O(\Delta t^2)$, including the error from spatial integration (discretization). Furthermore, the overall error is $O(\Delta) + O(\Delta t^2)$, if there is any discontinuity in the structure (Silling and Askari 2005).

## 7.4   Numerical Stability

Although the explicit time integration scheme is straightforward, it is only conditionally stable. Therefore, a stability condition is necessary to obtain convergent results. A stability condition for the time step size, $\Delta t$, is derived based on the approach by Silling and Askari (2005). According to this approach, the standard von Neumann stability analysis can be performed by assuming a displacement variation of

$$u_{(k)}^{n} = \zeta^{n} e^{(\kappa k \sqrt{-1})}, \tag{7.11}$$

where $\kappa$ and $\zeta$ are positive real and complex numbers, respectively. The stability analysis requires that $|\zeta| \leq 1$ for all values of $\kappa$. Satisfaction of this condition is necessary so the waves do not grow unboundedly over time. By using an explicit central difference formula, Eq. 7.3 results in

$$
\rho_{(k)} \left( \frac{u_{(k)}^{n+1} - 2u_{(k)}^n + u_{(k)}^{n-1}}{\Delta t^2} \right) = \sum_j \frac{2ad\,\delta\left(\theta_{(k)}^n + \theta_{(j)}^n\right) + 4b\delta\left(u_{(j)}^n - u_{(k)}^n\right)}{\left|\boldsymbol{\xi}_{(k)(j)}\right|} v_{c(j)} V_{(j)},
$$

(7.12)

where

$$
\theta_{(k)}^n = d\delta \sum_\ell \frac{u_{(\ell)}^n - u_{(k)}^n}{\left|\boldsymbol{\xi}_{(\ell)(k)}\right|} \left(v_{c(\ell)} V_{(\ell)}\right)
$$

(7.13a)

and

$$
\theta_{(j)}^n = d\delta \sum_\ell \frac{u_{(\ell)}^n - u_{(j)}^n}{\left|\boldsymbol{\xi}_{(\ell)(j)}\right|} \left(v_{c(\ell)} V_{(\ell)}\right).
$$

(7.13b)

Substituting Eqs. 7.11 in 7.12 yields

$$
\rho_{(k)} \left( \frac{\zeta^{n+1} - 2\zeta^n + \zeta^{n-1}}{\Delta t^2} \right) e^{\left(\kappa k \sqrt{-1}\right)}
$$
$$
= \sum_j \left( 2ad\,\delta \frac{\left(\theta_{(k)}^n + \theta_{(j)}^n\right)}{\left|\boldsymbol{\xi}_{(k)(j)}\right|} + 4b\delta \frac{\zeta^n \left(e^{\left(\kappa j \sqrt{-1}\right)} - e^{\left(\kappa k \sqrt{-1}\right)}\right)}{\left|\boldsymbol{\xi}_{(k)(j)}\right|} \right) \left(v_{c(j)} V_{(j)}\right),
$$

(7.14)

where

$$
\theta_{(k)}^n = d\delta \sum_\ell \frac{\zeta^n e^{\left(\kappa \ell \sqrt{-1}\right)} - \zeta^n e^{\left(\kappa k \sqrt{-1}\right)}}{\left|\boldsymbol{\xi}_{(\ell)(k)}\right|} \left(v_{c(\ell)} V_{(\ell)}\right)
$$

(7.15a)

and

$$
\theta_{(j)}^n = d\delta \sum_\ell \frac{\zeta^n e^{\left(\kappa \ell \sqrt{-1}\right)} - \zeta^n e^{\left(\kappa j \sqrt{-1}\right)}}{\left|\boldsymbol{\xi}_{(\ell)(j)}\right|} \left(v_{c(\ell)} V_{(\ell)}\right).
$$

(7.15b)

Rearranging Eq. 7.14 results in

$$
\rho_{(k)} \left( \frac{\zeta^2 - 2\zeta + 1}{\Delta t^2} \right)
$$

$$
= \sum_j \left( 2ad\,\delta \frac{\left( \bar{\theta}^n_{(k)} + \bar{\theta}^n_{(j)} \right)}{\left| \boldsymbol{\xi}_{(k)(j)} \right|} + 4b\delta \frac{\left( e^{\left( \kappa(j-k)\sqrt{-1} \right)} - 1 \right)}{\left| \boldsymbol{\xi}_{(k)(j)} \right|} \right) \zeta \left( v_{c(j)} V_{(j)} \right), \tag{7.16}
$$

where

$$
\bar{\theta}^n_{(k)} = d\delta \sum_\ell \frac{\left( e^{\left( \kappa(\ell-k)\sqrt{-1} \right)} - 1 \right)}{\left| \boldsymbol{\xi}_{(\ell)(k)} \right|} \left( v_{c(\ell)} V_{(\ell)} \right) \tag{7.17a}
$$

and

$$
\bar{\theta}^n_{(j)} = d\delta \sum_\ell \frac{\left( e^{\left( \kappa(\ell-j)\sqrt{-1} \right)} - 1 \right)}{\left| \boldsymbol{\xi}_{(\ell)(j)} \right|} \left( v_{c(\ell)} V_{(\ell)} \right). \tag{7.17b}
$$

Since exponential terms can be written in terms of sine and cosine functions, and sine is an odd function, Eq. 7.16 can be rewritten as

$$
\rho_{(k)} \left( \frac{\zeta^2 - 2\zeta + 1}{\Delta t^2} \right)
$$

$$
= \sum_j \left( 2ad\,\delta \frac{\left( \bar{\theta}^n_{(k)} + \bar{\theta}^n_{(j)} \right)}{\left| \boldsymbol{\xi}_{(k)(j)} \right|} + 4b\delta \frac{(\cos(\kappa(j-k)) - 1)}{\left| \boldsymbol{\xi}_{(k)(j)} \right|} \right) \zeta \left( v_{c(j)} V_{(j)} \right), \tag{7.18}
$$

where

$$
\bar{\theta}^n_{(k)} = d\delta \sum_\ell \frac{\cos(\kappa(\ell - k)) - 1}{\left| \boldsymbol{\xi}_{(\ell)(k)} \right|} \left( v_{c(\ell)} V_{(\ell)} \right) \tag{7.19a}
$$

and

$$
\bar{\theta}^n_{(j)} = d\delta \sum_\ell \frac{\cos(\kappa(\ell - j)) - 1}{\left| \boldsymbol{\xi}_{(\ell)(j)} \right|} \left( v_{c(\ell)} V_{(\ell)} \right). \tag{7.19b}
$$

By defining

$$M_\kappa = -\frac{1}{2} \sum_j \left( 2ad\,\delta \frac{\left( \bar{\theta}_{(k)}^n + \bar{\theta}_{(j)}^n \right)}{\left| \boldsymbol{\xi}_{(k)(j)} \right|} + 4b\delta \frac{(\cos(\kappa(j-k))) - 1)}{\left| \boldsymbol{\xi}_{(k)(j)} \right|} \right) \left( v_{c(j)} V_{(j)} \right), \quad (7.20)$$

Equation 7.18 takes the form

$$\zeta^2 - 2\left( 1 - \frac{M_\kappa \Delta t^2}{\rho_{(k)}} \right) \zeta + 1 = 0. \tag{7.21}$$

The solution to the quadratic equation results in

$$\zeta = 1 - \frac{M_\kappa \Delta t^2}{\rho_{(k)}} \pm \sqrt{\left( 1 - \frac{M_\kappa \Delta t^2}{\rho_{(k)}} \right)^2 - 1}. \tag{7.22}$$

Enforcing the condition $|\zeta| \le 1$ yields

$$\Delta t < \sqrt{2\rho_{(k)}/M_\kappa}, \text{ for all } \kappa \text{ values.} \tag{7.23}$$

In order for this condition to be valid for all $\kappa$ values implies

$$M_\kappa \le \sum_j \left( 2ad\,\delta \frac{\left( d\delta \sum_\ell \left( \frac{1}{|\boldsymbol{\xi}_{(\ell)(k)}|} + \frac{1}{|\boldsymbol{\xi}_{(\ell)(j)}|} \right) V_{(\ell)} \right)}{\left| \boldsymbol{\xi}_{(k)(j)} \right|} + \frac{4b\delta}{\left| \boldsymbol{\xi}_{(k)(j)} \right|} \right) \left( v_{c(j)} V_{(j)} \right). \tag{7.24}$$

By using Eqs. 7.23 and 7.24, the stability criterion on the time step size can be expressed as

$$\Delta t < \sqrt{\frac{2\rho_{(k)}}{\sum_j \left( 2ad\,\delta \frac{\left( d\delta \sum_\ell \left( \frac{1}{|\boldsymbol{\xi}_{(\ell)(k)}|} + \frac{1}{|\boldsymbol{\xi}_{(\ell)(j)}|} \right) V_{(\ell)} \right)}{\left| \boldsymbol{\xi}_{(k)(j)} \right|} + \frac{4b\delta}{\left| \boldsymbol{\xi}_{(k)(j)} \right|} \right) \left( v_{c(j)} V_{(j)} \right)}}. \tag{7.25}$$

The use of a safety factor that has a value of less than 1 is recommended as it makes the analysis more stable in case of some type of nonlinearity in the structure.

It is also worth noting that the stable time step size is dependent on the horizon size rather than the grid size because of the dependency of the PD material parameters on the horizon (Silling and Askari 2005).

## 7.5   Adaptive Dynamic Relaxation

Although the equation of motion of the peridynamic theory is in dynamic form, it can still be applicable to solve quasi-static or static problems by using a dynamic relaxation technique. As explained by Kilic and Madenci (2010), the dynamic relaxation method is based on the fact that the static solution is the steady-state part of the transient response of the solution. By introducing an artificial damping to the system, the solution is guided to the steady-state solution as fast as possible. However, it is not always possible to determine the most effective damping coefficient. Therefore, the damping coefficient is determined at each time step by using the Adaptive Dynamic Relaxation (ADR) scheme introduced by Underwood (1983).

According to the ADR method, the PD equation of motion is written as a set of ordinary differential equations for all material points in the system by introducing new fictitious inertia and damping terms

$$\mathbf{D\ddot{U}}(\mathbf{X}, t) + c\mathbf{D\dot{U}}(\mathbf{X}, t) = \mathbf{F}(\mathbf{U}, \mathbf{U}', \mathbf{X}, \mathbf{X}'), \qquad (7.26)$$

where $\mathbf{D}$ is the fictitious diagonal density matrix and $c$ is the damping coefficient whose values are determined by Greschgorin's theorem (Underwood 1983) and Rayleigh's quotient, respectively. The vectors $\mathbf{X}$ and $\mathbf{U}$ contain the initial position and displacement of the collocation (material) points, respectively, and they can be expressed as

$$\mathbf{X}^T = \left\{ \mathbf{x}_{(1)}, \mathbf{x}_{(2)}, \ldots, \mathbf{x}_{(M)} \right\} \qquad (7.27a)$$

and

$$\mathbf{U}^T = \left\{ \mathbf{u}\left(\mathbf{x}_{(1)}, t\right), \mathbf{u}\left(\mathbf{x}_{(2)}, t\right), \ldots, \mathbf{u}\left(\mathbf{x}_{(M)}, t\right) \right\}, \qquad (7.27b)$$

where $M$ is the total number of material points in the structure. Finally, the vector $\mathbf{F}$ is composed of PD interaction and body forces and its $i^{\text{th}}$ component can be expressed as

$$\mathbf{F}_{(i)} = \sum_{j=1}^{N} \left( \mathbf{t}_{(i)(j)} - \mathbf{t}_{(j)(i)} \right) \left( v_{cj} V_{(j)} \right) + \mathbf{b}_{(i)}. \qquad (7.28)$$

By utilizing central-difference explicit integration, displacements and velocities for the next time step can be obtained as

$$\dot{\mathbf{U}}^{n+1/2} = \frac{\left((2 - c^n \Delta t)\dot{\mathbf{U}}^{n-1/2} + 2\Delta t \mathbf{D}^{-1}\mathbf{F}^n\right)}{(2 + c^n \Delta t)} \qquad (7.29a)$$

and

$$\mathbf{U}^{n+1} = \mathbf{U}^n + \Delta t\, \dot{\mathbf{U}}^{n+1/2}, \qquad (7.29b)$$

where $n$ indicates the $n^{th}$ iteration. Although Eq. 7.29a cannot be used to start the iteration process due to an unknown velocity field at $t^{-1/2}$, it can be assumed that $\mathbf{U}^0 \neq 0$ and $\dot{\mathbf{U}} = 0$. Therefore, the integration can be started by

$$\dot{\mathbf{U}}^{1/2} = \frac{\Delta t\, \mathbf{D}^{-1}\mathbf{F}^0}{2}. \qquad (7.30)$$

Note that the only physical term in this algorithm is the force vector, $\mathbf{F}$. The density matrix, $\mathbf{D}$, damping coefficient, $c$, and time step size , $\Delta t$, do not have to be physical quantities. Thus, their values can be chosen to obtain faster convergence.

In dynamic relaxation, a time step size of 1 ($\Delta t = 1$) is a convenient choice. The diagonal elements of the density matrix, $\mathbf{D}$, can be chosen based on Greschgorin's theorem and can be expressed as

$$\lambda_{ii} \geq \frac{1}{4}\Delta t^2 \sum_j |K_{ij}|, \qquad (7.31)$$

in which $K_{ij}$ is the stiffness matrix of the system under consideration. The inequality sign ensures stability of the central-difference explicit integration; the derivation of this stability condition is given by Underwood (1983). Although this approach achieves near-optimal values, these values are coordinate frame dependent because they depend on absolute values of the global stiffness matrix as stated in the context of the finite element method of Lovie and Metzger (1999). Therefore, an alternative way can be followed by choosing the values based on the minimum element dimension to make the frame invariant, as suggested by Sauve and Metzger (1997). This approach seems to reduce overshooting as compared to Greschgorin's theorem.

Hence, the present solutions of the PD equations also utilize a frame-invariant density matrix. The construction of the stiffness matrix requires determination of the derivative of PD interaction forces with respect to the relative displacement vector, $\boldsymbol{\eta}$. Since the PD interaction forces given in Eq. 7.3 are nonlinear functions of $\boldsymbol{\eta}$, it is not always possible to determine its derivative. However, elements of the stiffness matrix can be calculated by using a small displacement assumption as

$$
\begin{aligned}
\sum_j \left|K_{ij}\right| &= \sum_{j=1}^{N} \frac{\partial\left(\mathbf{t}_{(i)(j)} - \mathbf{t}_{(j)(i)}\right)}{\partial\left(\left|\mathbf{u}_{(j)} - \mathbf{u}_{(i)}\right|\right)} \cdot \mathbf{e} \\
&= \sum_{j=1}^{N} \frac{\left|\boldsymbol{\xi}_{(i)(j)} \cdot \mathbf{e}\right|}{\left|\boldsymbol{\xi}_{(i)(j)}\right|} \frac{4\delta}{\left|\boldsymbol{\xi}_{(i)(j)}\right|} \left(\frac{1}{2} \frac{ad^2\delta}{\left|\boldsymbol{\xi}_{(i)(j)}\right|} \left(v_{c(i)}V_{(i)} + v_{c(j)}V_{(j)}\right) + b\right),
\end{aligned}
\tag{7.32}
$$

in which $\mathbf{e}$ is the unit vector along the $x$-, $y$-, or $z$-direction. Note that the summation given in Eq. 7.32 can be employed to determine the elements of the stiffness matrix and it is frame invariant.

As described by Underwood (1983), the damping coefficient can be determined by using the lowest frequency of the system. The lowest frequency can be obtained by utilizing Rayleigh's quotient, which is given as

$$
\omega = \sqrt{\frac{\mathbf{U}^T \mathbf{K} \mathbf{U}}{\mathbf{U}^T \mathbf{D} \mathbf{U}}}.
\tag{7.33}
$$

However, the elements of the density matrix given in Eq. 7.31 may have large numerical values, which make the denominator in Eq. 7.33 numerically difficult to compute. In order to overcome this problem, Eq. 7.26 can be written in a different form at the $n^{th}$ iteration:

$$
\ddot{\mathbf{U}}^n(\mathbf{X}, t^n) + c^n \dot{\mathbf{U}}^n(\mathbf{X}, t^n) = \mathbf{D}^{-1} \mathbf{F}^n\left(\mathbf{U}^n, \mathbf{U}'^n, \mathbf{X}, \mathbf{X}'\right).
\tag{7.34}
$$

The damping coefficient in Eq. 7.34 can be expressed by using Eq. 7.33 as

$$
c^n = 2\sqrt{\left(\left(\mathbf{U}^n\right)^T {}^1\mathbf{K}^n \mathbf{U}^n\right) \Big/ \left(\left(\mathbf{U}^n\right)^T \mathbf{U}^n\right)},
\tag{7.35}
$$

in which $^1\mathbf{K}^n$ is the diagonal "local" stiffness matrix, which is given as

$$^1K_{ii}^n = -\left(F_i^n/\lambda_{ii} - F_i^{n-1}/\lambda_{ii}\right)\Big/\left(\Delta t\,\ddot{u}_i^{n-1/2}\right). \tag{7.36}$$

## 7.6 Numerical Convergence

The spacing between material points (grid size), $\Delta$, and the horizon size, $\delta$, influence the computational process. It is important to determine the optimum values of these parameters in order to achieve high accuracy with sufficiently small amount of computational time.

As explained in Silling and Askari (2005), the horizon size can be chosen based on the characteristic length dimensions. If dimensions are on the order of the nanoscale, then the horizon may represent the maximum distance of physical interactions between atoms or molecules. Therefore, it is important to specify its actual value for an accurate outcome of the analysis. For macroscale analysis, the horizon does not have a physical correspondence and its value can be chosen based on convenience. To determine the most optimum value of the horizon, a benchmark study of a one-dimensional bar with length $L$ subjected to an initial strain loading of $\partial u_x/\partial x = 0.001\,H(\Delta t - t)$ is considered. The spatial integration is performed by using a very fine grid, so the numerical error due to grid size is minimum. Six different horizon sizes are considered, $\delta = (1, 3, 5, 10, 25, 50)\,\Delta$. For each of these cases, the displacement versus time variation of a collocation point, which is located close to the center of the bar, is monitored and compared against the analytical solution given by Rao (2004). As demonstrated in Fig. 7.6a–f, the highest accuracy is achieved for the horizon sizes of $\delta = \Delta$ and $3\,\Delta$. The discrepancy between analytical and numerical solutions becomes larger when the horizon size increases due the excessive wave dispersion (Silling and Askari 2005).

Furthermore, the computational time increases substantially as the horizon size increases. It is recommended to choose a horizon size of $\delta = 3\,\Delta$ since $\delta = \Delta$ may cause grid dependence on crack propagation and not be able to capture crack branching behavior, as demonstrated in Fig. 7.7 for a square plate with a central crack subjected to a velocity boundary condition of $V_0 = 50\,\mathrm{m/s}$. The model with a horizon size of $\delta = 3\,\Delta$ captures the expected crack branching behavior due to a very high velocity boundary condition, whereas the model with a horizon size of $\delta = \Delta$ can only capture the self-similar crack growth.

As mentioned in Sect. 7.1, the discretization error is on the order of $O(\Delta^2)$. Therefore, it is important to use a sufficient number of grid points to reduce the numerical error and at the same time achieve the desired numerical efficiency. By considering the vibration of a bar, it is possible to visualize the effect of grid size on

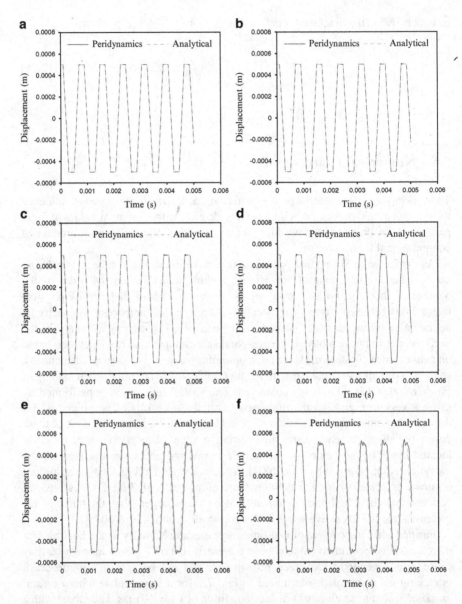

**Fig. 7.6** Variation of displacement with time at the center of the bar for horizon size values of (**a**) $\delta = \Delta$, (**b**) $\delta = 3\,\Delta$, (**c**) $\delta = 5\,\Delta$, (**d**) $\delta = 10\,\Delta$, (**e**) $\delta = 25\,\Delta$, and (**f**) $\delta = 50\,\Delta$

the accuracy for four different grid size values, $\Delta = L/10$, $L/100$, $L/1000$, and $L/10000$, as shown in Fig. 7.8a–d. The horizon size is specified as $\delta = 3\Delta$. Sufficient accuracy is obtained at a grid size value of $\Delta = L/1000$. Note that the error in the very coarse grid size case of $\Delta = L/10$ increases as the time progresses.

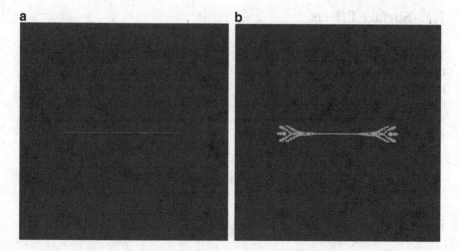

**Fig. 7.7** Damage distribution in a square plate with a central crack subjected to a velocity boundary condition of $V_0 = 50$ m/s for horizon values of (**a**) $\delta = \Delta$ and (**b**) $\delta = 3\,\Delta$

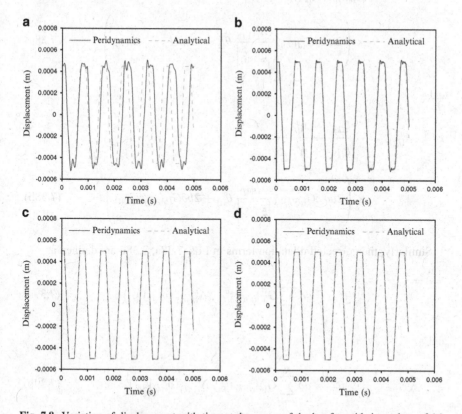

**Fig. 7.8** Variation of displacement with time at the center of the bar for grid size values of (**a**) $\Delta = L/10$, (**b**) $\Delta = L/100$, (**c**) $\Delta = L/1000$, and (**d**) $\Delta = L/10000$

## 7.7  Surface Effects

The lack of interactions due to free surfaces may cause inaccuracies, especially for the material points close to the surfaces. This problem can be largely overcome by introducing surface correction factors. Detailed information about the surface correction factors and their determination procedure are given in Chap. 4. The surface corrections can be directly invoked in the equation of motion, Eq. 7.3, by rewriting it in a slightly different form as

$$\rho_{(k)} \ddot{\mathbf{u}}^n_{(k)} = \sum_{j=1}^{N} \left( \bar{\mathbf{t}}^n_{(k)(j)} - \bar{\mathbf{t}}^n_{(j)(k)} \right) \left( v_{c(j)} V_{(j)} \right) + \mathbf{b}^n_{(k)}, \tag{7.37}$$

where the corrected PD interaction forces can be expressed as

$$\bar{\mathbf{t}}^n_{(k)(j)} = \frac{\boldsymbol{\xi}_{(k)(j)} + \boldsymbol{\eta}^n_{(k)(j)}}{\left| \boldsymbol{\xi}_{(k)(j)} + \boldsymbol{\eta}^n_{(k)(j)} \right|}$$

$$\times \left( 2ad\,\delta G_{(d)(k)(j)} \frac{\Lambda^n_{(k)(j)}}{\left| \boldsymbol{\xi}_{(k)(j)} \right|} \bar{\theta}^n_{(k)} + 2b\delta G_{(b)(k)(j)} s_{(k)(j)} \right) \tag{7.38a}$$

and

$$\bar{\mathbf{t}}^n_{(j)(k)} = -\frac{\boldsymbol{\xi}_{(k)(j)} + \boldsymbol{\eta}^n_{(k)(j)}}{\left| \boldsymbol{\xi}_{(k)(j)} + \boldsymbol{\eta}^n_{(k)(j)} \right|}$$

$$\times \left( 2ad\,\delta G_{(d)(k)(j)} \frac{\Lambda^n_{(k)(j)}}{\left| \boldsymbol{\xi}_{(k)(j)} \right|} \bar{\theta}^n_{(j)} + 2b\delta G_{(b)(k)(j)} s_{(k)(j)} \right). \tag{7.38b}$$

Similarly, the corrected dilatation terms in Eqs. 7.38a, 7.38b are defined as

$$\bar{\theta}^n_{(k)} = d\delta \sum_{\ell=1}^{N} G_{(d)(k)(\ell)} s^n_{(k)(\ell)} \Lambda^n_{(k)(\ell)} \left( v_{c(\ell)} V_{(\ell)} \right) \tag{7.39a}$$

and

$$\bar{\theta}^n_{(j)} = d\delta \sum_{\ell=1}^{N} G_{(d)(j)(\ell)} s^n_{(j)(\ell)} \Lambda^n_{(j)(\ell)} \left( v_{c(\ell)} V_{(\ell)} \right). \tag{7.39b}$$

Note that surface correction factors are consistently required in the time integration process. Therefore, they should be computed prior to the start of time integration. Since the determination of the correction factors requires a test loading condition on the actual structure, it is important to initialize the displacement and velocity values of collocation (material) points before starting the time integration process.

## 7.8  Application of Initial and Boundary Conditions

The PD equation of motion yields the acceleration of the collocation points. The displacement and velocity of collocation points can be obtained by integrating the acceleration; it requires the initial condition values of these quantities. Therefore, all of the collocation points should be subjected to initial displacement and velocity conditions. Various ways of specifying the initial conditions are explained in detail in Chap. 2. The initial conditions can be specified either in the form of displacement and velocity values on all material points as given in Eqs. 2.23a, b or in terms of displacement and velocity gradients as given in Eqs. 2.25a, b.

As also explained in Chap. 2, the displacement and velocity constraints can be applied in the peridynamic theory by following a different approach than in classical continuum mechanics. The constraint conditions can be imposed to material points inside a fictitious boundary region, $\mathcal{R}_c$, as demonstrated in Fig. 7.9, with a width equivalent to the horizon size, $\delta$. Displacement and velocity constraints can be applied by using Eqs. 2.26 and 2.28, respectively. On the other hand, the external loads can be applied as body loads through a material layer of $\mathcal{R}_\ell$, with a width of $\Delta$, as shown in Fig. 7.9. The magnitude of body force applied to collocation points inside this region can be obtained by using Eqs. 2.34a, b, depending on the nature of the applied loading condition, i.e., distributed pressure or point force.

## 7.9  Pre-existing Crack and No-Fail Zone

In many practical applications, cracks may initially exist in the structure and be located at various sites of the structure. The PD approach to create these initial cracks is rather straightforward. Any interaction between two material points

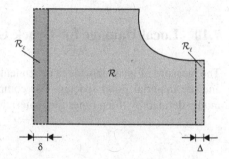

**Fig. 7.9** Boundary regions for (**a**) displacement and velocity constraints and (**b**) external loads

**Fig. 7.10** Termination
of PD interactions that pass
through a crack surface

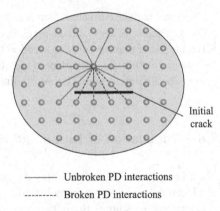

Initial
crack

——————  Unbroken PD interactions

- - - - - - - -  Broken PD interactions

**Fig. 7.11** No-fail zones

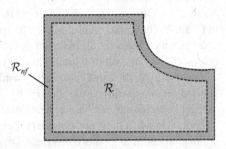

passing through a crack surface is terminated permanently, as shown in Fig. 7.10. Therefore, an entire set of terminated interactions represents a crack surface. If multiple cracks exist in the structure, the same procedure can be repeated for each crack surface.

For some applications under extreme loading conditions, unexpected failure may occur between collocation points located close to the external boundaries. In such cases, a region with a suitable width can be chosen as a "no fail zone," $\mathcal{R}_{nf}$, as shown in Fig. 7.11. The interactions associated with the collocation points located in this region are not allowed to fail. The thickness of the "no fail zone" should be chosen in such a way that it will have no adverse effect on the overall fracture behavior of the structure.

## 7.10   Local Damage for Crack Growth

The measure of local damage is dependent on the relationship between the horizon and the material point spacing. For computational efficiency, a horizon is commonly defined by three times the material point spacing, $\Delta$, i.e., $\delta = 3\Delta$.

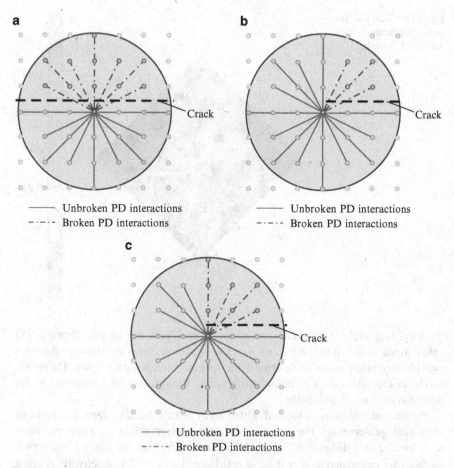

Fig. 7.12 Local damage at a material point (a) on the crack plane, (b) in front of the crack tip, and (c) behind the crack tip

As shown in Fig. 7.12a, the elimination of interaction between the material point $\mathbf{x}_{(j-)}$ and the others, $\mathbf{x}_{(k+)}$, above the dashed line representing the crack surface results in a local damage of $\varphi \approx 0.38$ for the material point $\mathbf{x}_{(j-)}$. Although computationally not feasible, the local damage at material point $\mathbf{x}_{(j-)}$ approaches one-half as the horizon approaches infinity. If the material point $\mathbf{x}_{(j-)}$ is located immediately ahead of the dashed line representing the crack surface, as shown in Fig. 7.12b, its interactions are still intact with the material points $\mathbf{x}_{(k+)}$ above the dashed line and directly aligned with $\mathbf{x}_{(j-)}$. Thus, the local damage at $\mathbf{x}_{(j-)}$ is calculated as $\varphi \approx 0.14$. If the material point $\mathbf{x}_{(j-)}$ is located immediately behind the dashed line representing the crack surface, as shown in Fig. 7.12c, its interactions are no longer intact with the material points $\mathbf{x}_{(k+)}$ above the dashed line and directly aligned with $\mathbf{x}_{(j-)}$. Thus, the local damage at $\mathbf{x}_{(j-)}$ is calculated as $\varphi \approx 0.24$.

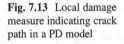

**Fig. 7.13** Local damage measure indicating crack path in a PD model

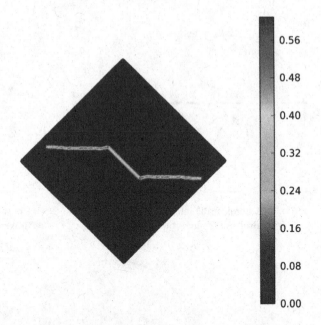

According to these local damage values, a crack path can be established in PD calculations with a horizon size of $\delta = 3\Delta$. However, the local damage does not provide any information to determine the specific broken interactions. Therefore, the local damage values at neighboring points should also be considered in the determination of a crack path.

As the material point is located farther away from a crack surface, its degree of local damage decreases. For example, the local damage values for a material point $\mathbf{x}_{(j^-)}$ located at a distance of $0.5\Delta$, $1.5\Delta$, and $2.5\Delta$ from the dashed line (crack surface) are calculated as $\varphi \approx 0.38$, $\varphi \approx 0.16$, and $\varphi \approx 0.02$, respectively. A crack surface results in discernible local damage values only for material points within a distance of $2\Delta$ away from the crack. Therefore, the local damage values can be used to identify crack path and tip with an error of less than $2\Delta$. Figure 7.13 shows the local damage in a plate with a crack in a peridynamic model. Both the path of the crack and the tip are clearly visible.

## 7.11  Spatial Partitioning

In the PD theory, the number of interactions is limited by defining a region called the horizon. The horizon makes the computations tractable; otherwise, the number of interactions that needs to be taken into account at each time step is $N^2$ for $N$ material points inside the body. This is especially very time consuming if the number of material points is large. According to the continuity assumption of the body, a material point must have the same neighbors during the deformation

**Fig. 7.14** Uniform grid and
interaction of collocation
points

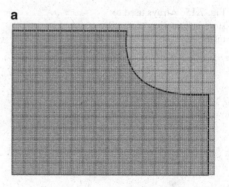

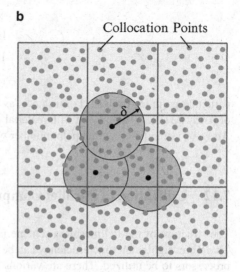

Collocation Points

process. Therefore, it is sufficient to determine the family members of a material
point within its horizon only once during the computation process.

While establishing the family members, it may be computationally advantageous to split the domain into equally sized cells, as shown in Fig. 7.14a. The size of
the cells should be larger than the horizon size. During the search process of the
family members, it is only necessary to examine the collocation points in the
neighboring cells, as shown in Fig. 7.14b.

Another important issue is following an efficient process for storing the family
members of the collocation points in order to overcome possible memory
limitations. For this purpose, two different arrays can be utilized. The first array
(see Array #1 in Fig. 7.15) can store all family members of material points
sequentially in a single column. The second array (see Array #2 in Fig. 7.15) can
be utilized as an indicator for the first array, so that the family members of a
particular material point can be easily extracted from the first array. Each element of
Array #2 corresponds to the location of the first material point within the family of a

**Fig. 7.15** Arrays used to store information on family members

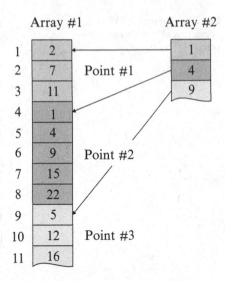

particular point in Array #1. For instance, as shown in Fig. 7.15, the second element of Array #2 (i.e., 4) associated with material point #2 indicates the fourth element of Array #1 as the first material point number within the family of material point #2.

## 7.12   Utilization of Parallel Computing and Load Balancing

The structure of the PD meshless scheme is very suitable for parallel computing. Therefore, significant time efficiency can be achieved, depending on the number of processors to be utilized. There are various tools available for parallel computing, such as central processing units (CPU) and graphics processing units (GPU). The most important aspect of parallel programming is the load balancing, so that full advantage of the parallel programming can be realized. Efficient load balancing can be obtained by distributing an approximately equal number of collocation points to each processor. Otherwise if a processor finishes its job earlier than others at the end of the time step, then it has to wait for other processors to finish their jobs to proceed to the next time step. The other important issue is to keep the number of PD interactions between collocation points that are assigned to different processors at a minimum level, since the computation of these interactions is carried out by a single processor to avoid a race condition. A race condition occurs when multiple processors try to access the same shared memory.

The computational domain can be divided into subunits and each of these subunits can be assigned to a specific processor by using binary space decomposition, as shown in Fig. 7.16 (Berger and Bokhari 1987). This method can handle variations in the collocation point concentration at different regions of the domain. The decomposition process continues in multiple steps, where each subunit is divided into two new rectangular subunits in every step. Based on the number of

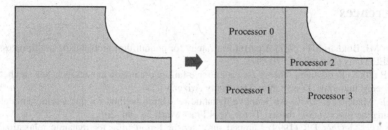

**Fig. 7.16**   Processor distribution

**Fig. 7.17**   Tree structure
to construct decomposition

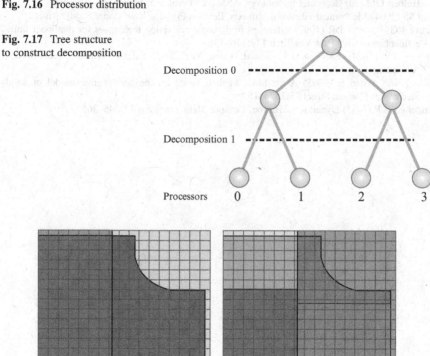

**Fig. 7.18**   Binary space decomposition

collocation points assigned to each subunit, a workload estimate can be calculated.
For instance, if there are $p$ number of available processors, where $p$ is not necessarily an even number, the domain is split into two, each having $s_1$ and $s_2$ collocation
points. Partitioning is performed such that the ratio of $s_1$ and $s_2$ is equal or close to
$(p/2)/(p - p/2)$. The partition direction is chosen to be the longest side of the
domain in order to reduce the number of interactions between subunits. Then, $p/2$
processors are assigned to the subunit having $s_1$ collocation points and $p - p/2$
processors are assigned to subunit with $s_2$ collocation points. Each subunit is then
divided into other subunits as long as the assigned number of processors is greater
than one. Figures 7.17 and 7.18 demonstrate a tree structure of a two-step binary
decomposition by using four processors and the subunits that are assigned to these
four processors, respectively.

# References

Berger MJ, Bokhari SH (1987) A partition strategy for nonuniform problems on multiprocessors. IEEE Trans Comp C-36:570–580

Kilic B (2008) Peridynamic theory for progressive failure prediction in homogeneous and heterogeneous materials. Dissertation, University Arizona

Kilic B, Madenci E (2010) An adaptive dynamic relaxation method for quasi-static simulations using the peridynamic theory. Theor Appl Fract Mech 53:194–201

Lovie TG, Metzger DR (1999) Lumped mass tensor formulation for dynamic relaxation. In: Hulbert GM (ed) Computer technology, ASME PVP vol 385, Boston, pp 255–260

Rao SS (2004) Mechanical vibrations, 4th edn. Pearson Prentice Hall, Upper Saddle River

Sauve RG, Metzger DR (1997) Advances in dynamic relaxation techniques for nonlinear finite element analysis. J Pres Ves Tech 117:170–176

Silling SA (2004) EMU user's manual, Code Ver. 2.6d. Sandia National Laboratories, Albuquerque

Silling SA, Askari E (2005) A meshfree method based on the peridynamic model of solid mechanics. Comput Struct 83:1526–1535

Underwood P (1983) Dynamic relaxation. Comput Meth Trans Anal 1:245–265

# Chapter 8
# Benchmark Problems

This chapter provides solutions to many benchmark problems and comparisons with those of classical continuum mechanics, i.e., analytical or finite element analysis. Failure is not allowed in the construction of peridynamic solutions. These benchmark problems concern primarily structures with simple geometries and under simple quasi-static and dynamic loads.

A one-dimensional bar initially stretched is released after a short period of time, and then the same bar is considered under quasi-static external tension. Next, a two-dimensional isotropic and a specially orthotropic plate are considered under uniaxial tension or uniform temperature change. The subsequent problems concern three-dimensional modeling of a block of material under tension, bending, or compression, and with a spherical cavity under radial extension. The peridynamic solutions to these problems are obtained by developing specific FORTRAN programs, as available on the website http://extras.springer.com.

## 8.1 Longitudinal Vibration of a Bar

A bar is subjected to an initial stretch for a short period of time, and then the stretch is removed. As illustrated in Fig. 8.1, the bar is clamped at the left end. The solution is obtained by specifying the geometric parameters, material properties, initial and boundary conditions, as well as the peridynamic discretization and time integration parameters as:

Geometric Parameters

Length of the bar: $L = 1$ m
Cross-sectional area: $A = h \times h = 1 \times 10^{-6} \, \text{m}^2$

E. Madenci and E. Oterkus, *Peridynamic Theory and Its Applications*,
DOI 10.1007/978-1-4614-8465-3_8, © Springer Science+Business Media New York 2014

**Fig. 8.1** Geometry of a bar
subjected to initial strain
and its discretization

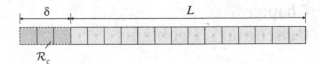

## Material Properties

Young's modulus: $E = 200$ GPa
Poisson's ratio: $\nu = 0.25$
Mass density: $\rho = 7850$ kg/m$^3$

## Boundary Conditions

$u_x(x = 0) = 0$

## Initial Conditions

Initial displacement gradient: $\partial u_x / \partial x = \varepsilon H(\Delta t - t)$ with $\varepsilon = 0.001$
Initial velocity: $\dot{u}_x(x, t) = 0$

## PD Discretization and Time Integration Parameters

Total number of material points in the x-direction: $1000 + 3$
Total number of material points in the y-direction: 1
Total number of material points in the z-direction: 1
Spacing between material points: $\Delta = 0.001$ m
Incremental volume of material points: $\Delta V = 1 \times 10^{-9}$ m$^3$
Volume of fictitious boundary region: $\Delta V_\delta = 3 \times 1 \times 1 \times \Delta V = 3 \times 10^{-9}$ m$^3$
Horizon: $\delta = 3.015\,\Delta$
Adaptive Dynamic Relaxation: OFF
Time step size: $\Delta t = 1.94598 \times 10^{-7}$ s
Total number of time steps: 26,000

*Numerical Results:* As given by Rao (2004), the analytical solution to this problem
can be easily constructed in the form

$$u_x(x,t) = \frac{8\varepsilon L}{\pi^2} \sum_{n=0}^{\infty} \frac{(-1)^n}{(2n+1)^2} \sin\left(\frac{(2n+1)\pi x}{2}\right) \cos\left(\sqrt{\frac{E}{\rho}}\frac{(2n+1)\pi}{2}t\right). \quad (8.1)$$

A material point located at $x = 0.4995$ m is monitored, and its displacement
variation with time is compared against the analytical solution. As shown in
Fig. 8.2, it is evident that the peridynamic (PD) simulation successfully captures
the expected longitudinal vibration.

**Fig. 8.2** Displacement of a
material (collocation) point
located at $x = 0.4995$ m
as a function of time

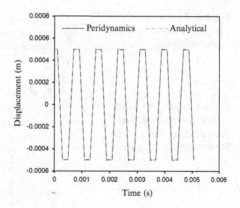

## 8.2   Bar Under Tension

The bar described in the previous case is now initially at rest, but subjected to a
quasi-static tension loading of $F = 200$ N at the free end (Fig. 8.3). The
peridynamic discretization parameters are the same as before, except for the applied
loading and time integration scheme. The external applied loading is introduced in
the form of a body force density in the boundary layer region.

Volume of boundary layer: $\Delta V_\Delta = 1 \times 1 \times 1 \times \Delta V = 1 \times 10^{-9} \, \text{m}^3$
Applied body force density: $b_x = F/\Delta V_\Delta = 2 \times 10^{11} \, \text{N/m}^3$
Adaptive Dynamic Relaxation: ON
Incremental time step size: $\Delta t = 1.0$ s
Total number of time steps: 10,000

*Numerical Results:* As observed in Fig. 8.4, the displacement of a collocation point
near the center of the bar appears to converge to a steady-state value after a time
step of 5,000. Hence, the displacement of collocation points along the bar at the end
of a time step of 10,000 is compared with the simple analytical solution, i.e.,

$$u_x = \frac{F}{AE}x = 0.001\,x. \tag{8.2}$$

As shown in Fig. 8.5, there is a close agreement between the PD predictions and
the analytical solution.

**Fig. 8.3** Geometry of
a bar under tension and
its discretization

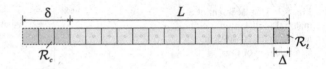

**Fig. 8.4** Displacement of
the collocation (material)
point near the center of the
bar as time step increases

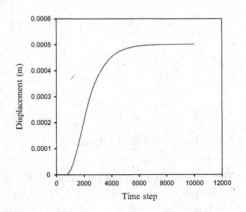

**Fig. 8.5** Comparison of
peridynamic prediction and
analytical solution for axial
displacements along the bar
at a time step value of
10,000

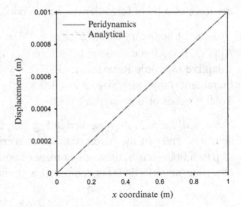

## 8.3   Isotropic Plate Under Uniaxial Tension or Uniform Temperature Change

A rectangular isotropic plate is subjected to uniaxial uniform tension or uniform temperature change, as shown Fig. 8.6. It is free of any displacement constraints. The applied tension is introduced in the form a body force density in the boundary layer region. The solution is obtained by specifying the geometric parameters, material properties, initial and boundary conditions, as well as the peridynamic discretization and time integration parameters as:

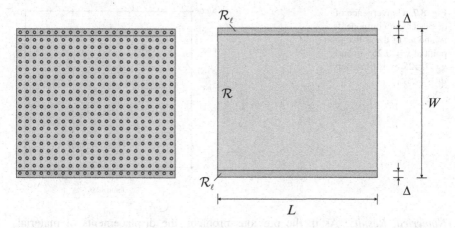

**Fig. 8.6** Geometry of a plate under uniaxial tension or uniform temperature change and its discretization

## Geometric Parameters

Length of the plate: $L = 1$ m
Width of the plate: $W = 0.5$ m
Thickness of the plate: $h = 0.01$ m

## Material Properties

Young's modulus: $E = 200$ GPa
Poisson's ratio: $\nu = 1/3$
Mass density: $\rho = 7850$ kg/m$^3$
Thermal expansion coefficient: $\alpha = 23 \times 10^{-6}/°$C

## Applied Loading

Uniaxial tension loading: $p_0 = 200$ MPa
Uniform temperature change: $\Delta T = 50°$C

## PD Discretization Parameters

Total number of material points in the x-direction: 100
Total number of material points in the y-direction: 50
Total number of material points in the z-direction: 1
Spacing between material points: $\Delta = 0.01$ m
Incremental volume of material points: $\Delta V = 1 \times 10^{-6}$ m$^3$
Boundary layer volume: $\Delta V_\Delta = 1 \times 50 \times 1 \times \Delta V = 50 \times 10^{-6}$ m$^3$
Applied body force density: $b_x = (p_0 W h)/\Delta V_\Delta = 2 \times 10^{10}$ N/m$^3$
Horizon: $\delta = 3.015\ \Delta$
Adaptive Dynamic Relaxation: ON
Incremental time step size: $\Delta t = 1.0$ s
Total number of time steps: 4,000

**Fig. 8.7** Convergence of displacement components, $u_x$ and $u_y$, of the collocation point at $x = 0.255$ m and $y = 0.125$ m as time step increases

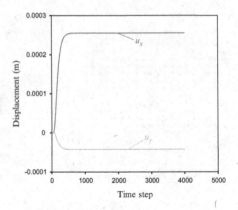

*Numerical Results:* As in the previous problem, the displacements of material points are monitored to ensure that the number of time steps used in the analysis is sufficient to reach the steady-state condition. As shown in Fig. 8.7, there is a rapid convergence of displacement components $u_x$ and $u_y$ to their steady-state values at $x = 0.255$ m and $y = 0.125$ m as the time step increases. Therefore, a total time step number of 4000 is considered sufficient to achieve a quasi-static solution.

The steady-state PD solutions for $u_x(x, y = 0)$ and $u_y(x = 0, y)$ are compared with the analytical solutions given as

$$u_x(x, y = 0) = \frac{p_0}{E} x \qquad (8.3a)$$

and

$$u_y(x = 0, y) = -\nu \frac{p_0}{E} y, \qquad (8.3b)$$

respectively. For both displacement components, there is a very good correlation between PD and analytical results, as shown in Fig. 8.8.

In the presence of only applied uniform temperature change of $\Delta T = 50°C$, the steady-state PD solutions for $u_x(x, y = 0)$ and $u_y(x = 0, y)$ are compared with the analytical solutions given as

$$u_x(x, y = 0) = \alpha (\Delta T) x \qquad (8.4a)$$

and

$$u_y(x = 0, y) = \alpha (\Delta T) y. \qquad (8.4b)$$

As shown in Fig. 8.9, there exists a remarkably close agreement between the analytical and peridynamic results.

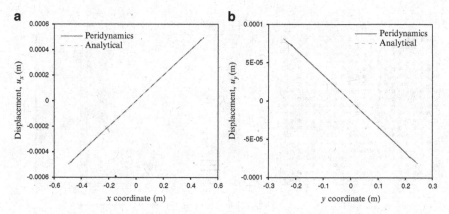

**Fig. 8.8** Displacement variations along the center lines under uniform tension: (**a**) $u_x(x, y = 0)$ and (**b**) $u_y(x = 0, y)$

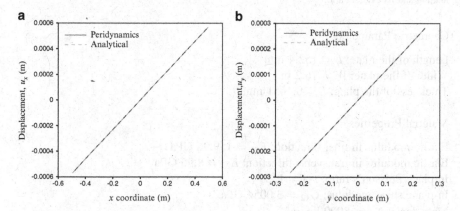

**Fig. 8.9** Displacement variations along the center lines under uniform temperature change: (**a**) $u_x(x, y = 0)$ and (**b**) $u_y(x = 0, y)$

## 8.4   Lamina Under Uniaxial Tension or Uniform Temperature Change

This problem is similar to the previous one except for the material properties. It has directional material properties representing a fiber-reinforced lamina. The fibers are aligned with the direction of tensile loading. A rectangular lamina is subjected to uniaxial uniform tension or uniform temperature change, as shown Fig. 8.10. It is free of any displacement constraints. The applied tension is introduced in the form of a body force density in the boundary layer region. The solution is obtained by specifying the geometric parameters, material properties, initial and boundary conditions, as well as the peridynamic discretization and time integration parameters as:

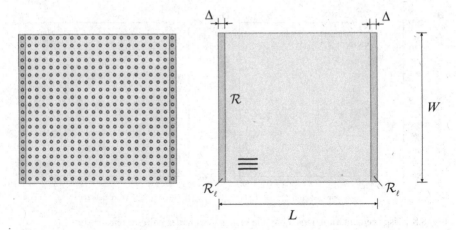

**Fig. 8.10** Geometry of a composite lamina under uniaxial tension or uniform temperature change loading and its discretization

Geometric Parameters

Length of the plate: $L = 152.4$ mm
Width of the plate: $W = 76.2$ m
Thickness of the plate: $h = 0.1651$ mm

Material Properties

Elastic modulus in fiber direction: $E_{11} = 159.96$ GPa
Elastic modulus in transverse direction: $E_{22} = 8.96$ GPa
In-plane Poisson's ratio: $\nu_{12} = 1/3$
In-plane shear modulus: $G_{12} = 3.0054$ GPa
Mass density: $\rho = 8000$ kg/m$^3$
Thermal expansion coefficient in fiber direction: $\alpha_1 = -1.52 \times 10^{-6}/°C$
Thermal expansion coefficient in transverse direction: $\alpha_2 = 34.3 \times 10^{-6}/°C$

Applied Loading:

Uniaxial tension loading: $p_0 = 159.96$ MPa
Uniform temperature change: $\Delta T = 50°C$

PD Discretization Parameters

Total number of material points in the x-direction: 240
Total number of material points in the y-direction: 120
Total number of material points in the z-direction: 1
Spacing between material points: $\Delta = 0.635$ mm
Incremental volume of material points: $\Delta V = 66.5724475 \times 10^{-12}$ m$^3$
Boundary layer volume: $\Delta V_\Delta = 1 \times 120 \times 1 \times \Delta V = 7.989 \times 10^{-9}$ m$^3$

**Fig. 8.11** Convergence of displacement components, $u_x$ and $u_y$, of the collocation point at $x = 38.4175$ mm and $y = 18.7325$ mm as time step increases

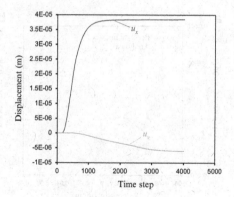

Applied body force density: $b_x = (p_0 Wh)/\Delta V_\Delta = 25.19 \times 10^{10}$ N/m$^3$
Horizon: $\delta = 3.015\ \Delta$
Adaptive Dynamic Relaxation: ON
Incremental time step size: $\Delta t = 1.0$ s
Total number of time steps: 4,000

*Numerical Results:* Based on a convergence study, 4,000 time steps are sufficient to reach the steady-state condition, as shown in Fig. 8.11, the same as in the isotropic plate case. Under uniaxial tensile loading, the steady-state solutions for displacement components along both the central $x$-axis and $y$-axis from PD analysis are compared with the analytical solutions given by

$$u_x(x, y = 0) = \frac{p_0}{E_{11}} x \qquad (8.5a)$$

and

$$u_y(x = 0, y) = -\nu_{12} \frac{p_0}{E_{11}} y. \qquad (8.5b)$$

A close agreement exists between the peridynamic and analytical solutions as presented in Fig. 8.12. After removing the tension loading, the lamina is subjected to a uniform temperature change of $\Delta T = 50°C$. The steady-state PD solutions for $u_x(x, y = 0)$ and $u_y(x = 0, y)$ are compared with the analytical solutions given as

$$u_x(x, y = 0) = \alpha_1 (\Delta T) x \qquad (8.6a)$$

and

$$u_y(x = 0, y) = \alpha_2 (\Delta T) y. \qquad (8.6b)$$

As demonstrated in Fig. 8.13, an excellent match exists between the peridynamic predictions and analytical solutions.

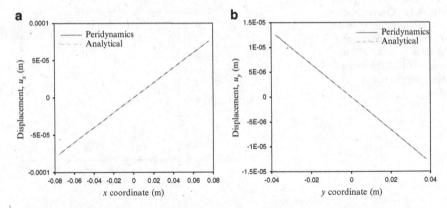

**Fig. 8.12** Displacement variations along the center lines: (**a**) $u_x(x, y = 0)$ and (**b**) $u_y(x = 0, y)$ under uniform tension

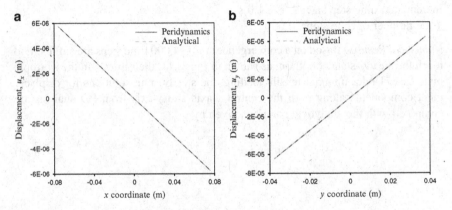

**Fig. 8.13** Displacement variations along the center lines: (**a**) $u_x(x, y = 0)$ and (**b**) $u_y(x = 0, y)$ under uniform temperature change

## 8.5   Block of Material Under Tension

A three-dimensional rectangular block is subjected to tensile loading at the free end. As illustrated in Fig. 8.14, it is fully clamped at the other end. The solution is obtained by specifying the geometric parameters, material properties, initial and boundary conditions, as well as the peridynamic discretization and time integration parameters as:

Geometric Parameters

Length of the block: $L = 1.0$ m
Width of the block: $W = 0.1$ m
Thickness of the block: $h = 0.1$ m

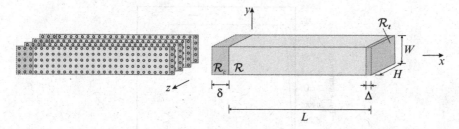

**Fig. 8.14** Geometry of a block under tension loading and its discretization

Material Properties

Young's modulus: $E = 200$ GPa
Poisson's ratio: $\nu = 0.25$
Mass density: $\rho = 7850$ kg/m$^3$

Boundary Conditions

Boundary condition at the left end: $u_x = u_y = u_z = 0$

Applied Loading

Uniaxial tensile load: $p_0 = 200$ MPa

PD Discretization Parameters

Total number of material points in the x-direction: $100 + 3$
Total number of material points in the y-direction: $10$
Total number of material points in the z-direction: $10$
Spacing between material points: $\Delta = 0.01$ m
Incremental volume of material points: $\Delta V = 1 \times 10^{-6}$ m$^3$
Volume of boundary layer: $\Delta V_\Delta = 1 \times 10 \times 10 \times \Delta V = 1 \times 10^{-4}$ m$^3$
Applied body force density: $b_x = (p_0 Wh)/\Delta V_\Delta = 2 \times 10^{10}$ N/m$^3$
Volume of fictitious boundary region: $\Delta V_\delta = 3 \times 10 \times 10 \times \Delta V = 3 \times 10^{-4}$ m$^3$
Horizon: $\delta = 3.015\,\Delta$
Adaptive Dynamic Relaxation: ON
Incremental time step size: $\Delta t = 1.0$ s
Total number of time steps: 4,000

*Numerical Results:* Based on the convergence study shown in Fig. 8.15, a total time step number of 1000 is sufficient to reach the steady-state condition. The steady-state solution is verified by comparing displacement components, shown in Fig. 8.16, along the central x-, y-, and z-axes with the analytical solutions given by

$$u_x(x, y = 0, z = 0) = \frac{p_0}{E}x, \tag{8.7a}$$

$$u_y(x = 0, y, z = 0) = -\nu\frac{p_0}{E}y, \tag{8.7b}$$

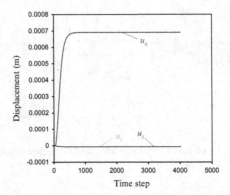

**Fig. 8.15** Convergence of displacement components, $u_x$, $u_y$, and $u_z$ at the collocation point of $x = 0.695$ m, $y = 0.025$ m, and $z = 0.025$ m as time step increases

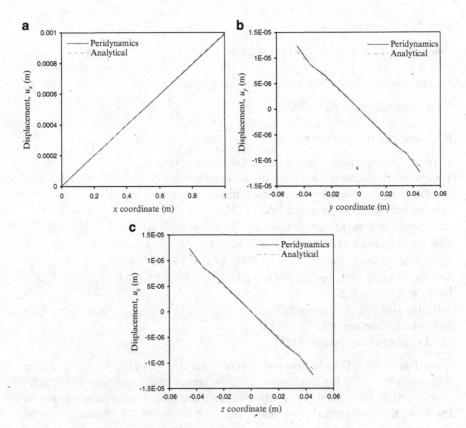

**Fig. 8.16** Displacement variations along the center lines under uniform tension: (**a**) $u_x(x, y = 0, z = 0)$, (**b**) $u_y(x = 0, y, z = 0)$, and (**c**) $u_z(x = 0, y = 0, z)$

and

$$u_z(x = 0, y = 0, z) = -\nu \frac{p_0}{E} z. \qquad (8.7c)$$

Although a coarse discretization is employed for the PD solution, a very good agreement is observed between the peridynamic and analytical solutions in Fig. 8.16.

## 8.6   Block of Material Under Transverse Loading

The rectangular block described in the preceding case is now subjected to a quasi-static transverse loading of $F = 5000$ N at the free end, as shown in Fig. 8.17. The peridynamic discretization parameters are the same as before except for the applied body force density, $b_y = F/\Delta V_\Delta = 5 \times 10^7$ N/m$^3$.

*Numerical Results:* Based on the convergence study shown in Fig. 8.18, a total time step number of 8000 is sufficient to reach the steady-state condition. The steady-state solution is verified by comparing the vertical displacement component, shown

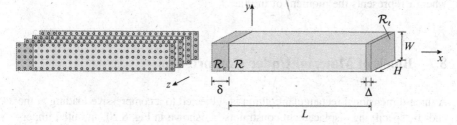

**Fig. 8.17**  Geometry of a rectangular block under transverse force and its discretization

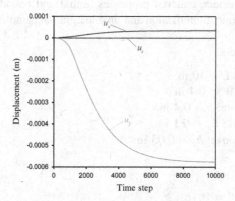

**Fig. 8.18**  Convergence of displacement components, $u_x$, $u_y$ and $u_z$ of the collocation point at $x = 0.695$ m, $y = 0.025$ m, and $z = 0.025$ m as time step increases

**Fig. 8.19** Displacement variations along the center line, $u_y(x, y = 0, z = 0)$ under transverse force

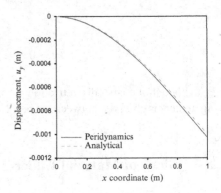

in Fig. 8.19, along the central $x$-axis with the finite element predictions. As presented in this figure, there is a good agreement between the peridynamic and analytical solution given by

$$u_y(x = 0, y, z = 0) = \frac{F}{6EI}(3L - x)x^2, \tag{8.8}$$

where $I$ represents the moment of inertia.

## 8.7  Block of Material Under Compression

A three-dimensional rectangular column is subjected to a compressive loading at the ends by specifying displacement constraints. As shown in Fig. 8.20, an initial imperfection is introduced in the form of a shallow groove at the upper surface near the center in order to trigger the out-of-plane displacement. The solution is obtained by specifying the geometric parameters, material properties, initial and boundary conditions, as well as the peridynamic discretization and time integration parameters as:

Geometric Parameters

Length of the plate: $L = 10$ in.
Width of the plate: $W = 0.4$ in.
Thickness of the plate: $h = 0.4$ in.
Length of the groove: $L_g = 0.1$ in.
Thickness of the groove: $h_g = 0.05$ in.

Material Properties

Young's modulus: $E = 10^7$ psi
Poisson's ratio: $\nu = 0.25$

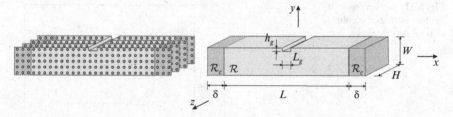

**Fig. 8.20** Geometry of a column under compression loading and its discretization

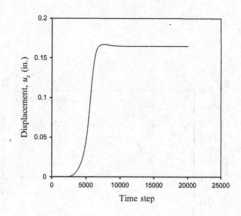

**Fig. 8.21** Convergence of transverse displacement component, $u_z$ of the collocation point at $x = 2.9$ 75 in., $y = 0.125$ in., and $z = 0.125$ in. as time step increases

Mass density: $\rho = 0.1$ lb/in$^3$

Boundary Conditions

Boundary condition at the left end: $u_x = 0.05$ in., $u_y = u_z = 0$
Boundary condition at the right end: $u_x = -0.05$ in., $u_y = u_z = 0$

PD Discretization Parameters

Total number of material points in the x-direction: $200 + 3 + 3$
Total number of material points in the y-direction: 8
Total number of material points in the z-direction: 8
Spacing between material points: $\Delta = 0.05$ in.
Incremental volume of material points: $\Delta V = 1.25 \times 10^{-4}$ in$^3$
Volume of fictitious boundary region: $\Delta V_\delta = 3 \times 8 \times 8 \times \Delta V = 2.4 \times 10^{-2}$ in$^3$
Horizon: $\delta = 3.015 \, \Delta$
Adaptive Dynamic Relaxation: ON
Incremental time step size: $\Delta t = 1.0$ s
Total number of time steps: 20,000

*Numerical Results:* As in the other cases, a convergence study was performed and, as demonstrated in Fig. 8.21, 15,000 time steps is sufficient to reach the steady-state

**Fig. 8.22** Convergence of
the force across the column
as time step increases

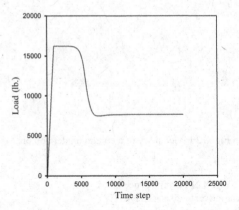

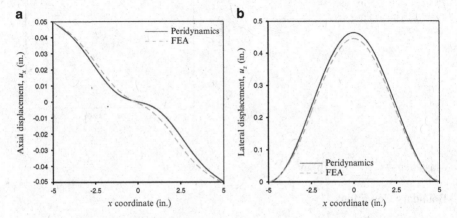

**Fig. 8.23** Variation of (**a**) axial displacement, $u_x$, and (**b**) transverse displacement, $u_z$, along central
$x$-axis

condition after the buckling occurs. As observed in Fig. 8.22, the peridynamic
prediction for column buckling load is about $P = 7650\,\text{N}$. The PD force is
calculated by summing all the forces in the axial direction passing through an
imaginary plane perpendicular to the loading direction located far from the loading
edges. The analytical prediction of $P = 8422\,\text{N}$ is obtained from

$$P = \frac{4\pi^2 EI}{L^2},\tag{8.9}$$

with $I$ being the moment of inertia. Even though the PD discretization is rather
coarse, these values are in reasonable agreement. Also, the peridynamic axial and
transverse displacements along the central $x$-axis are calculated and compared
against FEA solutions. As shown in Fig. 8.23, there is a very good agreement

**Fig. 8.24** Contour plot of transverse displacements, $u_z$, and the deformed shape of the column

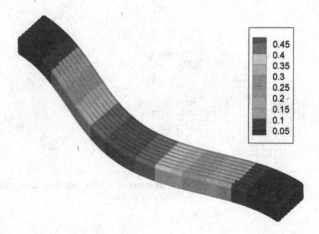

between the two approaches. Also, the contour plot of transverse displacement and the deformed shape of the column are presented in Fig. 8.24.

## 8.8 Block of Material with a Spherical Cavity Under Radial Extension

In this last problem, a spherical cavity inside a large block of material is subjected to radial extension, as shown in Fig. 8.25. The block is free of loading on its outer surfaces. The solution is obtained by specifying the geometric parameters, material properties, initial and boundary conditions, as well as the peridynamic discretization and time integration parameters as:

Geometric Parameters

Length of the block: $L = 1$ m
Width of the block: $W = 1$ m
Height of the block: $h = 1$ m
Radius of cavity: $a = 0.15$ m

Material Properties

Young's modulus: $E = 200$ GPa
Poisson's ratio: $\nu = 0.25$
Mass density: $\rho = 7850$ kg/m$^3$

Boundary Conditions

Radial displacement on the surface of the cavity: $u_r = u^* = 0.001$ m

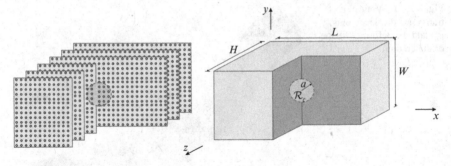

**Fig. 8.25** Geometry of a block of material with a spherical cavity under radial extension and its discretization

**Fig. 8.26** Convergence of displacement components, $u_x$, $u_y$, and $u_z$ of the collocation point at $x = 0.475$ m, $y = 0.375$ m, and $z = 0.275$ m as time step increases

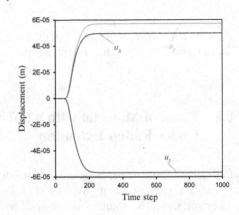

PD Discretization Parameters

Total number of material points in the x-direction: 81
Total number of material points in the y-direction: 81
Total number of material points in the z-direction: 81
Spacing between material points: $\Delta = 0.0125$ m
Incremental volume of material points: $\Delta V = 1.953125 \times 10^{-6}$ m$^3$
Volume of fictitious boundary region: $\Delta V_\delta = 4/3 \ \pi a^3 = 1.4137 \times 10^{-2}$ m$^3$
Horizon: $\delta = 3.015 \ \Delta$
Adaptive Dynamic Relaxation: ON
Incremental time step size: $\Delta t = 1.0$ s
Total number of time steps: 1,000

*Numerical Results:* After reaching the steady-state condition after 1,000 time steps, as in Fig. 8.26, the peridynamic simulation results are compared with the analytical solution given by

**Fig. 8.27** Radial
displacement variation
along $x$-axis

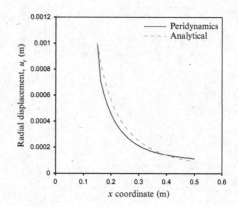

$$u_r = \frac{a^2}{r^2} u^*. \qquad (8.10)$$

As presented in Fig. 8.27, there is a good agreement between the two solutions.

# Reference

Rao SS (2004) Mechanical vibrations, 4th edn. Pearson Prentice Hall, Upper Saddle River

# Chapter 9
# Nonimpact Problems

In Chap. 8, peridynamic solutions of many benchmark problems were presented and compared with the classical theory in the absence of failure prediction. This chapter presents solutions to various problems while considering failure initiation and propagation. When available and suitable, the peridynamic (PD) predictions are compared with the finite element analysis (FEA) solutions.

An isotropic plate with a hole is slowly stretched along its horizontal boundaries. Its solution is straightforward when failure prediction is not a concern; however, it poses a challenge from a failure analysis point of view. Based on the Linear Elastic Fracture Mechanics (LEFM) concept, the traditional finite elements fail to address crack initiation and growth when there is no pre-existing crack in the structure. This problem demonstrates the capability of peridynamics when addressing crack initiation and its propagation. Next, an isotropic plate with a pre-existing crack is stretched rather fast along its horizontal boundaries. The peridynamic solution to this problem captures the effect of rate of loading (stretching) on the evolution of dynamic crack growth. In order to include the presence of thermal loading, a bimaterial strip with mismatch in thermal expansion coefficients is subjected to a uniform temperature change. Finally, an isotropic plate is subjected to a temperature gradient rather than a uniform temperature change. The peridynamic solutions to these problems are obtained by developing specific FORTRAN programs, which are available on the website http://extras.springer.com.

## 9.1 Plate with a Circular Cutout Under Quasi-Static Loading

As shown in Fig. 9.1, an isotropic plate with a circular cutout is subjected to a slow rate of stretch along its horizontal edges, representing quasi-static loading. There exist no initial cracks of any form in its domain. The solution is obtained by specifying the geometric parameters, material properties, initial and

E. Madenci and E. Oterkus, *Peridynamic Theory and Its Applications*,
DOI 10.1007/978-1-4614-8465-3_9, © Springer Science+Business Media New York 2014

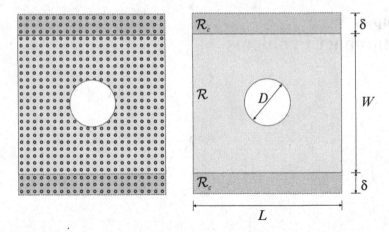

**Fig. 9.1** Geometry of a plate with a circular cutout under slow stretch and its discretization

boundary conditions as well as the peridynamic discretization and time integration parameters as:

### Geometric Parameters

Length of the plate: $L = 50$ mm
Width of the plate: $W = 50$ mm
Thickness of the plate: $h = 0.5$ mm
Diameter of the cutout: $D = 10$ mm

### Material Properties

Young's modulus: $E = 192$ GPa
Poisson's ratio: $\nu = 1/3$
Mass density: $\rho = 8000$ kg/m$^3$

### Boundary Conditions

$\dot{u}_y(x, \pm L/2, t) = \pm 2.7541 \times 10^{-7}$ m/s

### PD Discretization Parameters

Total number of material points in the x-direction: 100
Total number of material points in the y-direction: $100 + 3 + 3$
Total number of material points in the z-direction: 1
Spacing between material points: $\Delta = 0.0005$ m
Incremental volume of material points: $\Delta V = 1.25 \times 10^{-10}$ m$^3$
Volume of fictitious boundary layer: $\Delta V_\delta = 3 \times 100 \times 1 \times \Delta V = 3.75 \times 10^{-8}$ m$^3$
Horizon: $\delta = 3.015\,\Delta$
Critical stretch (failure off): $s_c = 1$
Critical stretch (failure on): $s_c = 0.02$

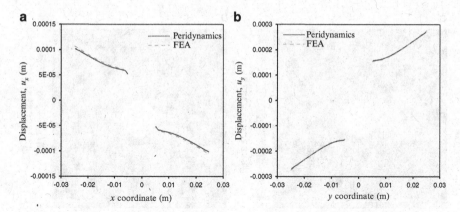

**Fig. 9.2** Variation of horizontal displacement (**a**) and vertical displacement (**b**) along the central axes at the end of 1,000 time steps when failure is not allowed

Adaptive Dynamic Relaxation: ON
Time step: $\Delta t = 1.0$ s
Total number of time steps: 1,000

*Numerical Results:* First, the displacement field due to the applied loading is obtained and compared against finite element predictions in the absence of failure. The variations of horizontal and vertical displacements along the central $x$-axis and $y$-axis, respectively, are shown in Fig. 9.2. A close agreement is observed between PD predictions and FEA results with ANSYS, a commercially available code. This indicates that the values of PD parameters such as grid size and horizon, and the volume of the boundary region, provide accurate results. After establishing the values of the PD parameters, failure among the material points is allowed by specifying a critical stretch value of $s_c = 0.02$, and the damage progression is examined at different time steps. Although there is no pre-existing crack in the plate, failure initiates in the form of a crack at the stress concentration sites. This is clearly an exceptional feature of the PD theory, unlike the other existing techniques that require pre-existing cracks. As shown in Fig. 9.3a, the damage initiates in the stress concentration sites at the end of 650 time steps. At the end of 700 time steps (Fig. 9.3b), the local damage value of some material points exceeds $\varphi = 0.38$, resulting in self-similar crack growth. Due to the low value of applied velocity along the boundary, representative of quasi-static loading, the crack continues to propagate toward the external vertical boundaries, as shown in Fig. 9.3c, d.

## 9.2   Plate with a Pre-existing Crack Under Velocity Boundary Conditions

The circular hole is replaced with a pre-existing crack, as shown in Fig. 9.4. Also, its horizontal edges are subjected to a very fast rate of stretch (velocity) in order to observe how the rate of loading affects the evolution of dynamic crack growth. The

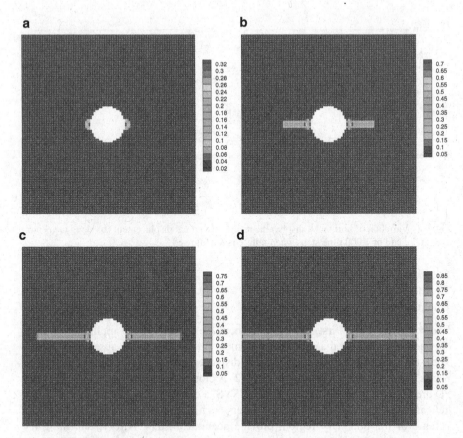

**Fig. 9.3** Damage plots for the plate with a circular cutout at the end of (**a**) 650 time steps, (**b**) 700 time steps, (**c**) 800 time steps, and (**d**) 1,000 time steps

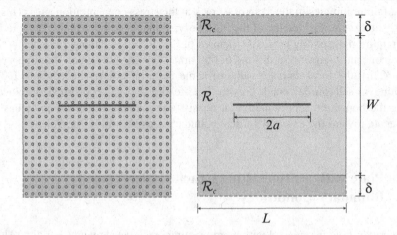

**Fig. 9.4** Geometry of a plate with a pre-existing crack under velocity boundary conditions and its discretization

plate properties and geometry are the same as before except for the thickness. The thickness and boundary conditions, as well as the peridynamic discretization and time integration parameters, are specified as:

Geometric Parameters

Thickness of the plate: $h = 0.0001$ m
Initial length of the pre-existing crack: $2a = 0.01$ m

Boundary Conditions

Case 1: $\dot{u}_y(x, \pm L/2, t) = \pm 20.0$ m/s
Case 2: $\dot{u}_y(x, \pm L/2, t) = \pm 50.0$ m/s

PD Discretization Parameters

Total number of material points in the x-direction: 500
Total number of material points in the y-direction: $500 + 3 + 3$
Total number of material points in the z-direction: 1
Spacing between material points: $\Delta = 0.0001$ m
Incremental volume of material points: $\Delta V = 1 \times 10^{-12}$ m$^3$
Volume of fictitious boundary layer: $\Delta V_\delta = 3 \times 100 \times 1 \times \Delta V = 3 \times 10^{-10}$ m$^3$
Horizon: $\delta = 3.015 \, \Delta$
Adaptive Dynamic Relaxation: OFF
Time step: $\Delta t = 1.3367 \times 10^{-8}$ s
Total number of time steps: 1,250
Critical stretch (failure off): $s_c = 1$
Critical stretch (failure on): $s_c = 0.04472$

*Numerical Results:* First, failure is not allowed (the interaction between the material points never ceases), and the crack opening displacement is computed as shown in Fig. 9.5. Unlike the elliptical crack opening displacement of classical continuum mechanics, the PD analysis predicts a cusp-like crack opening displacement near the crack tip. As explained by Silling (2000), the elliptical crack opening displacement is a mathematical requirement of the unbounded stresses (physically impossible) near the crack tip. The PD theory successfully captures a more physically meaningful crack opening shape. When failure is allowed by using a critical stretch value of $s_c = 0.04472$, a self-similar crack growth is observed at the end of 1,250 time steps, as shown in Fig. 9.6a. This growth is typical for a mode-I type of loading. The position of the crack tip or crack growth is determined based on the local damage value of any material point that exceeds $\varphi = 0.38$ along the x-axis. The growth of a crack as a function of time is shown in Fig. 9.6b, and the crack growth speed can be evaluated as 1,650 m/s. This crack speed is less than the Rayleigh wave speed of 2,800 m/s, which is considered to be the upper limit of the crack growth speed for a mode-I type of loading (Silling and Askari 2005). If the applied velocity boundary condition is increased from $V_0(t) = 20$ m/s to

**Fig. 9.5** Crack opening displacement near the crack tip at the end of 1,250 time steps when failure is not allowed

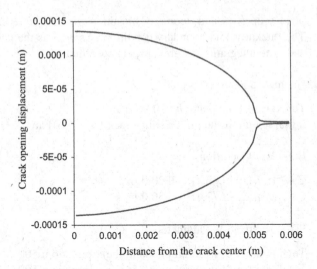

Distance from the crack center (m)

a                                    b

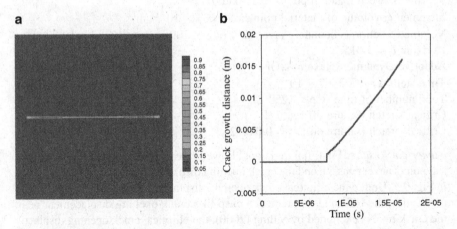

**Fig. 9.6** (a) Damage indicating self-similar crack growth at the end of 1,250 time steps under the velocity boundary condition of $V_0(t) = 20\,\text{m/s}$ and (b) crack growth as a function of time

$V_0(t) = 50\,\text{m/s}$, the crack growth characteristics change from self-similar to branching, as shown in Fig. 9.7. It is worth noting that the only parameter that was different between the two PD analyses while obtaining Figs. 9.6a and 9.7 is due to the applied velocity boundary condition. All other parameters remain the same. The PD theory captures a very complex phenomenon of crack branching without resorting to any external criteria that triggers branching.

**Fig. 9.7** Damage
indicating crack branching
at the end of 1,000 time
steps under a velocity
boundary condition of
$V_0(t) = 50 \, \text{m/s}$

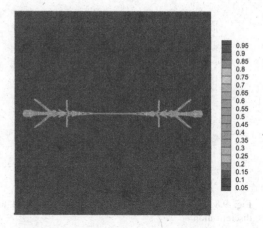

## 9.3   Bimaterial Strip Subjected to Uniform Temperature Change

A bimaterial strip is subjected to a uniform temperature change, as shown in Fig. 9.8. Both the top and bottom regions have the same length and thickness, but different widths and thermal expansion coefficients. Although the bimaterial strip is free of constraints and the temperature change is uniform, the mismatch between the thermal expansion coefficients causes bending deformation. The interface has the same properties as those of the top plate, and failure is not allowed. The solution is obtained by specifying the geometric parameters, material properties, initial and boundary conditions, as well as the peridynamic discretization and time integration parameters as:

Geometric Parameters

Length of plate: $L = 30$ mm
Thickness of plate: $h = 0.1$ mm
Width of bottom plate: $W_b = 1$ mm
Width of top plate: $W_t = 3$ mm

Material Properties

Young's modulus of bottom plate: $E_b = 128.0 \, \text{GPa}$
Poisson's ratio of bottom plate: $\nu_b = 1/3$
Thermal expansion coefficient of bottom plate: $\alpha_b = 16.6 \times 10^{-6}/^\circ\text{C}$
Young's modulus of top plate: $E_t = 5.1 \, \text{GPa}$
Poisson's ratio of top plate: $\nu_t = 1/3$
Thermal expansion coefficient of top plate: $\alpha_t = 50 \times 10^{-6}/^\circ\text{C}$

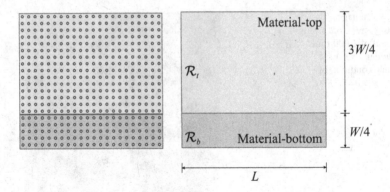

**Fig. 9.8** Geometry of a bimaterial strip subjected to uniform temperature change and its discretization

Applied Loading

Uniform temperature change: $\Delta T = 50°C$

PD Discretization Parameters

Total number of material points in x-direction: 300
Total number of material points in z-direction: 1
Total number of material points in bottom plate in y-direction: 10
Total number of material points in top plate in y-direction: 30
Spacing between material points, $\Delta = 0.1$ mm
Horizon: $\delta = 3.015 \; \Delta$
Adaptive Dynamic Relaxation: ON
Incremental time step size: $\Delta t = 1.0$ s
Total number of time steps: 20,000

*Numerical Results:* The PD predictions for displacement components, $u_x$ and $u_y$, along the interface between two plates are compared with those of FEA simulations. As observed in Fig. 9.9, the results from both approaches agree very well with each other. As expected, the bimaterial strip is curled down, presented in Fig. 9.9b, due to the mismatch between the coefficients of thermal expansion.

## 9.4   Rectangular Plate Subjected to Temperature Gradient

An isotropic plate is subjected to a nonuniform temperature change, as shown in Fig. 9.10. It is free of any constraints and, by specifying a unit critical stretch value, failure is not allowed. The solution is obtained by specifying the geometric parameters, material properties, initial and boundary conditions, as well as the peridynamic discretization and time integration parameters as:

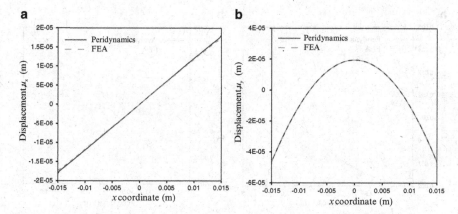

**Fig. 9.9** Variation of (**a**) $u_x$ displacement component, and (**b**) $u_y$ displacement component along the interface between two materials

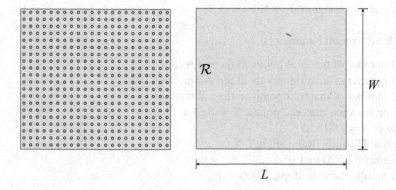

**Fig. 9.10** Geometry of a rectangular plate subjected to a temperature gradient and its discretization

## Geometric Parameters

Length of the plate: $L = 10$ in.
Width of the plate: $W = 4$ in.
Thickness of the plate: $h = 0.04$ in.

## Material Properties

Young's modulus: $E = 1 \times 10^7$ psi
Poisson's ratio: $\nu = 1/3$
Mass density: $\rho = 0.1$ lb/in$^3$
Thermal Expansion Coefficient: $\alpha = 24 \times 10^{-6}/^\circ C$

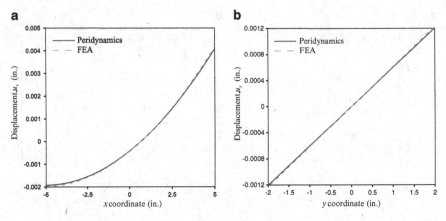

**Fig. 9.11** (a) $u_x$ displacement variation along central $x$-axis and (b) $u_y$ displacement variation

Applied Loading

Temperature change: $\Delta T = 5.0\,(x + 5.0)$

PD Discretization Parameters

Total number of material points in the x-direction: 250
Total number of material points in the y-direction: 100
Total number of material points in the z-direction: 1
Spacing between material points: $\Delta = 0.04$ in.
Horizon: $\delta = 3.015\ \Delta$
Adaptive Dynamic Relaxation: ON
Time step size: $\Delta t = 1.0$ s
Total number of time steps: 4,000

*Numerical Results:* The steady-state solution from PD is compared with the FEA predictions. The comparison of $u_x$ displacement values along the central $x$-axis and the $u_y$ displacement values along the central $y$-axis, shown in Fig. 9.11, indicates a close agreement.

# References

Silling SA (2000) Reformulation of elasticity theory for discontinuities and long-range forces. J Mech Phys Solids 48:175–209

Silling SA, Askari E (2005) A meshfree method based on the peridynamic model of solid mechanics. Comput Struct 83:1526–1535

# Chapter 10
# Impact Problems

This chapter concerns the peridynamic modeling of contact between two bodies due to an impact event. The impactor can be either rigid or deformable, and the target body is deformable. The interpenetration of bodies must be prevented between the bodies during the analysis. The treatment of contact due to a rigid impactor is different than that of a deformable impactor; Silling (2004) implemented two different techniques in the EMU code. The following sections will describe how interpenetration between the two bodies can be prevented while modeling contact due to a rigid or a deformable impactor. Also, applications are presented to well-known contact events such as the impact of two flexible bars, a rigid cylinder impacting a rectangular plate, and the Kalthoff and Winkler (1988) experiment. The peridynamic solutions to these problems are obtained by developing specific FORTRAN programs, which are available on the website http://extras. springer.com.

## 10.1  Impact Modeling

### 10.1.1  Rigid Impactor

The rigid impactor is not deformable at any instant, and it moves with its own velocity as a rigid body, as shown in Fig. 10.1a. The deformable target material is governed by the peridynamic equation of motion. After contact takes place between the impactor and target material, there is initially an interpenetration of material points, as illustrated in Fig. 10.1b. In order to reflect the physical reality, the material points inside the impactor are relocated to their new positions outside the impactor (see Fig. 10.1c). Their new locations are assigned in order to achieve the closest distance to the surface of the impactor. Hence, this process develops a contact surface between the impactor and material points at a particular time, $t$.

E. Madenci and E. Oterkus, *Peridynamic Theory and Its Applications*,
DOI 10.1007/978-1-4614-8465-3_10, © Springer Science+Business Media New York 2014

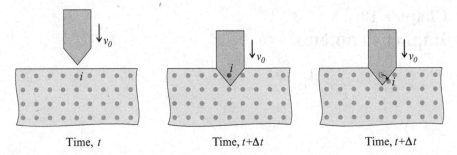

**Fig. 10.1** Relocation of material points inside a target material to represent contact with the impactor

The velocity of such a material point, $\mathbf{x}_{(k)}$, in its new location at the next time step, $t + \Delta t$, can be computed as

$$\bar{\mathbf{v}}_{(k)}^{t+\Delta t} = \frac{\bar{\mathbf{u}}_{(k)}^{t+\Delta t} - \mathbf{u}_{(k)}^{t}}{\Delta t}, \qquad (10.1)$$

where $\bar{\mathbf{u}}_{(k)}^{t+\Delta t}$ is the modified displacement vector at time $t + \Delta t$, $\mathbf{u}_{(k)}^{t}$ represents the displacement vector at time $t$, and $\Delta t$ corresponds to the time increment value.

At time $t + \Delta t$, the contribution of the material point, $\mathbf{x}_{(k)}$, to the reaction force from the target material to the impactor, $\mathbf{F}_{(k)}^{t+\Delta t}$, can be computed from

$$\mathbf{F}_{(k)}^{t+\Delta t} = -1 \times \rho_{(k)} \frac{\left( \bar{\mathbf{v}}_{(k)}^{t+\Delta t} - \mathbf{v}_{(k)}^{t+\Delta t} \right)}{\Delta t} V_{(k)}, \qquad (10.2)$$

where $\mathbf{v}_{(k)}^{t+\Delta t}$ is the velocity vector at time $t + \Delta t$ before relocating the material point $\mathbf{x}_{(k)}$, with $\rho_{(k)}$ and $V_{(k)}$ representing its density and volume, respectively. Summation of the contributions of all material points inside the impactor results in the total reaction force, $\mathbf{F}^{t+\Delta t}$, on the impactor at time $t + \Delta t$, and it can be expressed as

$$\mathbf{F}^{t+\Delta t} = \sum_{k=1} \mathbf{F}_{(k)}^{t+\Delta t} \lambda_{(k)}^{t+\Delta t}, \qquad (10.3)$$

where

$$\lambda_{(k)}^{t+\Delta t} = \begin{cases} 1 & \text{inside impactor} \\ 0 & \text{outside impactor} . \end{cases} \qquad (10.4)$$

## 10.1.2  Flexible Impactor

In the case of a flexible impactor, both the target material and the impactor are governed by the peridynamic equation of motion. However, when two bodies come close to each other within a critical distance, $r_{sh}$, they are forced to repel each other in order to define the contact between two bodies and also prevent sharing the same location by two or more material points, which is not acceptable from a continuum mechanics point of view.

This repelling short-range force, defined by Silling (2004), between material points can be expressed as

$$\mathbf{f}_{sh}\left(\mathbf{y}_{(j)}, \mathbf{y}_{(k)}\right) = \frac{\mathbf{y}_{(j)} - \mathbf{y}_{(k)}}{\left|\mathbf{y}_{(j)} - \mathbf{y}_{(k)}\right|} \min\left\{0, c_{sh}\left(\frac{\left|\mathbf{y}_{(j)} - \mathbf{y}_{(k)}\right|}{2r_{sh}} - 1\right)\right\}, \qquad (10.5)$$

where the short-range force constant, $c_{sh}$, and the critical distance, $r_{sh}$, can be chosen as

$$c_{sh} = 5c \qquad (10.6a)$$

and

$$r_{sh} = \frac{\Delta}{2}. \qquad (10.6b)$$

## 10.2  Validation

When failure initiation and growth is not permitted, the validity of the impact models is established by comparing peridynamic (PD) solutions with those of finite element analysis (FEA) using ANSYS. In the presence of fracture, the PD predictions are compared with observations such as those of the Kalthoff-Winkler experiment (Kalthoff and Winkler 1988). The first problem concerns the impact of two identical flexible bars. The solution to this problem is obtained by constructing a three-dimensional PD model. The second problem concerns a rigid disk impacting a rectangular plate on its edge. Its solution is obtained by constructing a two-dimensional PD model. The third problem simulates the Kalthoff-Winkler experiment with a three-dimensional PD model.

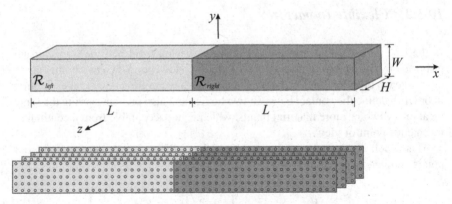

**Fig. 10.2** Impact of two identical flexible bars and PD discretization

## 10.2.1 Impact of Two Identical Flexible Bars

As shown in Fig. 10.2, the impact of two identical deformable bars is considered. Their velocities are equal but in opposite directions prior to their impact. Both are free of any displacement constraints and external loads. Also, failure is not allowed in order to compare the PD predictions with FEA results using ANSYS. The three-dimensional solution is constructed by specifying the geometric parameters, material properties, initial and boundary conditions, as well as the peridynamic discretization and time integration parameters as:

Geometric Parameters

Length of the identical bars: $L = 0.05$ m
Width of the identical bars: $W = 0.01$ m
Thickness of the identical bars: $h = 0.01$ m

Material Properties

Young's modulus: $E = 75$ GPa
Poisson's ratio: $\nu = 0.25$
Mass density: $\rho = 2700$ kg/m$^3$

Initial Conditions

Initial condition of the bars: $\dot{u}_x = \pm 10$ m/s

PD Discretization Parameters:

Total number of material points in the x-direction: 100
Total number of material points in the y-direction: 10
Total number of material points in the z-direction: 10

**Fig. 10.3** Comparison of displacement predictions in the $x$-direction from the PD and FEA at the centers $(\pm 0.025, 0.0, 0.0)$ of the *left* and *right* bars as time progresses

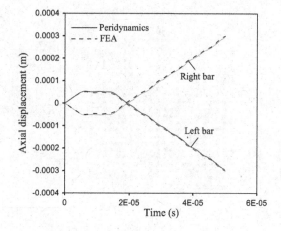

**Fig. 10.4** Comparison of the axial displacement predictions along the central $x$-axis from the PD and FEA at time step 535

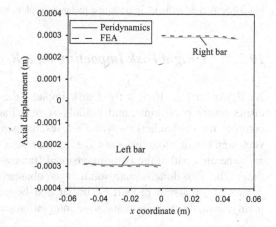

Spacing between material points: $\Delta = 0.001$ m

Incremental volume of material points: $\Delta V = 1 \times 10^{-9}$ m$^3$

Horizon: $\delta = 3.015\ \Delta$

Adaptive Dynamic Relaxation: OFF

Time step size: $\Delta t = 9.3184 \times 10^{-8}$ s

Total number of time steps: 535

*Numerical Results:* Figure 10.3 shows the axial displacement at the center of the bars. The displacement values change their sign and bars start to move in the opposite direction because the compressive waves generated after the initial impact of the bars propagate toward the free edges. These compressive waves are then transformed to tension waves when they reach the free edges, thus leading to the separation of the bars. This figure also shows the comparison of the PD and FEA predictions, and confirm the validity of the flexible impactor model proposed by Silling (2004). As shown in Fig. 10.4, the PD and FEA predictions for the axial displacement along the $x$-axis at the end of the analysis are also in close agreement.

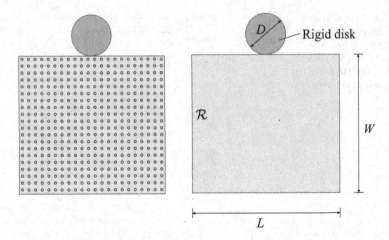

**Fig. 10.5** A rigid cylinder impacting a rectangular plate and PD discretization

## 10.2.2  A Rigid Disk Impacting on a Rectangular Plate

As shown in Fig. 10.5, a rigid disk impacts the edge of a plate that is free of displacement constraints, and initially at rest. Failure is not allowed in order to compare the PD predictions with FEA results using ANSYS. In the FEA model, a very high elastic modulus is assigned to represent the rigid impactor. Both models are generated within the two-dimensional framework to reduce the computational time. The two-dimensional solution is obtained by specifying the geometric parameters, material properties, initial and boundary conditions, as well as the peridynamic discretization and time integration parameters as:

Geometric Parameters

Length of the plate: $L = 0.2$ m
Width of the plate: $W = 0.1$ m
Thickness of the plate: $h = 0.009$ m

Material Properties

Young's modulus: $E = 191$ GPa
Poisson's ratio: $\nu = 1/3$
Mass density: $\rho = 8000$ kg/m$^3$

Impactor Properties:

Diameter of the impactor: $D = 0.05$ m
Thickness of the impactor: $H = 0.009$ m
Initial velocity of the impactor: $v_0 = 32$ m/s
Mass of the impactor: $m = 1.57$ kg

**Fig. 10.6** PD and FEA
predictions for the
displacement component
in the $y$-direction at the
center of the plate as
time progresses

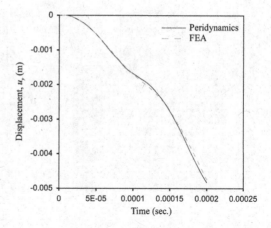

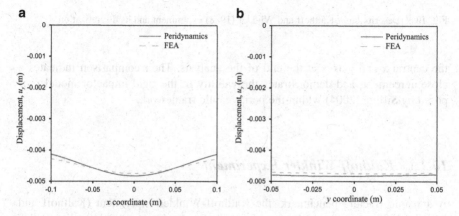

**Fig. 10.7** PD and FEA displacement predictions in the $y$-direction at a time step of 2000: (**a**) along
the central $x$-axis and (**b**) along the central $y$-axis

## PD Discretization Parameters

Total number of material points in the x-direction: 200
Total number of material points in the y-direction: 100
Total number of material points in the z-direction: 1
Spacing between material points: $\Delta = 0.001$ m
Incremental volume of material points: $\Delta V = 9 \times 10^{-9}$ m$^3$
Horizon: $\delta = 3.015 \, \Delta$
Adaptive Dynamic Relaxation: OFF
Time step size: $\Delta t = 1 \times 10^{-7}$ s
Total number of time steps: 2,000

*Numerical Results:* As shown in Fig. 10.6, a close agreement exists between the PD
and FEA predictions for displacement in the $y$-direction at the center of the plate as
time progresses. Figure 10.7 shows the PD and FEA displacement variations along

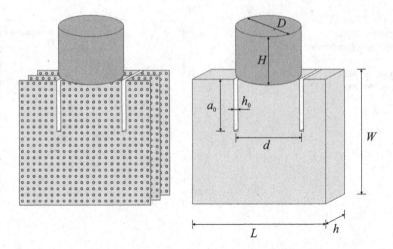

**Fig. 10.8** Description of Kalthoff and Winkler (1988) experiment and its discretization

the central $x$-and $y$-axes at the end of the analysis. Their comparison indicates a close agreement, and demonstrates the validity of the rigid impactor model proposed by Silling (2004) within the peridynamic framework.

### 10.2.3   Kalthoff-Winkler Experiment

A dynamic fracture benchmark, the Kalthoff-Winkler experiment (Kalthoff and Winkler 1988) concerns the impact of a steel plate having two notches (slits) with a cylindrical impactor, as depicted in Fig. 10.8. The slits are located symmetrically with respect to the central axis. The impactor is assumed rigid. The steel plate is free of displacement constraints, and initially at rest. Silling (2003) previously constructed the PD solution to this benchmark problem, and presents more detailed results and discussion. The solution is obtained by specifying the geometric parameters, material properties, initial and boundary conditions, as well as the peridynamic discretization and time integration parameters as:

Geometric Parameters

Length of the plate: $L = 0.2$ m
Width of the plate: $W = 0.1$ m
Thickness of the plate: $h = 0.009$ m
Distance between notches: $d = 0.05$ m
Notch length: $a_0 = 0.05$ m
Notch width: $h_0 = 0.0015$ m

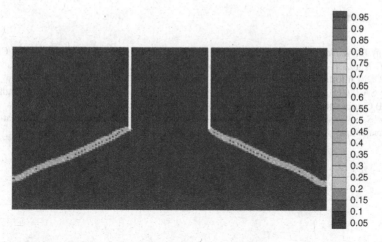

**Fig. 10.9** Contour plot for damage (crack propagation path)

Material Properties

Young's modulus: $E = 191$ GPa
Poisson's ratio: $\nu = 0.25$
Mass density: $\rho = 8000$ kg/m$^3$

Impactor Properties

Diameter of the impactor: $D = 0.05$ m
Height of the impactor: $H = 0.05$ m
Initial velocity of the impactor: $v_0 = 32$ m/s
Mass of the impactor: $m = 1.57$ kg

PD Discretization Parameters:

Total number of material points in the x-direction: 201
Total number of material points in the y-direction: 101
Total number of material points in the z-direction: 9
Spacing between material points: $\Delta = 0.001$ m
Incremental volume of material points: $\Delta V = 1 \times 10^{-9}$ m$^3$
Horizon: $\delta = 3.015\ \Delta$
Critical stretch: $s_c = 0.01$
Adaptive Dynamic Relaxation: OFF
Time step size: $\Delta t = 8.7 \times 10^{-8}$ s
Total number of time steps: 1,350

*Numerical Results:* Figure 10.9 shows the contour plots for damage patterns (crack growth). The PD predictions of 68° for crack propagation angles from the vertical axis are in excellent agreement with the experimental measurements (Kalthoff and Winkler 1988).

# References

Kalthoff JF, Winkler S (1988) Failure mode transition at high rates of shear loading. In: Chiem CY, Kunze H-D, Meyer LW (eds) Impact loading and dynamic behavior of materials, vol 1. DGM Informationsgesellschaft Verlag, Oberursel, pp 185–195

Silling SA (2003) Dynamic fracture modeling with a meshfree peridynamic code. In: Bathe KJ (ed) Computational fluid and solid mechanics. Elsevier, Oxford, pp 641–644

Silling SA (2004) EMU user's manual, Code Ver. 2.6d. Sandia National Laboratories, Albuquerque

# Chapter 11
# Coupling of the Peridynamic Theory and Finite Element Method

The PD theory provides deformation, as well as damage initiation and growth, without resorting to external criteria since material failure is invoked in the material response. However, it is computationally more demanding compared to the finite element method. Furthermore, the finite element method is very effective for modeling problems without damage. Hence, it is desirable to couple the PD theory and FEM to take advantage of their salient features if the regions of potential failure sites are identified prior to the analysis. Then, the regions in which failure is expected can be modeled by using the PD theory and the rest can be analyzed by using FEM.

A simple coupling approach is submodeling, demonstrated by Oterkus et al. (2012) and Agwai et al. (2012); it involves FEM for global analysis and PD theory for submodeling in order to perform failure prediction. The primary assumption in submodeling is that the structural details of the submodel do not significantly affect the global model. Also, the boundaries of the submodel should be sufficiently far away from local features so that St. Venant's principle holds for a valid submodeling analysis. The solution obtained from the global model along the boundary of the domain of interest is applied as displacement boundary conditions on the submodel. The global model should be refined enough to enable accurate calculation of the displacement on the boundary of the submodeling region. Also, different time discretizations of the displacement boundary condition should be considered because the time-dependent nature of boundary conditions may affect the results in submodeling.

Another straightforward way of coupling was suggested by Macek and Silling (2007) where the PD interactions are represented by using traditional truss elements and an embedded element technique for the overlap region. Lall et al. (2010) also utilized this approach to study shock and vibration reliability of electronics.

Recently, Liu and Hong (2012) introduced interface elements between FEM and PD regions. A finite number of peridynamic points are embedded inside the interface element to transfer information between PD and FEM regions. The peridynamic forces exerted on these embedded material points are distributed as

E. Madenci and E. Oterkus, *Peridynamic Theory and Its Applications*, 191
DOI 10.1007/978-1-4614-8465-3_11, © Springer Science+Business Media New York 2014

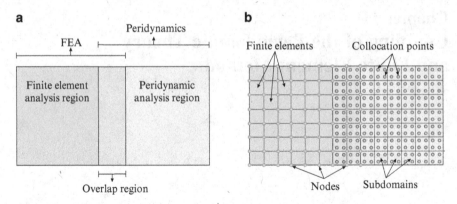

**Fig. 11.1** Schematic for coupling of the finite element method and peridynamics: (**a**) finite element (FEA) and peridynamic regions; (**b**) discretization

nodal forces to the interface element based on two particular schemes. In the first scheme, coupling forces are distributed to all nodes of the interface element. However, in the second scheme, the coupling forces are only distributed to nodes that are located on the interface plane between FEM and PD regions. The displacement of the embedded material points are not computed through a PD equation of motion. Instead, they are determined by utilizing the nodal displacements of interface elements and their shape functions.

Also, Lubineau et al. (2012) coupled local and nonlocal theories by introducing a transition (morphing) strategy. The definition of the morphing functions relies on the energy equivalence, and the transition region affects only constitutive parameters. The influence of local and nonlocal theories is captured by defining a function that automatically converges to full local and nonlocal formulations along their respective boundaries. In a recent study, Seleson et al. (2013) proposed a force-based blended model that coupled PD theory and classical elasticity by using nonlocal weights composed of integrals of blending functions. They also generalized this approach to couple peridynamics and higher-order gradient models of any order.

In addition to these techniques, Kilic and Madenci (2010) introduced a direct coupling of FEM and PD theory using an overlap region, shown in Fig. 11.1a, in which equations of both PD and FEM are solved simultaneously. The PD region is discretized with material points and the finite element region with traditional elements (Fig. 11.1b). Both the PD and FE equations are satisfied in the overlap region. Furthermore, the displacement and velocity fields are determined using finite element equations in the overlap region. These fields are then utilized to compute the body force densities using the PD theory. Finally, these body force densities serve as external forces for finite elements in the overlap region.

## 11.1 Direct Coupling

The direct coupling of PD theory and FEM presented herein concerns steady-state or quasi-static solutions. However, the PD equation of motion, Eq. 7.1, includes dynamic terms that need to be eliminated. Thus, the adaptive dynamic relaxation method, described in Chap. 7, is utilized to obtain a steady-state solution. The damping coefficient is changed adaptively in each time step. The dynamic relaxation method is based on the fact that the static solution is the steady-state part of the transient response of the solution.

In order to achieve direct coupling, the discrete PD equation of motion, Eq. 7.1, is rewritten as

$$\left\{ \begin{matrix} \ddot{\mathbf{U}}_p^n \\ \underline{\ddot{\mathbf{U}}}_p^n \end{matrix} \right\} + c^n \left\{ \begin{matrix} \dot{\mathbf{U}}_p^n \\ \underline{\dot{\mathbf{U}}}_p^n \end{matrix} \right\} = \begin{bmatrix} \mathbf{D}^{-1} & 0 \\ 0 & \underline{\mathbf{D}}^{-1} \end{bmatrix} \left\{ \begin{matrix} \mathbf{F}_p^n \\ \underline{\mathbf{F}}_p^n \end{matrix} \right\}, \tag{11.1}$$

in which $\mathbf{U}$ is a vector that contains displacements at the PD material points and the vector $\mathbf{F}$ is the summation of internal and external forces. The subscript $p$ denotes the variables associated with the PD region, and single and double underscores denote the variables located outside and inside the overlap region, respectively. The parameter $c^n$ represents the damping coefficient at the $n$th time increment. The coefficients of the fictitious diagonal density matrix, $\mathbf{D}$, are determined through Greschgorin's theorem (Underwood 1983). A detailed description of these parameters is presented in Chap. 7.

In order to achieve coupling of FEM with the PD theory, the direct assembly of finite element equations without constructing the global stiffness matrix is utilized so that the FE equations can be expressed as

$$\left\{ \begin{matrix} \ddot{\mathbf{U}}_f^n \\ \underline{\ddot{\mathbf{U}}}_f^n \end{matrix} \right\} + c^n \left\{ \begin{matrix} \dot{\mathbf{U}}_f^n \\ \underline{\dot{\mathbf{U}}}_f^n \end{matrix} \right\} = \begin{bmatrix} \mathbf{M}^{-1} & 0 \\ 0 & \underline{\mathbf{M}}^{-1} \end{bmatrix} \left\{ \begin{matrix} \mathbf{F}_f^n \\ \underline{\mathbf{F}}_f^n \end{matrix} \right\}, \tag{11.2}$$

in which subscript $f$ denotes the variables associated with the finite element region, and $\mathbf{M}$ is the diagonal mass matrix. The components of the mass matrix can be approximated as

$$\mathbf{M} = \mathbf{I}\tilde{\mathbf{m}}, \tag{11.3}$$

in which $\mathbf{I}$ is the identity matrix. The mass vector, $\tilde{\mathbf{m}}$, is constructed as

$$\tilde{\mathbf{m}} = \underset{e}{\mathbf{A}} \hat{\mathbf{m}}^{(e)}, \tag{11.4}$$

where $\mathbf{A}$ is the assembly operator and the operations are strictly performed as additions (Belytschko 1983). The components of vector $\hat{\mathbf{m}}^{(e)}$ can be written as

$$\hat{m}_i^{(e)} = \sum_{j=1}^{8} \left| k_{ij}^{(e)} \right|, \tag{11.5}$$

in which $k_{ij}^{(e)}$ indicates the components of the element stiffness matrix given by Zienkiewicz (1977).

The force vector $\mathbf{F}^n$ at the *nth* time increment can be expressed as

$$\mathbf{F}^n = \mathbf{f}^{ext}(t^n) - \mathbf{f}^{int}(\mathbf{u}^n), \tag{11.6}$$

where $t$ is time and $\mathbf{f}^{ext}$ is the vector of external forces.

The internal forces resulting from the deformation of the elements can be assembled into a global array of internal forces by using the convention of Belytschko (1983) as

$$\mathbf{f}^{int} = \mathbf{A}\, \mathbf{f}^{(e)}, \tag{11.7}$$

where $\mathbf{f}^{(e)}$ is the element force vector.

The element force vector is expressed as

$$\mathbf{f}^{(e)} = \mathbf{k}^{(e)} \mathbf{u}^{(e)}, \tag{11.8}$$

in which $\mathbf{k}^{(e)}$ is the element stiffness matrix described by Zienkiewicz (1977) and $\mathbf{u}^{(e)}$ is the vector representing the nodal displacements of the *eth* element.

The vector $\mathbf{u}_p$ representing displacements of a PD material point located inside the *eth* element can be obtained from

$$\mathbf{u}_p = \sum_{i=1}^{8} N_i \mathbf{u}_i^{(e)}, \tag{11.9}$$

where $N_i$ are the shape functions given by Zienkiewicz (1977). The vector $\mathbf{u}_i^{(e)}$ is the *ith* nodal displacements of the *eth* element and is extracted from the global solution vector, $\underline{\mathbf{U}}_f$, denoting nodal FE displacements. Determination of the vector $\mathbf{u}_p$ leads to the computation of vector $\underline{\mathbf{U}}_p$. The force density vector $\underline{\mathbf{F}}_p^n$ can then be computed by utilizing the force density vector $\mathbf{F}_p$ associated with the PD material point $\mathbf{x}_p$ inside the *eth* element (subdomain) as given by Eq. 7.1

$$\mathbf{F}_p = \mathbf{b}(\mathbf{x}_p, t) + \sum_{e=1}^{N} \sum_{j=1}^{N_e} w_{(j)} \left[ \mathbf{t}\big(\mathbf{u}(\mathbf{x}_{(j)}, t) - \mathbf{u}(\mathbf{x}_p, t), \mathbf{x}_{(j)} - \mathbf{x}_p, t\big) \right.$$
$$\left. - \mathbf{t}\big(\mathbf{u}(\mathbf{x}_p, t) - \mathbf{u}(\mathbf{x}_{(j)}, t), \mathbf{x}_p - \mathbf{x}_{(j)}, t\big) \right] V_{(j)}, \tag{11.10}$$

where $N$ is the number of elements within the horizon and $N_e$ is the number of collocation points in the $e^{th}$ element. The position vector $\mathbf{x}_{(j)}$ represents the location of the $j^{th}$ collocation (integration) point.

Since, for a quasi-static problem, the equilibrium should be satisfied for all material points, i.e., $\mathbf{F}_p = \mathbf{0}$, the body load exerted on material point $\mathbf{x}_p$ can then be calculated as

$$
\begin{aligned}
\mathbf{b}(\mathbf{x}_p, t) = -\sum_{e=1}^{N}\sum_{j=1}^{N_e} w_{(j)} & \left[ \mathbf{t}\big(\mathbf{u}(\mathbf{x}_{(j)}, t) - \mathbf{u}(\mathbf{x}_p, t), \mathbf{x}_{(j)} - \mathbf{x}_p, t\big) \right. \\
& \left. - \mathbf{t}\big(\mathbf{u}(\mathbf{x}_p, t) - \mathbf{u}(\mathbf{x}_{(j)}, t), \mathbf{x}_p - \mathbf{x}_{(j)}, t\big) \right] V_{(j)} .
\end{aligned}
\tag{11.11}
$$

The total body load associated with the element where the material point $\mathbf{x}_p$ is located can be computed as

$$
\mathbf{g}^{(e)} = \sum_{j=1}^{N_e} \mathbf{b}(\mathbf{x}_p, t).
\tag{11.12}
$$

Furthermore, the calculated total body load can be lumped into the finite element nodes as

$$
\mathbf{f}_I^{(e)} = \int_{V_e} dV_e N_I \rho\, \mathbf{g}^{(e)},
\tag{11.13}
$$

in which $\rho$ is the mass density of the *eth* element and $I$ indicates the *Ith* node of the *eth* element. Hence, $\mathbf{f}_I^{(e)}$ indicates the external force acting on the *Ith* node. The body force density is only known at the PD material points, which serve as integration points for the *eth* element in Eq. 11.13. Furthermore, $\underline{\underline{\mathbf{F}}}_f^n$ is constructed by assembling the nodal forces given by Eq. 11.13.

Finally, the coupled system of equations can be expressed as

$$
\ddot{\underline{\mathbf{U}}}^n + c^n \dot{\underline{\mathbf{U}}}^n = \mathbf{M}^{-1}\underline{\mathbf{F}}^n,
\tag{11.14}
$$

in which $\dot{\underline{\mathbf{U}}}^n$ and $\ddot{\underline{\mathbf{U}}}^n$ are the first and second time derivatives of the displacements, respectively, and can be expressed as

$$
\dot{\underline{\mathbf{U}}}^n = \left\{ \dot{\underline{\mathbf{U}}}_p^n \quad \dot{\underline{\mathbf{U}}}_f^n \quad \dot{\underline{\underline{\mathbf{U}}}}_f^n \right\}^T,
\tag{11.15a}
$$

$$
\ddot{\underline{\mathbf{U}}}^n = \left\{ \ddot{\underline{\mathbf{U}}}_p^n \quad \ddot{\underline{\mathbf{U}}}_f^n \quad \ddot{\underline{\underline{\mathbf{U}}}}_f^n \right\}^T.
\tag{11.15b}
$$

The matrix $\mathbf{M}$ can be written as

$$\mathbf{M} = \begin{bmatrix} \underline{\mathbf{D}} & 0 & 0 \\ 0 & \mathbf{M} & 0 \\ 0 & 0 & \underline{\mathbf{M}} \end{bmatrix}. \tag{11.16}$$

The vector $\mathbf{F}$ is given as

$$\mathbf{F}^n = \left\{ \underline{\mathbf{F}}_p^h \quad \underline{\mathbf{F}}_f^n \quad \underline{\underline{\mathbf{F}}}_f^n \right\}^T. \tag{11.17}$$

As suggested by Underwood (1983), the damping coefficient $c^n$ can be determined as

$$c^n = 2\sqrt{\left((\mathbf{U}^n)^T \, {}^1\mathbf{K}^n \mathbf{U}^n\right) \Big/ \left((\mathbf{U}^n)^T \, \mathbf{U}^n\right)}, \tag{11.18}$$

in which ${}^1\mathbf{K}^n$ is the diagonal "local" stiffness matrix expressed as (Underwood 1983)

$$ {}^1 K_{ii}^n = -\left( \underline{F}_i^n \Big/ \underline{m}_{ii} - \underline{F}_i^{n-1} \Big/ \underline{m}_{ii} \right) \Big/ \dot{U}_i^{n-1/2}. \tag{11.19}$$

The time integration is performed by utilizing the central-difference explicit integration, with a time step size of unity, as

$$\underline{\dot{\mathbf{U}}}^{n+1/2} = \frac{(2 - c^n)\underline{\dot{\mathbf{U}}}^{n-1/2} + 2\underline{\mathbf{M}}^{-1}\mathbf{F}^n}{(2 + c^n)}, \tag{11.20a}$$

$$\underline{\mathbf{U}}^{n+1} = \mathbf{U}^n + \underline{\dot{\mathbf{U}}}^{n+1/2}. \tag{11.20b}$$

However, the integration algorithm given by Eq. 11.20a, b cannot be used to start the integration due to an unknown velocity field at $t^{-1/2}$, but integration can be started by assuming that $\mathbf{U}^0 \neq 0$ and $\dot{\mathbf{U}}^0 = 0$, which yields

$$\underline{\dot{\mathbf{U}}}^{1/2} = \mathbf{M}^{-1}\mathbf{F}^{1/2}/2. \tag{11.21}$$

Finally, the steps in coupling FEM with PD can be summarized as:

1. Utilize displacement and velocity fields which are known at time steps i, where $i \leq n$.
2. Compute displacement of collocation points within the overlap region using nodal displacements within the overlap region.

3. Compute force densities associated with collocation points within the overlap region.
4. Apply force densities as body force to finite elements within the overlap region.
5. Integrate to find displacements and velocities at time step $(n+1)$.
6. Repeat previous steps to reach desired number of time steps.

## 11.2   Validation of Direct Coupling

The validity of the direct coupling approach is demonstrated by considering a bar and a plate with a hole, both of which are under tension. In the case of a bar, there exists only one overlap region between PD and FEM solution domains. In the case of a plate, the region of the hole where failure is expected to occur is modeled with the PD theory, and the regions far away from the hole are modeled with FEM, resulting in two overlap regions.

### 11.2.1   Bar Subjected to Tensile Loading

The isotropic bar under tension at both ends is divided into two regions for modeling with FEM and PD theory, as illustrated in Fig. 11.2.

Geometric Parameters

Length of the beam: $L = 10$ in. (FEM region, $L_f = 5$ in.; PD region, $L_p = 5$ in.)
Cross-sectional area: $A = h \times h = 0.16$ in$^2$

Material Properties

Young's modulus: $E = 10^7$ psi
Poisson's ratio: $\nu = 0.25$
Mass density: $\rho = 0.1$ lbs/in$^3$

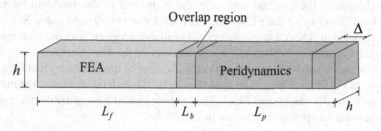

**Fig. 11.2**  Dimensions of the bar

**Fig. 11.3** Comparison
of horizontal displacements
of the bar

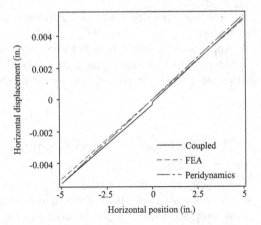

Boundary Conditions

Free of displacement constraints

Applied Loading

Uniaxial tensile force: $F = 1600$ lbs.

PD Discretization Parameters

Total number of material points in the x-direction: 200
Total number of material points in the y-direction: 8
Total number of material points in the z-direction: 8
Spacing between material points: $\Delta = 0.05$ in.
Incremental volume of material points: $\Delta V = 125 \times 10^{-6}$ in$^3$
Boundary layer volume: $\Delta V_\Delta = 1 \times 8 \times 8 \times 125 \times 10^{-6}$ in$^3 = 8 \times 10^{-3}$ in$^3$
Applied body force density: $b_x = F/\Delta V_\Delta = 2 \times 10^5$ lb/in$^3$
Overlap region: $L_b = 0.125$ in.
Horizon: $\delta = 3\Delta$
Adaptive Dynamic Relaxation: ON
Incremental time step size: $\Delta t = 1$ s

In addition to the coupled approach, the entire bar is also modeled by using
either the PD theory or the FEM. The FE model was constructed using SOLID45
brick elements of ANSYS, a commercially available program. The uniaxial tension
is applied as surface tractions at the end surfaces of the bar. A comparison of the
displacements from the coupled approach with those of the PD theory and the FEM
is shown in Fig. 11.3. There is an approximately 5 % difference among the models
using only the PD theory and FEM. The contour plot of horizontal displacements
from the coupled approach is shown in Fig. 11.4.

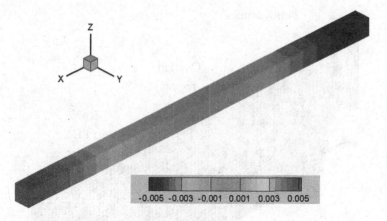

**Fig. 11.4** Horizontal displacement contour plot of the bar

**Fig. 11.5** Dimensions of the plate with a circular cutout

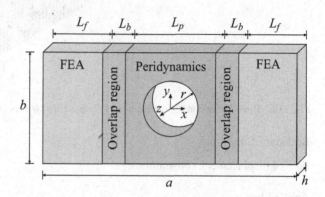

## 11.2.2 Plate with a Hole Subjected to Tensile Loading

The isotropic plate with a hole under tension at both ends is divided into three different regions for coupled modeling of the FEM and PD theory, as illustrated in Fig. 11.5. The extent of the PD and FEM regions is defined by $L_p$ and $L_f$, respectively.

Geometric Parameters

Length of the plate: $a = 9$ in. ($L_f = 2.5$ in. and $L_p = 4$ in.)
Width of the plate: $b = 3$ in.
Thickness of plate, $h = 0.2$ in.
Hole radius: $r = 0.5$ in.

Material Properties

Young's modulus: $E = 10^7$ psi
Poisson's ratio: $\nu = 0.25$
Mass density: $\rho = 0.1$ lbs/in$^3$

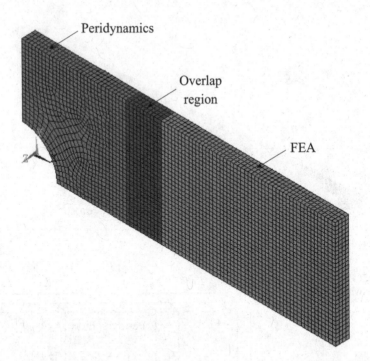

**Fig. 11.6** Three dimensional discretization of the plate for coupled analysis

Boundary Conditions

Free of displacement constraints

Applied Loading

Uniaxial tensile force: $F = 6000$ lbs.

PD Discretization Parameters

Total number of material points in the x-direction: 180
Total number of material points in the y-direction: 60
Total number of material points in the z-direction: 4
Spacing between material points: $\Delta = 0.05$ in.
Incremental volume of material points: $\Delta V = 125 \times 10^{-6}$ in$^3$
Overlap region: $L_b = 0.125$ in.
Boundary layer volume: $\Delta V_\Delta = 1 \times 4 \times 60 \times 125 \times 10^{-6}$ in$^3 = 0.03$ in$^3$
Applied body force density: $b_x = F/\Delta V_\Delta = 2 \times 10^5$ lb/in$^3$
Horizon: $\delta = 3\Delta$
Adaptive Dynamic Relaxation: ON
Incremental time step size: $\Delta t = 1$ s

The three-dimensional model is constructed by discretizing the domain, as shown in Fig. 11.6. The validity of the coupled approach is established by

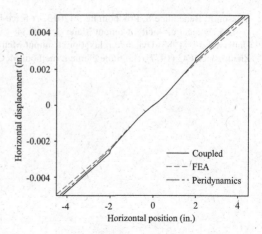

**Fig. 11.7** Horizontal displacements along the lower edge of plate

comparing the steady-state displacements from the PD theory and FEM using ANSYS, a commercially available program. Both the PD and FE models are constructed by utilizing the same discretization as that of coupled model shown in Fig. 11.6. The FE model was constructed using the SOLID45 brick elements of ANSYS. Figure 11.7 shows the horizontal displacements along the bottom line of the plate. The comparison of horizontal displacements indicates a close agreement among the coupled analysis, peridynamic theory, and finite element method.

# References

Agwai A, Guven I, Madenci E (2012) Drop-shock failure prediction in electronic packages by using peridynamic theory. IEEE Trans Adv Packag 2(3):439–447

Belytschko T (1983) An overview of semidiscretization and time integration procedures. Comput Meth Trans Anal 1:1–65

Kilic B, Madenci E (2010) Coupling of peridynamic theory and finite element method. J Mech Mater Struct 5:707–733

Lall P, Shantaram S, Panchagade D (2010) Peridynamic-models using finite elements for shock and vibration reliability of leadfree electronics. In: Proceedings of the 12th intersociety conference, thermal and thermomechanical phenomena in electronic systems (ITHERM), Las Vegas. IEEE, Piscataway, p 859, 2–5 June 2010

Liu W, Hong J (2012) A coupling approach of discretized peridynamics with finite element method. Comput Meth Appl Mech Eng 245–246:163–175

Lubineau G, Azdoud Y, Han F, Rey C, Askari A (2012) A morphing strategy to couple non-local to local continuum mechanics. J Mech Phys Solids 60:1088–1102

Macek RW, Silling SA (2007) Peridynamics via finite element analysis. Finite Elem Anal Des 43 (15):1169–1178

Oterkus E, Madenci E, Weckner O, Silling S, Bogert P, Tessler A (2012) Combined finite element and peridynamic analyses for predicting failure in a stiffened composite curved panel with a central slot. Comp Struct 94:839–850

Seleson P, Beneddine S, Prudhomme S (2013) A force-based coupling scheme for peridynamics and classical elasticity. Comput Mater Sci. 66:34–49

Underwood P (1983) Dynamic relaxation. Comput Meth Trans Anal 1:245–265

Zienkiewicz OC (1977) The finite element method. McGraw-Hill, London

# Chapter 12
# Peridynamic Thermal Diffusion

The peridynamic (PD) theory can be applied to other physical fields such as thermal diffusion, neutronic diffusion, vacancy diffusion, and electrical potential distribution. This paves the way for fully coupling various field equations and deformation within the framework of peridynamics using the same computational domain.

## 12.1 Basics

In heat conduction, the thermal energy is transported through phonons, lattice vibrations, and electrons. Usually, electrons are the vehicles through which thermal energy is transported in metals while phonons are the heat carriers in insulators and semiconductors. This process of thermal energy transfer is inherently nonlocal because the carriers arrive at one point, having brought thermal energy from another. The mean free path of the heat carriers is the average distance a carrier travels before its excess energy is lost. As the heat carriers' mean free path becomes comparable to the characteristic lengths, the nonlocality needs to be taken into account in the continuum model.

Although heat transfer and temperature are closely related, they are of a different nature. Temperature has only a magnitude, and heat transfer has a direction as well as a magnitude. Temperature difference between the material points in a medium is the driving force for any type of heat transfer. In a body, heat flows in the direction of decreasing temperature. Physical experiments show that the rate of heat flow is proportional to the gradient of the temperature, and the proportionality constant, $k$, represents thermal conductivity of the material. This observation, referred to as Fourier's law of heat conduction, is expressed as

$$\mathbf{q} = -k\,\nabla\Theta, \tag{12.1}$$

where $\mathbf{q}$ is the heat flux vector, $k$ is the thermal conductivity, and $\nabla\Theta$ is the gradient of temperature. The minus sign ensures that heat flows in the direction of decreasing

E. Madenci and E. Oterkus, *Peridynamic Theory and Its Applications*, DOI 10.1007/978-1-4614-8465-3_12, © Springer Science+Business Media New York 2014

temperature. The rate of heat entering through the bounding surface, $S$, with unit normal, $\mathbf{n}$, can be obtained from

$$\dot{Q} = - \int_S \mathbf{q}.\mathbf{n} dS, \qquad (12.2)$$

in which the minus sign ensures that heat flow is into the body. If the rate of heat, $\dot{Q}$, is positive, it indicates a heat gain. Otherwise, it is a heat loss. This formulation employing Fourier's law as the local constitutive relation has been used success-fully to model macroscale heat conduction.

## 12.2   Nonlocal Thermal Diffusion

Nonlocality often becomes important at low temperatures, as exhibited in cryogen-ics systems, since the heat carriers have a longer mean free path at lower temperatures. It has been found that nonlocality should also be accounted for in problems in which the temperature gradients are steep. This is because the penetra-tion depth, the length characterizing the temperature gradient, becomes short, even becoming the same order of magnitude as the mean free path of the carrier. In such instances, it is necessary to consider the nonlocality of the heat transport in a continuum model. With the miniaturization of devices, the short geometric length scales also necessitated the inclusion of nonlocal effects in microscale and nano-scale models (Tien and Chen 1994).

Several nonlocal heat conduction theories have been proposed in the last few decades. In the early 1980s, Luciani et al. (1983) developed a nonlocal theory to better represent electron heat transport down a steep temperature gradient by introducing a nonlocal expression for the heat flux. The nonlocal model was in better agreement with probabilistic simulations (Fokker-Planck simulations) than the local models. Later, Mahan and Claro (1988) proposed a nonlocal relation between the heat current, determined from Boltzmann's equation, and the tempera-ture gradient. In the 1990s, Sobolev (1994) introduced a model in which both space and time nonlocality are taken into account in the strong form, i.e., integral form, of the energy balance, Gibbs, and entropy balance equations. Lebon and Grmela (1996), proposed a weakly nonlocal model (weakly nonlocal models are typically based on gradient formulations). The model was based on nonequilibrium thermo-dynamics, for which an extra variable is added to the basic state variables to account for nonlocality. Subsequently, they extended their model to include nonlinearity (Grmela and Lebon 1998). More recently, the development of nonlocal heat conduction equations has been motivated by the miniaturization of devices. A number of researchers have put forth nonlocal models with the objective of capturing heat transport in microscale and nanoscale devices. One example of this is the ballistic-diffusive heat equation by Chen (2002), which was derived from the Boltzmann's equation, and it accounts for nonlocality in heat transport. Another example is by Alvarez and Jou (2007). They developed their model by including nonlocal (and memory/lag) effects in irreversible thermodynamics. Tzou and Guo (2010) constructed their model by incorporating a nonlocal (and lag) term into the Fourier law.

An area of interest is determining the temperature field in the presence of emerging discontinuities. One class of problems that contains a discontinuity is the heat transfer process that involves phase changes such as solidification and melting (Özişik 1980). This process is commonly referred to as the Stefan problem, and there are a number of technologically important problems that involve heat transfer with phase change. Examples of these include ablation of space vehicles during reentry and casting of metals. Another heat conduction problem with an emerging discontinuity is the rewetting problem from the nuclear industry. Rewetting in a nuclear reactor is employed to restore temperatures to a safe range following accidental dry-out or loss of coolant. Emergency cooling is introduced to the system via an upward moving water front or by spraying from the top of the reactor (Duffey and Porthous 1973; Dorfman 2004). A moving discontinuity occurs in the heat-generating solid at the quench front due to the sudden change in the heat transfer condition at the solid surface.

A peridynamic approach to heat conduction is advantageous because it not only accounts for nonlocality but it also allows for the determination of the temperature field in spite of discontinuities. The peridynamic heat conduction is a continuum model; it is not a discrete model. As such, the phonon and electron motions are not explicitly modeled. Initial successful attempts have recently been made to develop heat conduction equations in the peridynamic framework. Gerstle et al. (2008) developed a PD model for electromigration that accounts for heat conduction in a one-dimensional body. Additionally, Bobaru and Duangpanya (2010, 2012) introduced a multi-dimensional peridynamic heat conduction equation, and considered domains with discontinuities such as insulated cracks. Both studies adopted the bond-based PD approach. Later, Agwai (2011) derived a state-based PD heat conduction equation, which is described in the subsequent section.

## 12.3  State-Based Peridynamic Thermal Diffusion

Within the peridynamic framework, the interaction between material points is nonlocal. For thermal diffusion, the nonlocal interaction between material points is due to the exchange of heat energy. Therefore, a material point will exchange heat with points within its neighborhood defined by the horizon. In the Lagrangian formalism, the governing heat conduction equation corresponds to the Euler-Lagrange equation. The Euler-Lagrange equation based on the Lagrangian, $L$, is given in the following form (Moiseiwitsch 2004):

$$\frac{d}{dt}\left(\frac{\partial L}{\partial \dot{\Theta}}\right) - \frac{\partial L}{\partial \Theta} = 0, \tag{12.3a}$$

with

$$L = \int_V \mathcal{L}dV, \tag{12.3b}$$

in which $\Theta$ is the temperature and $\mathcal{L}$ is the Lagrangian density. The Lagrangian density of a peridynamic material point can be defined as

$$\mathcal{L} = Z + \rho \hat{s}\Theta, \tag{12.4}$$

where $Z$ is thermal potential and is a function of all the temperatures of the points with which $x$ interacts, $\rho$ is the density, and $\hat{s}$ is the heat source per unit mass, which includes the rate of heat generation per unit volume and the internal energy storage. There is a thermal potential associated with each material point, and the term $Z_{(i)}$ represents the thermal potential of material point $x_{(i)}$. The microthermal potential, $z_{(i)(j)}$, is the thermal potential due to the interaction (exchange of heat energy) between material points $x_{(i)}$ and $x_{(j)}$. The microthermal potential is related to heat energy exchange, which depends on the temperature difference between the material points. Therefore, the microthermal potential is dependent on the temperature difference between pairs of material points. More specifically, the microthermal potential, $z_{(i)(j)}$, depends on the temperature difference between point $i$ and all other material points that interact with point $x_{(i)}$. Note that the microthermal potential $z_{(j)(i)} \neq z_{(i)(j)}$, as $z_{(j)(i)}$ depends on the state of material points that interact with material point $x_{(j)}$. The microthermal potential is denoted as follows:

$$z_{(i)(j)} = z_{(i)(j)}\left(\Theta_{(1^i)} - \Theta_{(i)}, \Theta_{(2^i)} - \Theta_{(i)}, \cdots\right), \tag{12.5a}$$

$$z_{(j)(i)} = z_{(j)(i)}\left(\Theta_{(1^j)} - \Theta_{(j)}, \Theta_{(2^j)} - \Theta_{(j)}, \cdots\right), \tag{12.5b}$$

where $\Theta_{(i)}$ is the temperature at point $x_{(i)}$, $\Theta_{(1^i)}$ is the temperature of the first material point that interacts with point $x_{(i)}$, and, similarly, $\Theta_{(j)}$ is the temperature at point $x_{(j)}$ while $\Theta_{(1^j)}$ is the temperature of the first material point that interacts with point $x_{(j)}$.

The thermal potential of point $x_{(i)}$, $Z_{(i)}$ is defined as

$$Z_{(i)} = \frac{1}{2}\sum_{j=1}^{\infty}\frac{1}{2}\left(z_{(i)(j)}\left(\Theta_{(1^i)} - \Theta_{(i)}, \Theta_{(2^i)} - \Theta_{(i)}, \cdots\right)\right.$$
$$\left. + z_{(j)(i)}\left(\Theta_{(1^j)} - \Theta_{(j)}, \Theta_{(2^j)} - \Theta_{(j)}, \cdots\right)\right)V_{(j)}, \tag{12.6}$$

where $V_{(j)}$ is the volume associated with material point $\mathbf{x}_{(j)}$. Basically, this equation indicates that the thermal potential at a point is the summation over all the microthermal potential associated with that point. The microthermal potential and therefore thermal potential are both functions of temperature. The Euler-Lagrange equation, Eq. 12.3a, for material point $\mathbf{x}_{(k)}$ becomes

$$\frac{d}{dt}\left(\frac{\partial L}{\partial \dot{\Theta}_{(k)}}\right) - \frac{\partial L}{\partial \Theta_{(k)}} = 0, \tag{12.7a}$$

in which

$$L = \sum_{i=1}^{\infty} \mathfrak{L}_{(i)} V_{(i)}, \tag{12.7b}$$

with

$$\mathfrak{L}_{(i)} = Z_{(i)} + \rho \hat{s}_{(i)} \Theta_{(i)}. \tag{12.7c}$$

Consequently, invoking Eq. 12.6 into Eq. 12.7b results in the Lagrangian function as

$$L = \sum_{i=1}^{\infty} \left\{ \frac{1}{2} \sum_{j=1}^{\infty} \frac{1}{2} \left[ \begin{array}{l} z_{(i)(j)}\left(\Theta_{(1^i)} - \Theta_{(i)}, \Theta_{(2^i)} - \Theta_{(i)}, \cdots\right) \\ + z_{(j)(i)}\left(\Theta_{(1^j)} - \Theta_{(j)}, \Theta_{(2^j)} - \Theta_{(j)}, \cdots\right) \end{array} \right] V_{(j)} \right. \\ \left. + \rho \hat{s}_{(i)} \Theta_{(i)} \right\} V_{(i)}, \tag{12.8a}$$

which can be written in an expanded form by showing only the terms associated with the material point $\mathbf{x}_{(k)}$:

$$L = \cdots \frac{1}{2} \sum_{j=1}^{\infty} \left\{ \frac{1}{2} \left[ z_{(k)(j)}\left(\Theta_{(1^k)} - \Theta_{(k)}, \Theta_{(2^k)} - \Theta_{(k)}, \cdots\right) \right. \right.$$

$$\left. \left. + z_{(j)(k)}\left(\Theta_{(1^j)} - \Theta_{(j)}, \Theta_{(2^j)} - \Theta_{(j)}, \cdots\right) \right] V_{(j)} \right\} V_{(k)} \cdots$$

$$\cdots + \frac{1}{2} \sum_{i=1}^{\infty} \left\{ \frac{1}{2} \left[ z_{(i)(k)}\left(\Theta_{(1^i)} - \Theta_{(i)}, \Theta_{(2^i)} - \Theta_{(i)}, \cdots\right) \right. \right. \tag{12.8b}$$

$$\left. \left. + z_{(k)(i)}\left(\Theta_{(1^k)} - \Theta_{(k)}, \Theta_{(2^k)} - \Theta_{(k)}, \cdots\right) \right] V_{(k)} \right\} V_{(i)} \cdots$$

$$\cdots + \left(\rho \hat{s}_{(k)} \Theta_{(k)}\right) V_{(k)} \cdots$$

or

$$L = \cdots \sum_{j=1}^{\infty} \left\{ \frac{1}{2} \left[ z_{(k)(j)} \left( \Theta_{(1^k)} - \Theta_{(k)}, \ \Theta_{(2^k)} - \Theta_{(k)}, \cdots \right) \right. \right.$$

$$+ z_{(j)(k)} \left( \Theta_{(1^j)} - \Theta_{(j)}, \ \Theta_{(2^j)} - \Theta_{(j)}, \cdots \right) \right] V_{(j)} \Big\} V_{(k)} \cdots$$

$$\cdots + \left( \rho \hat{s}_{(k)} \Theta_{(k)} \right) V_{(k)} \cdots .$$

(12.8c)

With this representation, the Euler-Lagrange equation, Eq. 12.7a, becomes

$$\left( \sum_{j=1}^{\infty} \frac{1}{2} \left( \sum_{i=1}^{\infty} \frac{\partial z_{(k)(i)}}{\partial \left( \Theta_{(j)} - \Theta_{(k)} \right)} V_{(i)} \right) \frac{\partial \left( \Theta_{(j)} - \Theta_{(k)} \right)}{\partial \Theta_{(k)}} \right.$$

$$+ \sum_{j=1}^{\infty} \frac{1}{2} \left( \sum_{i=1}^{\infty} \frac{\partial z_{(i)(k)}}{\partial \left( \Theta_{(k)} - \Theta_{(j)} \right)} V_{(i)} \right) \frac{\partial \left( \Theta_{(k)} - \Theta_{(j)} \right)}{\partial \Theta_{(k)}} V_{(j)} \right) V_{(k)} + \rho \hat{s}_{(k)} V_{(k)} = 0$$

(12.9a)

or

$$- \sum_{j=1}^{\infty} \frac{1}{2} \left( \sum_{i=1}^{\infty} \frac{\partial z_{(k)(i)}}{\partial \left( \Theta_{(j)} - \Theta_{(k)} \right)} V_{(i)} \right) + \sum_{j=1}^{\infty} \left( \sum_{i=1}^{\infty} \frac{\partial z_{(i)(k)}}{\partial \left( \Theta_{(k)} - \Theta_{(j)} \right)} V_{(i)} \right) + \rho \hat{s}_{(k)} = 0,$$

(12.9b)

in which the terms $\sum_{i=1}^{\infty} V_{(i)} \partial z_{(k)(i)} / \partial \left( \Theta_{(j)} - \Theta_{(k)} \right)$ and $\sum_{i=1}^{\infty} V_{(i)} \partial z_{(i)(k)} / \partial \left( \Theta_{(k)} - \Theta_{(j)} \right)$ can be thought of as the heat flow density from material point $\mathbf{x}_{(j)}$ to material point $\mathbf{x}_{(k)}$ and the heat flow density from material point $\mathbf{x}_{(k)}$ to $\mathbf{x}_{(j)}$, respectively. Based on this interpretation, $\mathcal{H}_{(k)(j)}$ and $\mathcal{H}_{(j)(k)}$ are introduced and defined as

$$\mathcal{H}_{(k)(j)} = \frac{1}{2} \frac{1}{V_{(j)}} \left( \sum_{i=1}^{\infty} \frac{\partial z_{(k)(i)}}{\partial \left( \Theta_{(j)} - \Theta_{(k)} \right)} V_{(i)} \right) \quad \text{and} \quad \mathcal{H}_{(j)(k)} = \frac{1}{2} \frac{1}{V_{(j)}} \left( \sum_{i=1}^{\infty} \frac{\partial z_{(i)(k)}}{\partial \left( \Theta_{(k)} - \Theta_{(j)} \right)} V_{(i)} \right).$$

(12.10)

Using these definitions allows Eq. 12.9b to be rewritten as follows:

$$\sum_{j=1}^{\infty} \left( -\mathcal{H}_{(k)(j)} + \mathcal{H}_{(j)(k)} \right) V_{(j)} + \rho \hat{s}_{(k)} = 0.$$

(12.11)

A PD state can be thought of as an infinite dimensional array that contains certain information about all the interactions associated with a particular material point. All of the heat flow density associated with each interaction assembled in an infinite-dimensional array is referred to as the heat flow *scalar* state, $-h(\mathbf{x}, t)$, where $t$ is the time. The assembled heat flow state for material points $\mathbf{x}_{(k)}$ and $\mathbf{x}_{(j)}$ may be represented as

$$\underline{h}\big(\mathbf{x}_{(k)},t\big) = \left\{ \begin{array}{c} \vdots \\ \mathcal{H}_{(k)(j)} \\ \vdots \end{array} \right\} \quad \text{and} \quad \underline{h}\big(\mathbf{x}_{(j)},t\big) = \left\{ \begin{array}{c} \vdots \\ \mathcal{H}_{(j)(k)} \\ \vdots \end{array} \right\}. \tag{12.12}$$

The heat flow state associates each pair of interacting material points with a heat flow density, and enables the expressions for heat flow densities $\mathcal{H}_{(k)(j)}$ and $\mathcal{H}_{(j)(k)}$ as

$$\mathcal{H}_{(k)(j)} = \underline{h}\big(\mathbf{x}_{(k)},t\big)\big\langle \mathbf{x}_{(j)} - \mathbf{x}_{(k)}\big\rangle \quad \text{and} \quad \mathcal{H}_{(j)(k)} = \underline{h}\big(\mathbf{x}_{(j)},t\big)\big\langle \mathbf{x}_{(k)} - \mathbf{x}_{(j)}\big\rangle, \tag{12.13}$$

where the angled brackets include the interacting material points. The microthermal potentials may also be assembled in a state, which is called the microthermal potential *scalar* state, $\underline{z}(\mathbf{x},t)$, permitting the following representation:

$$z_{(k)(j)} = \underline{z}\big(\mathbf{x}_{(k)},t\big)\big\langle \mathbf{x}_{(j)} - \mathbf{x}_{(k)}\big\rangle \quad \text{and} \quad z_{(j)(k)} = \underline{z}\big(\mathbf{x}_{(j)},t\big)\big\langle \mathbf{x}_{(k)} - \mathbf{x}_{(j)}\big\rangle. \tag{12.14}$$

Applying the state notation, Eq. 12.11 can be rewritten as

$$\sum_{j=1}^{\infty} \big(\underline{h}\big(\mathbf{x}_{(k)},t\big)\big\langle \mathbf{x}_{(j)} - \mathbf{x}_{(k)}\big\rangle - \underline{h}\big(\mathbf{x}_{(j)},t\big)\big\langle \mathbf{x}_{(k)} - \mathbf{x}_{(j)}\big\rangle \big) V_{(j)} - \rho \widehat{s}_{(k)} = 0. \tag{12.15}$$

Transforming the summation to integration over the material points within the horizon as given by

$$\sum_{j=1}^{\infty} (\cdot)\, V_{(j)} \rightarrow \int_{H} (\cdot)\, dV_{\mathbf{x}'} \tag{12.16}$$

permits Eq. 12.15 to be recast as

$$\int_{H} (\underline{h}(\mathbf{x},t)\langle \mathbf{x}' - \mathbf{x}\rangle - \underline{h}(\mathbf{x}',t)\langle \mathbf{x} - \mathbf{x}'\rangle) dV_{\mathbf{x}'} - \rho \widehat{s} = 0, \tag{12.17}$$

where $\underline{h}(\mathbf{x},t)\langle \mathbf{x}' - \mathbf{x}\rangle = 0$ for $\mathbf{x}' \notin H$, and the domain of integration, $H$, is defined by the horizon of the material point, $\mathbf{x}$, that interacts with other material points in its own family.

For convenience, the following notation is adopted:

$$\underline{h}(\mathbf{x},t) = \underline{h}, \qquad \underline{h}(\mathbf{x}',t) = \underline{h}'. \tag{12.18}$$

Also, the temperature *scalar* state, $\underline{\tau}$, is defined as

$$\underline{\tau}(\mathbf{x},t)\langle \mathbf{x}' - \mathbf{x}\rangle = \Theta(\mathbf{x}',t) - \Theta(\mathbf{x},t). \tag{12.19}$$

The temperature state simply contains the temperature difference associated with each interaction of a particular material point. Since, the microthermal potential is dependent on the temperature difference of all the interactions associated with the material point, it may be written as a function of the temperature state

$$\underline{z} = \underline{z}(\underline{\tau}). \tag{12.20}$$

Therefore, the heat flow state can also be written as a function of the temperature state

$$\underline{h} = \underline{h}(\underline{\tau}). \tag{12.21}$$

As outlined by Bathe (1996), the heat conduction equation should explicitly include the rate at which heat energy is stored when the heat flow changes over a short period of time. This rate of internal energy storage density, $\dot{e}_s$, is a negative energy source and is given by

$$\dot{e}_s = c_v \frac{\partial \Theta}{\partial t}, \tag{12.22}$$

for which $c_v$ is the specific heat capacity.

Therefore, the source term in Eq. 12.15 is then replaced by $\hat{s} = \dot{e}_s - s_b$, where $s_b$ is the heat source due to volumetric heat generation per unit mass. Invoking Eq. 12.22 into Eq. 12.15 leads to the transient form of the state-based peridynamic thermal diffusion equation

$$\rho c_v \dot{\Theta}(\mathbf{x}, t) = \int_H \underline{h}(\mathbf{x}, t)\langle \mathbf{x}' - \mathbf{x}\rangle - \underline{h}(\mathbf{x}', t)\langle \mathbf{x} - \mathbf{x}'\rangle dV' + h_s(\mathbf{x}, t), \tag{12.23}$$

in which $h_s(\mathbf{x}, t) = \rho s_b(\mathbf{x}, t)$ is the heat source due to volumetric heat generation. The resulting equation is an integro-differential equation in time and space. It contains differentiation with respect to time, and integration in the spatial domain. It does not contain any spatial derivatives of temperature; thus, the PD thermal equation is valid everywhere whether or not discontinuities exist in the domain. Construction of its solution involves time and spatial integrations while being subject to conditions on the boundary of the domain, $\mathcal{R}$, and initial conditions.

## 12.4   Relationship Between Heat Flux and Peridynamic Heat Flow States

The heat flow *scalar* state, $\underline{h}$, contains the heat flow densities associated with all the interactions. The heat flow density, $\underline{h}(\mathbf{x},t)\langle\mathbf{x}' - \mathbf{x}\rangle$, has units of heat flow rate (rate of heat energy change) per volume squared. The integral in Eq. 12.23

$$\int_H \underline{h}(\mathbf{x},t)\langle\mathbf{x}' - \mathbf{x}\rangle - \underline{h}(\mathbf{x}',t)\langle\mathbf{x} - \mathbf{x}'\rangle dV' \tag{12.24}$$

is similar to the divergence of heat flux, $\nabla \cdot \mathbf{q}$, and it has units of heat flow rate per volume. Therefore, the peridynamic heat flow state can be related to the heat flux, $\mathbf{q}$.

Multiplying the PD heat conduction equation, Eq. 12.23, by a temperature variation of $\Delta\Theta$ and integrating over the entire domain results in

$$\int_V \rho c_v \dot{\Theta}\Delta\Theta dV = \int_V\int_H [\underline{h}(\mathbf{x},t)\langle\mathbf{x}' - \mathbf{x}\rangle - \underline{h}(\mathbf{x}',t)\langle\mathbf{x} - \mathbf{x}'\rangle]\Delta\Theta dV' dV$$
$$+ \int_V h_s(\mathbf{x},t)\Delta\Theta dVt . \tag{12.25}$$

Moving the last term on the right-hand side of Eq. 12.25, the heat generation term, to the left-hand side, and changing the integration from $H$ to $V$ due to the fact that

$$\underline{h}(\mathbf{x},t)\langle\mathbf{x}' - \mathbf{x}\rangle = \underline{h}(\mathbf{x}',t)\langle\mathbf{x} - \mathbf{x}'\rangle = 0 \text{ for } \mathbf{x}' \notin H, \tag{12.26}$$

leads to the following form of the equation:

$$\int_V [\rho c_v\dot{\Theta} - h_s(\mathbf{x},t)]\Delta\Theta dV = \int_V\int_V [\underline{h}(\mathbf{x},t)\langle\mathbf{x}' - \mathbf{x}\rangle - \underline{h}(\mathbf{x}',t)\langle\mathbf{x} - \mathbf{x}'\rangle]\Delta\Theta dV' dV . \tag{12.27}$$

If the parameters $\mathbf{x}$ and $\mathbf{x}'$ in the second integral on the right-hand side of Eq. 12.27 are exchanged, the second integral becomes

$$\int_V\int_V \underline{h}(\mathbf{x}',t)\langle\mathbf{x} - \mathbf{x}'\rangle\Delta\Theta dV' dV = \int_V\int_V \underline{h}(\mathbf{x},t)\langle\mathbf{x}' - \mathbf{x}\rangle\Delta\Theta' dV dV'. \tag{12.28}$$

Substituting from Eq. 12.28 into Eq. 12.27, leads to

$$\int\limits_{V} \left[ \rho c_v \dot{\Theta} - h_s(\mathbf{x},t) \right] \Delta\Theta dV = \int\limits_{V} \int\limits_{V} \underline{h}(\mathbf{x},t) \langle \mathbf{x}' - \mathbf{x} \rangle (\Delta\Theta - \Delta\Theta') dV' dV. \quad (12.29)$$

Invoking the variation of the temperature *scalar* state, $\Delta\underline{\tau}$, from Eq. 12.19 into Eq. 12.29 results in

$$\int\limits_{V} \left[ \rho c_v \dot{\Theta} - h_s(\mathbf{x},t) \right] \Delta\Theta dV = \int\limits_{V} \Delta Z dV, \quad (12.30)$$

where $\Delta Z$ corresponds to the variation of the PD thermal potential at $\mathbf{x}$ due to its interactions with all other material points:

$$\Delta Z = - \int\limits_{V} (\underline{h}(\mathbf{x},t) \langle \mathbf{x}' - \mathbf{x} \rangle)(\Delta\underline{\tau} \langle \mathbf{x}' - \mathbf{x} \rangle) dV'. \quad (12.31)$$

Considering only the material points within the horizon, Eq. 12.31 can be rewritten as

$$\Delta Z = - \int\limits_{H} (\underline{h}(\mathbf{x},t) \langle \mathbf{x}' - \mathbf{x} \rangle)(\Delta\underline{\tau} \langle \mathbf{x}' - \mathbf{x} \rangle) dV'. \quad (12.32)$$

Based on the classical formulation, the corresponding variation of thermal potential can be written as

$$\Delta\hat{Z}(\bar{\mathbf{G}}) = \frac{1}{2} \left( \Delta\bar{\mathbf{G}} \cdot k\bar{\mathbf{G}} + \bar{\mathbf{G}} \cdot k\Delta\bar{\mathbf{G}} \right) = k\bar{\mathbf{G}} \cdot \Delta\bar{\mathbf{G}}, \quad (12.33a)$$

with $\hat{Z}(\bar{\mathbf{G}})$ given by

$$\hat{Z}(\bar{\mathbf{G}}) = \frac{1}{2} \bar{\mathbf{G}} \cdot k\bar{\mathbf{G}}, \quad (12.33b)$$

where $k$ is the thermal conductivity and $\bar{\mathbf{G}} = \nabla\Theta$. After invoking the Fourier relation, $\mathbf{q} = -k\bar{\mathbf{G}}$, the variation of classical thermal potential can be rewritten as

$$\Delta\hat{Z}(\bar{\mathbf{G}}) = -\mathbf{q} \cdot \Delta\bar{\mathbf{G}}. \quad (12.34)$$

By applying the definition of scalar reduction given in the Appendix, the temperature gradient can be approximated as

$$\Delta\bar{\mathbf{G}} = \frac{1}{m} \Delta\underline{\tau} * \underline{\mathbf{X}} = \frac{1}{m} \int\limits_{H} \underline{w} \langle \mathbf{x}' - \mathbf{x} \rangle \Delta\underline{\tau} \langle \mathbf{x}' - \mathbf{x} \rangle \otimes \underline{\mathbf{X}} \langle \mathbf{x}' - \mathbf{x} \rangle dV', \quad (12.35)$$

in which $\Delta\underline{\tau}$ is a *scalar* state, thus not requiring the dyadic, $\otimes$, operation. It reduces to

$$\Delta\bar{\mathbf{G}} = \frac{1}{m} \int_H \underline{w}\langle\mathbf{x}'-\mathbf{x}\rangle\underline{\mathbf{X}}\langle\mathbf{x}' - \mathbf{x}\rangle\Delta\underline{\tau}\langle\mathbf{x}' - \mathbf{x}\rangle dV', \qquad (12.36)$$

where $\underline{w}$ is a scalar state representing the influence function, and $m$ is the scalar weighted volume defined in the Appendix.

Its substitution into Eq. 12.34 leads to the following:

$$\Delta\hat{Z} = -\frac{1}{m} \int_H \mathbf{q}^T\underline{w}\langle\mathbf{x}' - \mathbf{x}\rangle\underline{\mathbf{X}}\langle\mathbf{x}' - \mathbf{x}\rangle\Delta\underline{\tau}\langle\mathbf{x}' - \mathbf{x}\rangle dV'. \qquad (12.37)$$

Assuming that the variation of the PD thermal potential, $\Delta Z$, and classical thermal potential, $\Delta\hat{Z}$, are equal, $\Delta Z = \Delta\hat{Z}$, and comparing Eq. 12.31 to Eq. 12.37, it follows that

$$\underline{h}(\mathbf{x},t)\langle\mathbf{x}' - \mathbf{x}\rangle = \frac{1}{m}\mathbf{q}^T\underline{w}\langle\mathbf{x}'-\mathbf{x}\rangle\underline{\mathbf{X}}\langle\mathbf{x}' - \mathbf{x}\rangle, \qquad (12.38)$$

and this expression relates the heat flow state to the heat flux.

## 12.5 Initial and Boundary Conditions

The PD thermal equation does not contain any spatial derivatives; thus, boundary conditions are, in general, not necessary for the solution of an integro-differential equation. However, such conditions on temperature can be imposed in a "fictitious material layer" along the boundary of a nonzero volume.

Heat flux does not directly appear in the PD thermal diffusion equation. Therefore, the application of heat flux is also different from that of the classical heat conduction theory. The difference can be illustrated by considering a region, $\Omega$, that is in thermal equilibrium. If this region is fictitiously divided into two domains, $\Omega^-$ and $\Omega^+$ as shown in Fig. 12.1, there must be rates of heat flow $\dot{Q}^+$ and $\dot{Q}^-$ entering through the cross-sectional surfaces, $\partial\Omega$, of domain $\Omega^+$ and $\Omega^-$.

According to classical heat conduction theory, the heat flow rates, $\dot{Q}^+$ and $\dot{Q}^-$, can be determined by integrating the normal component of the heat flux over the cross-sectional area, $\partial\Omega$, of domains $\Omega^+$ and $\Omega^-$ as

$$\dot{Q}^+ = - \int_{\partial\Omega} \mathbf{q}^+ \cdot \mathbf{n}^+ dS \qquad (12.39a)$$

**Fig. 12.1** Boundary conditions: (**a**) heat fluxes through the cross-sectional area, (**b**) heat flow rate in classical heat conduction theory, (**c**) heat flow density of a material point in domain $\Omega^+$ with other material points in domain $\Omega^-$, (**d**) heat flux density from domain $\Omega^+$ due to domain $\Omega^-$

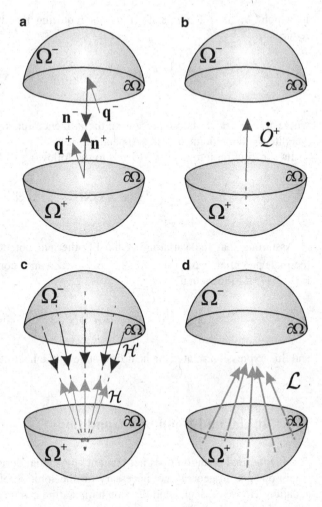

and

$$\dot{Q}^- = -\int_{\partial\Omega} \mathbf{q}^- \cdot \mathbf{n}^- d\mathcal{S}, \qquad (12.39b)$$

in which $\mathbf{q}^+$ and $\mathbf{q}^-$ are the heat fluxes across the surfaces with unit normal vectors, $\mathbf{n}^+$ and $\mathbf{n}^-$, of domains $\Omega^+$ and $\Omega^-$, as shown in Fig. 12.1a, b.

In the case of the PD theory, the material points located in domain $\Omega^+$ interact with the other material points in domain $\Omega^-$ (Fig. 12.1c). Thus, the heat flow rate, $\dot{Q}^+$, can be computed by volume integration of the heat flux densities (Fig. 12.1d) over domain $\Omega^+$ as

$$\dot{Q}^+ = \int_{\Omega^+} \mathcal{L}(\mathbf{x})dV, \qquad (12.40a)$$

in which $\mathcal{L}(\mathbf{x})$, acting on a material point in domain $\Omega^+$, is determined by

$$\mathcal{L}(x) = \int_{\Omega^-} [\underline{h}(\mathbf{x}, t)\langle \mathbf{x}' - \mathbf{x} \rangle - \underline{h}(\mathbf{x}', t)\langle \mathbf{x} - \mathbf{x}' \rangle]dV. \qquad (12.40b)$$

Note that if the volume $\Omega^-$ is void, the volume integration in Eq. 12.40b vanishes. Hence, the heat flux cannot be applied as a boundary condition since its volume integrations result in a zero value. Therefore, the heat flux can be applied as rate of volumetric heat generation in a "real material layer" along the boundary of a nonzero volume.

### 12.5.1 Initial Conditions

Time integration requires the application of initial temperature values at each material point in the domain, $\mathcal{R}$, as shown in Fig. 12.2, and they can be specified as

$$\Theta(\mathbf{x}, t = 0) = \Theta^*(\mathbf{x}). \qquad (12.41)$$

### 12.5.2 Boundary Conditions

Boundary conditions can be imposed as temperature, heat flux, convection, and radiation. As shown in Fig. 12.2, the prescribed boundary temperature is imposed in a layer of a fictitious region, $\mathcal{R}_t$, along the boundary of the actual material surface, $\mathcal{S}_t$, of the actual material region, $\mathcal{R}$. Based on numerical experiments, the extent of the fictitious boundary layer must be equal to the horizon, $\delta$, in order to ensure that the prescribed temperatures are sufficiently reflected in the actual material region. The prescribed heat flux, convection, and radiation are imposed in boundary layer regions, $\mathcal{R}_f$, $\mathcal{R}_c$, and $\mathcal{R}_r$, respectively, with depth, $\Delta$, along the boundary of the material region, $\mathcal{R}$, as shown in Fig. 12.2.

#### 12.5.2.1 Temperature

As shown in Fig. 12.3a, the prescribed boundary temperature, $\Theta^*(\mathbf{x}^*, t)$, can be imposed in a layer of a fictitious region, $\mathcal{R}_t$, along the boundary of the actual material surface, $\mathcal{S}_t$, as

**Fig. 12.2** Boundary layers for imposing temperature, heat flux, convection, and radiation

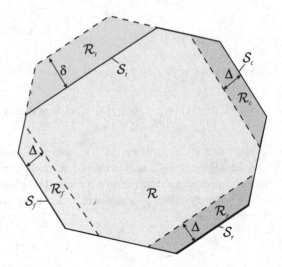

$$\Theta(\mathbf{y}, t + \Delta t) = 2\Theta^*(\mathbf{x}^*, t + \Delta t) - \Theta(\mathbf{z}, t), \quad \mathbf{x}^* \in \mathcal{S}_t, \quad \mathbf{y} \in \mathcal{R}_t, \quad \mathbf{z} \in \mathcal{R}, \quad (12.42)$$

in which $\mathbf{z}$ represents the position of a material point in $\mathcal{R}$, and $\mathbf{x}^*$ represents the location of a point on the surface, $\mathcal{S}_t$. Their relative position is such that the distance, $d = |\mathbf{x}^* - \mathbf{z}|$, between them is the shortest. The location of the image material point in $\mathcal{R}_t$ is obtained from $\mathbf{y} = \mathbf{z} + 2d\mathbf{n}$, with $\mathbf{n} = (\mathbf{x}^* - \mathbf{z})/|\mathbf{x}^* - \mathbf{z}|$. The implementation of the prescribed temperature boundary condition is demonstrated in Fig. 12.3b. For the case of $\Theta^*(\mathbf{x}^*, t) = 0$, this representation enforces the temperature variation in the fictitious region to become the negative mirror image of the temperature variation near the boundary surface in the actual material, as shown in Fig. 12.3c.

### 12.5.2.2   Heat Flux

Application of this type of boundary condition is accomplished by first calculating the rate of heat entering through the bounding surface by using Eq. 12.2, converting the heat flow rate, $\dot{Q}$, to a heat generation per unit volume, and then specifying this volumetric heat generation to collocations points in the boundary region. Assuming the cross-sectional area is constant for each material point, conversion is achieved by

$$\tilde{Q} = \frac{\dot{Q}}{V_f} = \frac{-\int_{\mathcal{S}_f} \mathbf{q}.\mathbf{n}d\mathcal{S}}{V_f} = -\frac{\mathbf{q}\cdot\mathbf{n}\,\mathcal{S}_f}{\mathcal{S}_f\Delta} = -\frac{\mathbf{q}\cdot\mathbf{n}}{\Delta}, \qquad (12.43)$$

**Fig. 12.3** (**a**) Material
point and its image in a
fictitious domain for
applying, (**b**) a constant
temperature condition and
(**c**) a zero temperature
condition

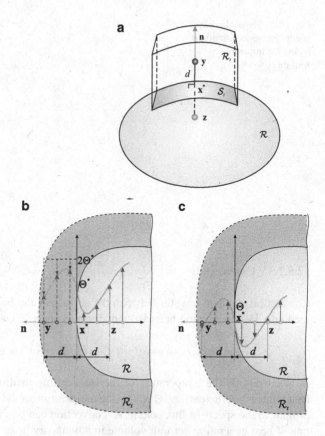

where $\tilde{Q}$ is the volumetric heat generation, $\mathbf{q}$ is the heat flux, $\mathcal{S}_f$ is the area over
which the heat flux is applied, and $V_f$ is the volume of the boundary region.

Specified flux, $\mathbf{q}^*(\mathbf{x}, t)$, over the surface $\mathcal{S}_f$, shown in Fig. 12.2, can be applied as
the rate of volumetric heat generation in a boundary layer, $\mathcal{R}_f$, as

$$h_s(\mathbf{x}, t) = -\frac{1}{\Delta}\mathbf{q}^*(\mathbf{x}, t) \cdot \mathbf{n}, \text{ for } \mathbf{x} \in \mathcal{R}_f. \qquad (12.44)$$

If there exists no specified flux, $\mathbf{q}^*(\mathbf{x}, t) = 0$, volumetric heat generation, $\tilde{Q}$
calculated from Eq. 12.43 vanishes. Thus, the implementation of a zero flux
boundary condition can be viewed as imposing a zero-valued volumetric heat
generation. Alternative to this implementation, zero flux can be achieved by
assigning the mirror image of the temperature values near the boundary in the
actual domain to the material points in the fictitious region, as shown in Fig. 12.4.

**Fig. 12.4** Material point
and its image in a fictitious
region for imposing
zero flux

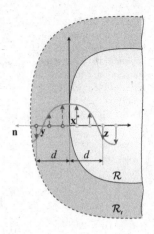

### 12.5.2.3  Convection

Convection is a heat transfer between the surface of the body and the surrounding
medium. The convection boundary condition is specified as

$$\mathbf{q}(\mathbf{x}, t) \cdot \mathbf{n} = h(\Theta(\mathbf{x}, t) - \Theta_\infty), \quad \text{for} \quad \mathbf{x} \in \mathcal{S}_c, \tag{12.45}$$

in which $\Theta_\infty$ is the temperature of the surrounding medium, $h$ is the convective
heat transfer coefficient, and $\Theta(\mathbf{x}, t)$ is the temperature of the body on the surface, $\mathcal{S}_c$.
Similar to the specified flux condition, convection can be imposed in the form of a
rate of heat generation per unit volume in a boundary layer region, $\mathcal{R}_c$, as

$$h_s(\mathbf{x}, t) = \frac{1}{\Delta} h(\Theta_\infty - \Theta(\mathbf{x}, t)), \quad \text{for} \quad \mathbf{x} \in \mathcal{R}_c. \tag{12.46}$$

### 12.5.2.4  Radiation

Radiation is a heat transfer between the surface of the body and the surrounding
medium. The radiation boundary condition can be written as

$$\mathbf{q}(\mathbf{x}, t) \cdot \mathbf{n} = \varepsilon\sigma\big(\Theta^4(\mathbf{x}, t) - \Theta_{ss}^4\big), \quad \text{for} \quad \mathbf{x} \in \mathcal{S}_r, \tag{12.47}$$

in which $\Theta_{ss}$ is the temperature of the surface surrounding the body, $\Theta(\mathbf{x}, t)$ is the
surface temperature of the body, $\sigma$ is the Stefan-Boltzman constant, and $\varepsilon$ is
emissivity of the boundary surface. Similar to the imposition of the convection
condition, radiation can also be imposed in the form of rate of heat generation per
unit volume in a boundary layer region, $\mathcal{R}_r$, as

$$h_s(\mathbf{x}, t) = \frac{1}{\Delta} \varepsilon \sigma (\Theta_{ss}^4 - \Theta^4(\mathbf{x}, t)), \quad \text{for} \quad \mathbf{x} \in \mathcal{R}_r. \tag{12.48}$$

## 12.6   Bond-Based Peridynamic Thermal Diffusion

If it is assumed that the heat flow density associated between two material points, $\mathbf{x}$ and $\mathbf{x}'$, is a function of the temperature difference only between these two points, then the following expression holds true:

$$\underline{h}(\mathbf{x}, t)\langle \mathbf{x}' - \mathbf{x} \rangle = -\underline{h}(\mathbf{x}', t)\langle \mathbf{x} - \mathbf{x}' \rangle. \tag{12.49}$$

This leads to the specialized bond-based PD thermal diffusion. In this specialized case, the heat flow density, $f_h(\mathbf{x}', \mathbf{x}, t)$, is defined as

$$f_h(\mathbf{x}', \mathbf{x}, t) = \underline{h}(\mathbf{x}, t)\langle \mathbf{x}' - \mathbf{x} \rangle - \underline{h}(\mathbf{x}', t)\langle \mathbf{x} - \mathbf{x}' \rangle = 2\underline{h}(\mathbf{x}, t)\langle \mathbf{x}' - \mathbf{x} \rangle, \tag{12.50}$$

so that the PD heat conduction equation can be written as

$$\rho c_v \dot{\Theta}(\mathbf{x}, t) = \int_H f_h(\Theta', \Theta, \mathbf{x}', \mathbf{x}, t) dV_{\mathbf{x}'} + \rho s_b(\mathbf{x}, t). \tag{12.51}$$

The term $f_h$, also referred to as the thermal response function, is the heat flow density function that governs only the interaction of material point $\mathbf{x}$ with $\mathbf{x}'$. In the case of bond-based PD thermal diffusion, the pairwise interactions are independent of each other, and the heat flow between a pair of material points does not depend on the temperature difference between other pairs of material points. The thermal response function, $f_h(\mathbf{x}', \mathbf{x})$ is zero for material points outside the horizon; i.e., $|\boldsymbol{\xi}| = |\mathbf{x}' - \mathbf{x}| > \delta$.

## 12.7   Thermal Response Function

The pairwise heat flow density can be related to the microthermal potential through

$$f_h = \frac{\partial z}{\partial \tau}. \tag{12.52}$$

The microthermal potential, $z$, represents the thermal potential between a pair of interacting points. The temperature difference between the material points $\mathbf{x}'$ and $\mathbf{x}$ at any time is given by

$$\tau(\mathbf{x}', \mathbf{x}, t) = \Theta(\mathbf{x}', t) - \Theta(\mathbf{x}, t).$$                (12.53)

The thermal potential at point $\mathbf{x}$ is then a summation over all microthermal potentials associated with this point, and is defined as

$$Z(\mathbf{x}, t) = \frac{1}{2} \int_H z(\mathbf{x}', \mathbf{x}, t) dV_{\mathbf{x}'}.$$                (12.54)

The pairwise heat flow density function, $f_h$, can be expressed as

$$f_h(\mathbf{x}', \mathbf{x}, t) = \kappa \frac{\tau(\mathbf{x}', \mathbf{x}, t)}{|\boldsymbol{\xi}|},$$                (12.55)

where $\kappa$ is the thermal microconductivity. The microthermal potential corresponding to the thermal response function, $f_h$, can be obtained as

$$z = \kappa \frac{\tau^2}{2|\boldsymbol{\xi}|}.$$                (12.56)

The microconductivity is a PD parameter that can be related to the standard conductivity for a specified horizon.


## 12.8   Peridynamic Microconductivity

The microconductivity can be determined by equating the peridynamic thermal potential to the classical thermal potential at a point arising from a simple linear temperature field. The expression for the microconductivity will differ depending on the form of the thermal response function. The form given in Eq. 12.55 differs from those introduced by Bobaru and Duangpanya (2010, 2012) and Gerstle et al. (2008). In the most general case, heat transfer through a medium is three dimensional. However, certain problems can be classified as two or one dimensional depending on the relative magnitudes of heat transfer rates in different directions.


### 12.8.1   One-Dimensional Analysis

For one-dimensional analysis, a simple linear temperature field of the form $\Theta(x) = x$ results in the PD temperature difference of

$$\tau = \Theta(x') - \Theta(x) = x' - x = \xi = |\xi|. \tag{12.57}$$

Invoking this temperature difference into Eq. 12.56 results in the PD microthermal potential as

$$z = \kappa \frac{\xi^2}{2|\xi|}, \tag{12.58}$$

where $|\xi| = |x' - x|$. Substituting for $z$ from Eq. 12.58 into Eq. 12.54 and performing the integration leads to the PD thermal potential as

$$Z = \frac{1}{2} \int_H z(\xi) dV_\xi = \frac{\kappa}{2} \int_0^\delta \left( \frac{\xi^2}{|\xi|} \right) A d\xi = \frac{\kappa \delta^2 A}{4}, \tag{12.59}$$

where $A$ is the cross-sectional area of the volume associated with the material point $x'$. The corresponding classical thermal potential from Eq. 12.33b is obtained as

$$\hat{Z} = \frac{1}{2}k. \tag{12.60}$$

Equating the peridynamic thermal potential in Eq. 12.59 to the classical thermal potential given in Eq. 12.60 and solving for $\kappa$ results in the PD microconductivity for one-dimensional analysis as

$$\kappa = \frac{2k}{A\delta^2}. \tag{12.61}$$

### 12.8.2   Two-Dimensional Analysis

For two-dimensional analysis, a simple linear temperature field of the form $\Theta(x,y) = (x + y)$ results in the PD temperature difference of

$$\tau = \Theta(x',y') - \Theta(x,y) = x' + y' \tag{12.62}$$

for the material point of interest, $x$, located at the origin $(x = 0, y = 0)$. Invoking this temperature difference into Eq. 12.56 results in the PD microthermal potential as

$$z = \kappa \frac{(x' + y')^2}{2|\xi|}, \tag{12.63}$$

where $|\xi| = \sqrt{x'^2 + y'^2}$. Substituting for $z$ from Eq. 12.63 into Eq. 12.54 and performing the integration over the horizon leads to the PD thermal potential as

$$Z(\mathbf{x},t) = \frac{1}{2} \int\limits_{0}^{2\pi} \int\limits_{0}^{\delta} \kappa \frac{(\xi Cos(\theta) + \xi Sin(\theta))^2}{2|\xi|} h\xi d\xi d\theta = \frac{\pi h \kappa \delta^3}{6}, \qquad (12.64)$$

in which polar coordinates, $(\xi, \theta)$, are utilized to perform the integration over a disk with thickness $h$ and radius $\delta$. The corresponding classical thermal potential from Eq. 12.33b is obtained as

$$\hat{Z} = k. \qquad (12.65)$$

Equating the PD thermal potential in Eq. 12.64 to the classical thermal potential given in Eq. 12.65 and solving for $\kappa$ results in the PD microconductivity for two-dimensional analysis as

$$\kappa = \frac{6k}{\pi h \delta^3}. \qquad (12.66)$$

## 12.8.3   Three-Dimensional Analysis

For three-dimensional analysis, a simple linear temperature field of the form $\Theta(x, y) = (x + y + z)$ results in the PD temperature difference of

$$\tau = \Theta(x', y', z') - \Theta(x, y, z) = (x' + y' + z') \qquad (12.67)$$

for the material point of interest, $\mathbf{x}$, located at the origin $(x = 0, y = 0, z = 0)$. Invoking this temperature difference into Eq. 12.56 results in the PD microthermal potential as

$$z = \kappa \frac{(x' + y' + z')^2}{2|\xi|}, \qquad (12.68)$$

where $|\xi| = \sqrt{x'^2 + y'^2 + z'^2}$. Substituting for $z$ from Eq. 12.68 into Eq. 12.54 and performing the integration over the horizon leads to the PD thermal potential as

$$Z(\mathbf{x},t) = \frac{1}{2} \int_0^{\delta} \int_0^{2\pi} \int_0^{\pi} \kappa \frac{(\xi Cos(\theta)Sin(\phi) + \xi Sin(\theta)Sin(\phi) + \xi Cos(\phi))^2}{2|\xi|}$$

$$\times Sin\phi d\phi d\theta \xi^2 d\xi = \frac{\pi\kappa\delta^4}{4}, \qquad (12.69)$$

in which spherical coordinates, $(\xi, \theta, \phi)$, are utilized to perform the integration over a sphere with radius $\delta$. The corresponding classical thermal potential from Eq. 12.33b is obtained as

$$\hat{Z} = \frac{3}{2}k. \qquad (12.70)$$

Equating the peridynamic thermal potential in Eq. 12.69 to the classical thermal potential given in Eq. 12.70 and solving for $\kappa$ results in the PD microconductivity for three-dimensional analysis as

$$\kappa = \frac{6k}{\pi\delta^4}. \qquad (12.71)$$

## 12.9 Numerical Procedure

Numerical techniques are employed in order to solve for the PD thermal diffusion equation. The region of interest is discretized into subdomains in which the temperature is assumed to be constant. Thus, each subdomain is represented as a single integration point located at its mass center with an associated volume and integration weight, $w_{(j)} = 1$. Subsequently, the integration in the governing equation, given in Eq. 12.51, is numerically performed as

$$\rho_{(i)} c_{v(i)} \dot{\Theta}_{(i)}^n = \sum_{j=1}^{N} f_h\left(\tau^n(\mathbf{x}_{(j)} - \mathbf{x}_{(i)})\right) V_{(j)} + h_{s(i)}^n, \qquad (12.72)$$

for which $n$ is the time step number, $i$ represents the point of interest, and $j$ represents the points within the horizon of $i$. The volume of the subdomain associated with the collocation point $\mathbf{x}_{(j)}$ is denoted by $V_{(j)}$. The time integration is accomplished using the forward difference time stepping scheme. When forward differencing is employed, the following equation is solved:

$$\Theta_{(i)}^{n+1} = \Theta_{(i)}^n + \frac{\Delta t}{\rho_{(i)}c_{v(i)}}\left\{\sum_{j=1}^{N} f_h\left(\tau^n\left(\mathbf{x}_{(j)} - \mathbf{x}_{(i)}\right)\right)V_{(j)} + h_{s(i)}^n\right\}, \qquad (12.73)$$

where $\Delta t$ is the time step size.

### 12.9.1  Discretization and Time Stepping

A one-dimensional region is considered to describe the details of the numerical scheme. The discretization of a one-dimensional region into subdomains is depicted in Fig. 12.5. Each subdomain has one integration point. The integration point represents a material point. The solution is constructed for material point $\mathbf{x}_{(i)}$. The material point $\mathbf{x}_{(i)}$ interacts with all points within its horizon, represented by $\mathbf{x}_{(j)}$. As shown in Fig. 12.6, material point $\mathbf{x}_{(i)}$ interacts with six other material points, $\mathbf{x}_{(j)}$ ($j = i-3, i-2, i-1, i+1, i+2,$ and $i+3$), in its horizon. Thus, the radius of the horizon is $\delta = 3\Delta$, where $\Delta = |x_{(i+1)} - x_{(i)}|$.

The discretized form of the PD thermal diffusion equation for material point $\mathbf{x}_{(i)}$ becomes

$$\rho_{(i)}c_{v(i)}\dot{\Theta}_{(i)}^n = \sum_{j=1}^{N} f_{h(i)(j)}^n V_{(j)} + h_{s(i)}^n, \qquad (12.74)$$

in which the thermal response function, represented by $f_{h(i)(j)}^n$, is determined at each time step for every interaction. The discretized equation for the thermal response function, $f_h$, is cast as

$$f_{h(i)(j)}^n = \kappa \frac{\tau_{(i)(j)}^n}{\left|\boldsymbol{\xi}_{(i)(j)}\right|}. \qquad (12.75)$$

The relative initial position is defined as $\boldsymbol{\xi}_{(i)(j)} = \mathbf{x}_{(j)} - \mathbf{x}_{(i)}$ while the relative temperature is defined as $\tau_{(i)(j)}^n = \Theta_{(j)}^n - \Theta_{(i)}^n$. The thermal interaction of material point $\mathbf{x}_{(i)}$ with the points within its horizon is illustrated in Fig. 12.6.

The discretized thermal diffusion equation can be expanded as

$$\rho_{(i)}c_{v(i)}\dot{\Theta}_{(i)}^n = f_{h(i)(i+1)}^n V_{(i+1)} + f_{h(i)(i+2)}^n V_{(i+2)} + f_{h(i)(i+3)}^n V_{(i+3)}$$
$$+ f_{h(i)(i-1)}^n V_{(i-1)} + f_{h(i)(i-2)}^n V_{(i-2)} + f_{h(i)(i-3)}^n V_{(i-3)} + h_{s(i)}^n. \qquad (12.76)$$

For marching in time, the forward differencing scheme is used. The time derivative of temperature at material point $\mathbf{x}_{(i)}$ is determined at the current time

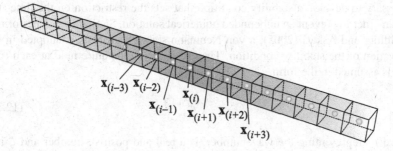

**Fig. 12.5**  Discretization of one-dimensional region with collocation points

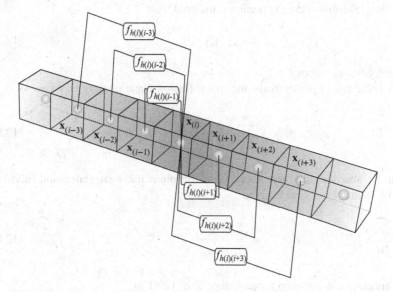

**Fig. 12.6**  Thermal interaction of points with the horizon of material point $\mathbf{x}_{(i)}$

step, $n$, from Eq. 12.76. By employing time integration via the forward differencing technique, the temperature at the next time step, $(n + 1)$, is determined. This algorithm may be expressed as

$$\Theta_{(i)}^{n+1} = \Theta_{(i)}^{n} + \Delta t \dot{\Theta}_{(i)}^{n}. \tag{12.77}$$

## 12.9.2  Numerical Stability

The forward differencing method utilized for the numerical time integration of the peridynamic thermal diffusion equation is conditionally stable. Therefore, it is necessary to develop a stability condition that sets the restriction on the time step size in order to prevent an unbounded numerical solution. Similar to that performed by Silling and Askari (2005), a von Neumann stability analysis is adopted in the derivation of the stability condition. Therefore, the temperature field at each time step is assumed in the form

$$\Theta_{(i)}^n = \zeta^n e^{\Gamma i \sqrt{-1}}, \tag{12.78}$$

where $\Gamma$, representing the wavenumber, is a real and positive number and $\zeta$ is a complex number. The condition on the time step size ensures that the solution does not grow in an unbounded manner over time. In order for the solution to be bounded over time, the following expression must hold true:

$$|\zeta| \leq 1 \tag{12.79}$$

for every wavenumber $\Gamma$.

The discretized peridynamic thermal diffusion equation may be recast as

$$\rho_{(i)} c_{v(i)} \dot{\Theta}_{(i)}^n = \sum_{j=1}^{N} \frac{\kappa}{|\boldsymbol{\xi}_{(i)(j)}|} \left( \Theta_{(j)}^n - \Theta_{(i)}^n \right) V_{(j)} + h_{s(i)}^n. \tag{12.80}$$

In the absence of a heat source due to volumetric heat generation, invoking Eq. 12.78 into Eq. 12.80 leads to

$$\frac{\rho_{(i)} c_{v(i)}}{\Delta t} \left( \zeta^{n+1} - \zeta^n \right) e^{\Gamma i \sqrt{-1}} = \sum_{j=1}^{N} \frac{\kappa}{|\boldsymbol{\xi}_{(i)(j)}|} \left( \zeta^n e^{\Gamma(j-i)\sqrt{-1}} - \zeta^n \right) e^{\Gamma i \sqrt{-1}} V_{(j)}. \tag{12.81}$$

Canceling out common terms reduces Eq. 12.81 to

$$\frac{\rho_{(i)} c_{v(i)}}{\Delta t} (\zeta - 1) = \sum_{j=1}^{N} \frac{\kappa}{|\boldsymbol{\xi}_{(i)(j)}|} \left( e^{\Gamma(j-i)\sqrt{-1}} - 1 \right) V_{(j)}$$

$$= \sum_{j=1}^{N} \frac{\kappa}{|\boldsymbol{\xi}_{(i)(j)}|} (\cos \Gamma(j - i) - 1) V_{(j)}. \tag{12.82}$$

This equation can be recast as

$$\frac{\rho_{(i)} c_{v(i)}}{\Delta t} (\zeta - 1) = -M_\Gamma, \tag{12.83}$$

in which $M_\Gamma$ is defined by

$$M_\Gamma = \sum_{j=1}^{N} \frac{\kappa}{|\boldsymbol{\xi}_{(i)(j)}|} (1 - \cos \Gamma(j-i)) V_{(j)}. \tag{12.84}$$

Solving for $\zeta$ in Eq. 12.83 gives

$$\zeta = 1 - \frac{\Delta t}{\rho_{(i)} c_{v(i)}} M_\Gamma. \tag{12.85}$$

Enforcing the condition $|\zeta| \leq 1$ results in the following constraint:

$$0 \leq \frac{\Delta t}{\rho_{(i)} c_{v(i)}} M_\Gamma \leq 2. \tag{12.86}$$

The restriction on the time step size is determined as

$$\Delta t < \frac{2\rho_{(i)} c_{v(i)}}{M_\Gamma}. \tag{12.87}$$

For the condition $|\zeta| \leq 1$ to be valid for all wavenumbers, $\Gamma$, Eq. 12.84 leads to the condition of

$$M_\Gamma \leq \sum_{j=1}^{N} 2 \frac{\kappa}{|\boldsymbol{\xi}_{(i)(j)}|} V_{(j)}. \tag{12.88}$$

Substituting Eq. 12.88 into Eq. 12.87 leads to the stability condition as

$$\Delta t < \frac{\rho_{(i)} c_{v(i)}}{\sum_{j=1}^{N} \frac{\kappa}{|\boldsymbol{\xi}_{(i)(j)}|} V_{(j)}}. \tag{12.89}$$

Due to the dependence of $\kappa$ on the horizon, the stability condition given in Eq. 12.89 is dependent on $\delta$.

## 12.10   Surface Effects

The PD microconductivity parameter, $\kappa$, that appears in the thermal response function, $f_h$, is determined by computing the thermal potential of a material point whose horizon is completely embedded in the material. The value of this parameter depends on the domain of integration defined by the horizon. Therefore, the value of $\kappa$ requires correction if the material point is close to free surfaces or material

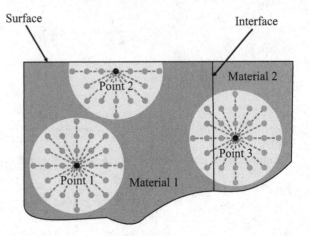

**Fig. 12.7** Surface effects in the domain of interest

interfaces (Fig. 12.7). Since the presence of free surfaces is problem dependent, it is impractical to resolve this issue analytically. The correction of the material parameters is achieved by numerically integrating the PD thermal potential at each material point inside the body for simple temperature distribution and comparing it to its counterpart obtained from the classical thermal potential.

The simple temperature distribution can be linear in form, and the corresponding thermal potential, $Z_\infty$, of a point completely embedded in the material is calculated using Eq. 12.33b. Subsequently, the PD thermal potential due to the applied linear temperature distribution is computed for each material point through numerical integration over its horizon from

$$Z_{(i)} = \frac{1}{2}\int_H z(\xi)dV_\xi = \frac{1}{2}\sum_{j=1}^N z_{(i)(j)}V_{(j)},\tag{12.90}$$

in which the micropotential, $z_{(i)(j)}$, between material points $\mathbf{x}_{(i)}$ and $\mathbf{x}_{(j)}$ depends on the material microconductivity.

As shown in Fig. 12.8, the material point $\mathbf{x}_{(i)}$ may interact with material points $\mathbf{x}_{(j)}$ and $\mathbf{x}_{(m)}$. Material points $\mathbf{x}_{(i)}$ and $\mathbf{x}_{(j)}$ are embedded in material 1, and $\mathbf{x}_{(m)}$ is embedded in material 2. Thus, the microconductivity between points $\mathbf{x}_{(i)}$ and $\mathbf{x}_{(j)}$ is $\kappa_{(i)(j)}$, and it differs from $\kappa_{(i)(m)}$ between material points $\mathbf{x}_{(i)}$ and $\mathbf{x}_{(m)}$. Because the material points $\mathbf{x}_{(i)}$ and $\mathbf{x}_{(m)}$ are embedded in two different materials, their microconductivity, $\kappa_{(i)(m)}$, can be expressed in terms of an equivalent thermal conductivity as

$$k_{(i)(m)} = \frac{\ell_1 + \ell_2}{\frac{\ell_1}{k_1}+\frac{\ell_2}{k_2}},\tag{12.91}$$

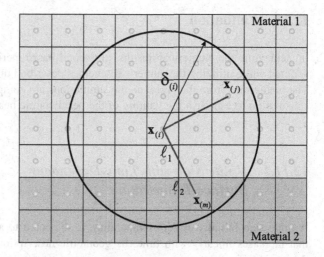

**Fig. 12.8** Material point $\mathbf{x}_{(i)}$ close to an interface

in which $\ell_1$ represents the segment of the distance between material points $\mathbf{x}_{(i)}$ and $\mathbf{x}_{(m)}$ in material 1 whose thermal conductivity is $k_1$, and $\ell_2$ represents the segment in material 2 whose thermal conductivity is $k_2$.

The thermal potential of material point $\mathbf{x}_{(i)}$ is denoted by $Z_{(i)}$. The correction factor is determined for each material point in the domain as

$$g_{(i)} = \frac{Z_\infty}{Z_{(i)}}. \tag{12.92}$$

Therefore, the discretized thermal diffusion equation including the correction factor for point $\mathbf{x}_{(i)}$ becomes

$$\rho_{(i)} c_{v(i)} \dot{\Theta}^n_{(i)} = \sum_{j=1}^N g_{(i)(j)} f^n_{h(i)(j)} V_{(j)} + \rho_{(i)} s^n_{b(i)}, \tag{12.93}$$

where $g_{(i)(j)} = g_{(i)} + g_{(j)}/2$. Finally, the discretized equation of motion for material point $\mathbf{x}_{(i)}$, including surface and volume correction, $v_c$, is rewritten as

$$\rho_{(i)} c_{v(i)} \dot{\Theta}^n_{(i)} = \sum_{j=1}^N g_{(i)(j)} f^n_{h(i)(j)} \left( v_{c(j)} V_{(j)} \right) + \rho_{(i)} s^n_{b(i)}. \tag{12.94}$$

Also, the thermal response functions between material points $\mathbf{x}_{(i)}$ and $\mathbf{x}_{(j)}$ and $\mathbf{x}_{(i)}$ and $\mathbf{x}_{(m)}$ are modified to reflect the change in microconductivity as

$$f^n_{h(i)(m)} = \kappa_{(i)(m)} \frac{\tau^n_{(i)(m)}}{\left| \boldsymbol{\xi}_{(i)(m)} \right|} \quad \text{and} \quad f^n_{h(i)(j)} = \kappa_{(i)(j)} \frac{\tau^n_{(i)(j)}}{\left| \boldsymbol{\xi}_{(i)(j)} \right|}. \tag{12.95}$$

## 12.11  Validation

In achieving the numerical results, the bond-based peridynamics approach is adopted while utilizing the numerical schemes described in the preceding sections. The predictions from the peridynamic simulations are compared against the classical solution to establish the validity of the peridynamic heat transfer analysis.

### 12.11.1  *Finite Slab with Time-Dependent Surface Temperature*

A finite slab initially at zero temperature is subjected to a boundary temperature that increases linearly with time. Its geometric description and discretization are depicted in Fig. 12.9.

Geometric Parameters

Slab thickness: $L = 0.01$ m

Material Properties

Specific heat capacity: $c_v = 64$ J/kgK
Thermal conductivity: $k = 233$ W/mK
Mass density: $\rho = 260$ kg/m$^3$

Initial Conditions

$\Theta(x, 0) = 0°$C, $0 \le x \le L$

Boundary Conditions

$\Theta(0, t) = 0$, $\Theta(L, t) = At$ with $A = 500$, $0 \le t < \infty$

PD Discretization Parameters

Total number of material points in the $x$-direction: 100
Spacing between material points: $\Delta = 0.0001$ m

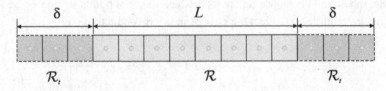

**Fig. 12.9** Discretization of the finite slab and the fictitious boundary regions for temperatures

**Fig. 12.10** Temperature variations from peridynamics and classical analytical solutions

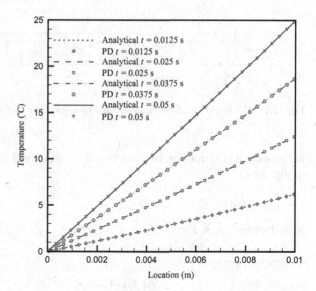

Incremental volume of material points: $\Delta V = 1 \times 10^{-12}$ m$^3$
Volume of fictitious boundary layer: $V_\delta = (3) \times \Delta V = 3 \times 10^{-12}$ m$^3$
Horizon: $\delta = 3.015\Delta$
Time step size: $\Delta t = 10^{-6}$ s

The classical analytical solution (Jiji 2009) can be expressed as

$$\Theta(x,t) = A\frac{x}{L} + A\frac{\rho c_v 2L^2}{k\pi^3}$$
$$\times \sum_{n=1}^{\infty} \frac{(-1)^n}{n^3}\sin\left(\frac{n\pi}{L}x\right)\left[1 - \exp\left(-\frac{k}{\rho c_v}\left(\frac{n\pi}{L}\right)^2 t\right)\right]. \qquad (12.96)$$

The temperature variation is predicted at $t = 0.0125$ s, $t = 0.025$ s, $t = 0.0375$ s, and $t = 0.05$ s. Both analytical and PD predictions are shown in Fig. 12.10, and they are in close agreement. Because the temperature on the right boundary increases as a function of time, the rate of heat transfer from the right boundary also increases, as expected.

## 12.11.2 Slab with Convection Boundary Condition

A plate of thickness $L$, initially at temperature $\Theta(x,0) = F(x)$, dissipates heat by convection for times $t > 0$ from its surfaces into an environment at $\Theta_\infty = 0°C$. The plate initially has a linear temperature profile, and two surfaces are subjected

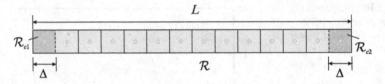

**Fig. 12.11** Discretization of the finite slab and boundary regions for convection

to convective heat transfer. Its geometric description and discretization are depicted in Fig. 12.11.

Geometric Parameters

Slab thickness: $L = 1$ m

Material Properties

Specific heat capacity: $c_v = 64$ J/kgK
Thermal conductivity: $k = 233$ W/mK
Mass density: $\rho = 260$ kg/m$^3$

Initial Conditions

$\Theta(x, 0) = F(x), \quad 0 \le x \le L, \text{ with } F(x) = x$

Boundary Conditions

$-k\partial\Theta/\partial x = h_1(\Theta_\infty - \Theta), \quad t > 0, \text{ at } x = 0$
$k\partial\Theta/\partial x = h_2(\Theta_\infty - \Theta), \quad t > 0, \text{ at } x = L$
with $h_1 = 10\,W/m^2K, \; h_2 = 20\,W/m^2K, \; \Theta_\infty = 0°C$

PD Discretization Parameters

Total number of material points in the $x$-direction: 500
Spacing between material points: $\Delta = 0.002$ m
Incremental volume of material points: $\Delta V = 8 \times 10^{-9}$ m$^3$
Volume of boundary layer: $V_\Delta = 8 \times 10^{-9}$ m$^3$
Horizon: $\delta = 3.015\Delta$
Time step size: $\Delta t = 10^{-6}$ s
Rate of heat generation per unit volume at $x = 0$:
$h_{s1}(\mathbf{x}, t) = h_1(\Theta_\infty - \Theta(\mathbf{x}, t))/\Delta, \mathbf{x} \in \mathcal{R}_{c1}$
Rate of heat generation per unit volume at $x = L$:
$h_{s2}(\mathbf{x}, t) = h_2(\Theta_\infty - \Theta(\mathbf{x}, t))/\Delta, \mathbf{x} \in \mathcal{R}_{c2}$

**Fig. 12.12** Temperature
variations from
peridynamics and classical
analytical solutions

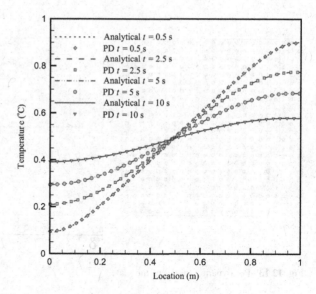

The classical analytical solution of the problem is given by Özişik (1980) as

$$\Theta(x,t) = \sum_{m=1}^{\infty} e^{-\frac{k}{\rho c_v}\beta_m^2 t} \frac{1}{N(\beta_m)} X(\beta_m, x) \int_0^L X(\beta_m, x') F(x') dx', \qquad (12.97)$$

in which $X(\beta_m, x)$ represents the eigenfunctions, $\beta_m$ represents the eigenvalues, and $N(\beta_m)$ represents the normalization integral. The eigenfunctions, eigenvalues, and normalization integral are as follows:

$$X(\beta_m, x) = \beta_m Cos(\beta_m x) + H_1 Sin(\beta_m x), \qquad (12.98a)$$

$$\tan(\beta_m L) = \frac{\beta_m (H_1 + H_2)}{\beta_m^2 - H_1 H_2}, \qquad (12.98b)$$

$$N(\beta_m) = \frac{1}{2}\left[ (\beta_m^2 + H_1^2)\left( L + \frac{H_2}{\beta_m^2 + H_2^2}\right) + H_1\right], \qquad (12.98c)$$

with $H_1 = h_1/k$ and $H_2 = h_2/k$. The temperature variation is predicted at $t = 0.5$ s, $t = 2.5$ s, $t = 5$ s, and $t = 10$ s. Both analytical and PD predictions are shown in Fig. 12.12, and they are in close agreement.

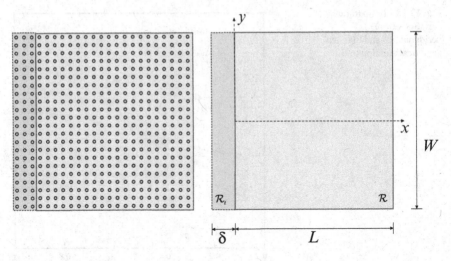

**Fig. 12.13** Peridynamic model of the plate

## 12.11.3  Plate Under Thermal Shock with Insulated Boundaries

A square plate of isotropic material under thermal shock with insulated boundaries, shown in Fig. 12.13, was first considered by Hosseini-Tehrani and Eslami (2000) using the Boundary Element Method (BEM).

Geometric Parameters

Length: $L = 10$ m
Width: $W = 10$ m
Thickness: $H = 1$ m

Material Properties

Specific heat capacity: $c_v = 1$ J/kgK
Thermal conductivity: $k = 1$ W/mK
Mass density: $\rho = 1$ kg/m$^3$

Initial Conditions

$\Theta(x, y, t = 0) = 0°C$

Boundary Conditions

$\Theta_x(x = 10, y) = 0, \quad t > 0$
$\Theta_y(x, y = \pm 5) = 0, \quad t > 0$
$\Theta(x = 0, t) = 5te^{-2t}, \quad t > 0$

**Fig. 12.14** Temperature variation from peridynamics and BEM at $y = 0$ (Hosseini-Tehrani and Eslami 2000)

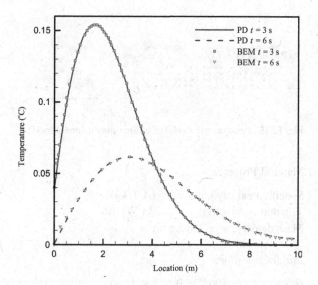

## PD Discretization Parameters

Total number of material points in the $x$-direction: 500
Total number of material points in the $y$-direction: 500
Spacing between material points: $\Delta = 0.02$ m
Incremental volume of material points: $\Delta V = 4 \times 10^{-4}$ m$^3$
Volume of fictitious boundary layer: $V_\delta = (3 \times 500) \times \Delta V = 0.6$ m$^3$
Horizon: $\delta = 3.015\Delta$
Time step size: $\Delta t = 5 \times 10^{-4}$ s

The temperature variations at $y = 0$ are predicted for $t = 3$ s and $t = 6$ s. Both BEM and PD predictions are shown in Fig. 12.14, and they are in close agreement.

## 12.11.4 Three-Dimensional Block with Temperature and Insulated Boundaries

A block of isotropic material is subjected to constant temperatures at both ends while its lateral surfaces are insulated. The schematic of the problem is described in Fig. 12.15.

## Geometric Parameters

Length: $L = 0.01$ m
Width: $W = 0.001$ m
Thickness: $H = 0.001$ m

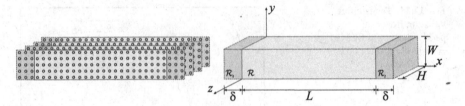

**Fig. 12.15** Peridynamic model of a three-dimensional block

## Material Properties

Specific heat capacity: $c_v = 64$ J/kgK
Thermal conductivity: $k = 233$ W/mK
Mass density: $\rho = 260$ kg/m$^3$

## Initial Conditions

$\Theta(x, y, z, 0) = 100°C, \ 0 \leq x \leq L, \ 0 \leq y \leq W, \ 0 \leq z \leq H$

## Boundary Conditions

$\Theta(0, y, z, t) = 0°C, \ \Theta(L, y, z, t) = 300°C, \ t > 0$
$\Theta_{,y}(x, 0, z, t) = 0, \ \Theta_{,y}(x, W, z, t) = 0, \ t > 0$
$\Theta_{,z}(x, y, 0, t) = 0, \ \Theta_{,z}(x, y, H, t) = 0, \ t > 0$

## PD Discretization Parameters

Total number of material points in the $x$-direction: 100
Total number of material points in the $y$-direction: 10
Total number of material points in the $z$-direction: 10
Spacing between material points: $\Delta = 0.0001$ m
Incremental volume of material points: $\Delta V = 1 \times 10^{-12}$ m$^3$
Volume of fictitious boundary layer: $V_\delta = (3 \times 10 \times 10) \times \Delta V = 3 \times 10^{-10}$ m$^3$
Horizon: $\delta = 3.015\Delta$
Time step size: $\Delta t = 10^{-7}$ s

Since the block is insulated on its lateral surfaces, the temperature profile along the block can be compared with the one-dimensional analytical solution of the problem given by

$$\Theta(x, t) = \Theta(0, t) - \frac{\Theta(0, t) - \Theta(L, t)}{L} x - \frac{2}{L} \sum_{n=1,3,5,\ldots}^{\infty} \sin\left(\frac{n\pi}{L} x\right)$$

$$\times \left[ \frac{L}{n\pi} (\Theta(0, t) - (-1)^n \Theta(L, t)) - \frac{100L}{n\pi} ((-1)^n - 1) \right] e^{-\frac{k}{\rho c_v}\left(\frac{n^2\pi^2}{L^2}\right)t}.$$

$$(12.99)$$

**Fig. 12.16** Temperature variations from peridynamics and classical analytical solutions

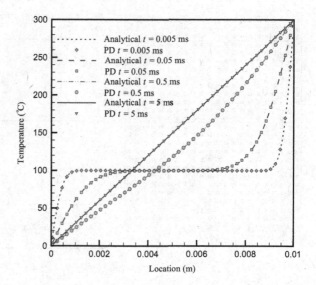

The temperature variation is predicted at $t = 5 \times 10^{-6}$s, $t = 5 \times 10^{-5}$s, $t = 5 \times 10^{-4}$s, and $t = 5 \times 10^{-3}$s. As the block reaches a steady-state condition, the temperature profile approaches a linear variation along the block. As observed in Fig. 12.16, the thermal response predicted by the peridynamic heat transfer model is in close agreement with the analytical solution.

## 12.11.5  Dissimilar Materials with an Insulated Crack

In order to verify the peridynamic model in solving for the heat transfer concerning dissimilar materials, a plate with two different materials having an insulated interface crack is considered, as shown in Fig. 12.17. The peridynamic predictions and their comparison with ANSYS are given in Fig. 12.18. As observed, there is a close agreement.

Geometric Parameters

Length: $L = 2$ cm
Width : $W = 2$ cm
Thickness: $H = 0.01$ cm
Crack length: $2a = 1.0$ cm

Material Properties

Specific heat capacity: $c_v = 1$ J/kgK
Thermal conductivity: $k = 1.14$ W/cmK
Mass density: $\rho = 1 \, \text{kg/cm}^3$

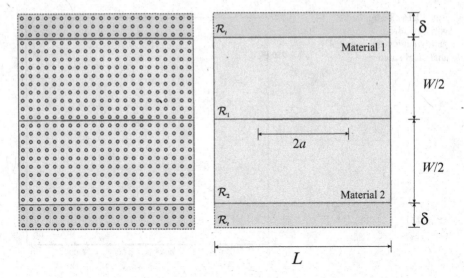

**Fig. 12.17** Peridynamic model of a plate with an insulated interface crack

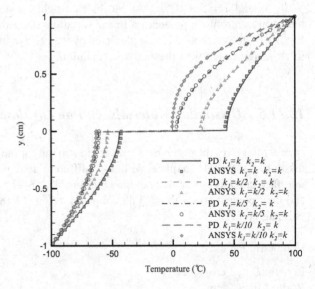

**Fig. 12.18** Temperature variations along $x = 0$, across the interface of the plates with thermal conductivity $k_1$ for the upper half and $k_2$ for the lower half at $t = 0.5$ s

### Initial Conditions

$$\Theta(x, y, z, 0) = 0, \quad -L/2 \leq x \leq L/2, \quad -W/2 \leq y \leq W/2$$

### Boundary Conditions

$$\Theta(x, W/2, t) = 100°C, \quad \Theta(x, -W/2, t) = -100°C, \quad t > 0$$
$$\Theta_{,x}(L/2, y, t) = 0, \quad \Theta_{,x}(-L/2, y, t) = 0, \quad t > 0$$

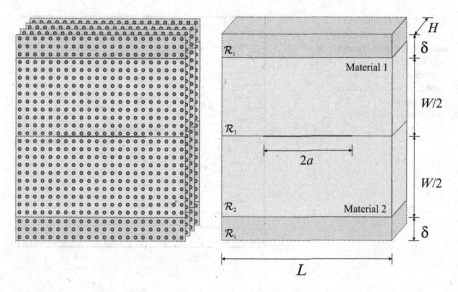

**Fig. 12.19** Three-dimensional peridynamic model of a plate with a crack

PD Discretization Parameters

Total number of material points in the $x$-direction: 200
Total number of material points in the $y$-direction: 200
Spacing between material points: $\Delta = 0.01$ cm
Incremental volume of material points: $\Delta V = 1 \times 10^{-6}$ cm$^3$
Volume of fictitious boundary layer: $V_\delta = (3 \times 200) \times \Delta V = 6 \times 10^{-4}$cm$^3$
Horizon: $\delta = 3.015 \times \Delta$
Time step size: $\Delta t = 10^{-4}$ s

In order to demonstrate the three-dimensional capability of the PD analysis, the plate geometry with an insulated crack is also discretized in the thickness direction, as shown in Fig. 12.19. The peridynamic results are compared with the two-dimensional predictions. As observed in Fig. 12.20, there exists a close agreement between the two models.

Geometric Parameters

Length: $L = 2$ cm
Width: $W = 2$ cm
Thickness: $H = 0.2$ cm
Crack length: $2a = 1.0$ cm

Material Properties

Specific heat capacity: $c_v = 1$ J/kgK
Thermal conductivity: $k = 1.14$ W/cmK
Mass density: $\rho = 1$ kg/cm$^3$

**Fig. 12.20** Temperature
field from two- and three-
dimensional peridynamic
analyses for $k_1 = k_2 = k$ at
$t = 0.5$ s (two-dimensional
model = *solid line*, three-
dimensional
model = *dashed line*)

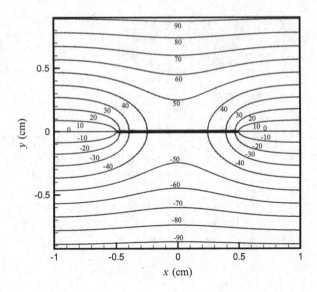

Initial Conditions

$$\Theta(x, y, z, 0) = 0 - L/2 \le x \le L/2, \quad -W/2 \le y \le W/2, \quad -H \le z \le 0$$

Boundary Conditions

$$\Theta(x, W/2, z, t) = 100°C, \quad \Theta(x, -W/2, z, t) = -100°C, \quad t > 0$$
$$\Theta_{,x}(L/2, y, z, t) = 0, \quad \Theta_{,x}(-L/2, y, z, t) = 0, \quad t > 0$$
$$\Theta_{,z}(x, y, 0, t) = 0, \quad \Theta_{,z}(x, y, -H, t) = 0, \quad t > 0$$

PD Discretization Parameters

Total number of material points in the x-direction: 100
Total number of material points in the y-direction: 100
Total number of material points in the z-direction: 10
Spacing between material points: $\Delta = 0.02$ cm
Incremental volume of material points: $\Delta V = 8 \times 10^{-6}$ cm$^3$
Volume of fictitious boundary layer: $V_\delta = (3 \times 100 \times 10) \times \Delta V$ cm$^3$
Horizon: $\delta = 3.015 \times \Delta$
Time step size: $\Delta t = 10^{-5}$ s

## 12.11.6  *Thick Plate with Two Inclined Insulated Cracks*

In order to further demonstrate the three-dimensional capability of the PD analysis,
a thick plate with two insulated inclined cracks is considered under two different
types of boundary conditions. The plate geometry is symmetric with respect to the

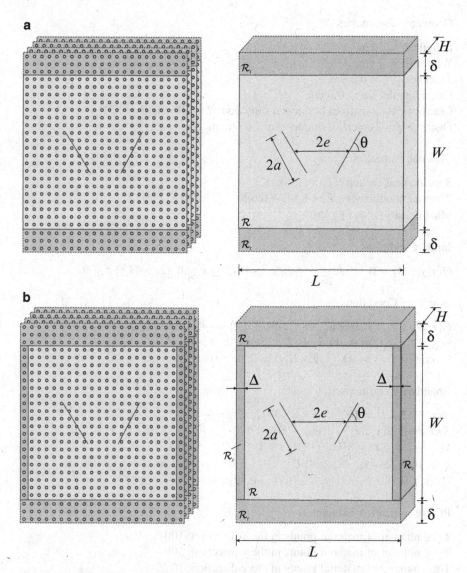

**Fig. 12.21** Peridynamic model of the thick plate: (**a**) boundary conditions type I; (**b**) boundary conditions type-II

vertical direction. For the first type of boundary condition, the plate is subjected to constant temperature at the top and bottom surfaces while the remaining surfaces are insulated. For the second type of boundary condition, the plate is subjected to constant temperature at the top and bottom surfaces and convective heat transfer on the left and right surfaces while the remaining surfaces are insulated. The discretization and PD model of the plate for these two different types of boundary conditions are shown in Fig. 12.21a, b.

Geometric Parameters

Length: $L = 2$ cm
Width: $W = 2$ cm
Thickness: $H = 0.2$ cm
Crack lengths: $2a = 0.6$ cm
Crack orientations from horizontal direction: $\theta = 60°$ and $\theta = 120°$
Distance between crack centers: $2e = 0.66$ cm

Material Properties

Specific heat capacity: $c_v = 1\,\text{J/kgK}$
Thermal conductivity: $k = 1.14\,\text{W/cmK}$
Mass density: $\rho = 1\,\text{kg/cm}^3$

Initial Conditions

$\Theta(x, y, z, 0) = 0 - L/2 \leq x \leq L/2, \ -W/2 \leq y \leq W/2, \ -H \leq z \leq 0$

Boundary Conditions-I

$\Theta(x, W/2, z, t) = 100°\text{C}, \ \Theta(x, -W/2, z, t) = -100°\text{C}, \ t > 0$
$\Theta_x(L/2, y, z, t) = 0, \ \Theta_x(-L/2, y, z, t) = 0, \ t > 0$
$\Theta_z(x, y, 0, t) = 0, \ \Theta_z(x, y, -H, t) = 0, \ t > 0$

Boundary Conditions-II

$\Theta(x, W/2, z, t) = 100°\text{C}, \ \Theta(x, -W/2, z, t) = -100°\text{C}, \ t > 0$
$-kT_x(-L/2, y, z, t) = h(\Theta_\infty - \Theta), \ t > 0$
$kT_x(L/2, y, z, t) = h(\Theta_\infty - \Theta), \ t > 0$
$h = 10\,\text{W/cm}^2\text{K}, \ \Theta_\infty = 0°\text{C}$
$\Theta_z(x, y, 0, t) = 0, \ \Theta_z(x, y, -H, t) = 0, \ t > 0$

PD Discretization Parameters

Total number of material points in the $x$-direction: 100
Total number of material points in the $y$-direction: 100
Total number of material points in the $z$-direction: 10
Spacing between material points: $\Delta = 0.02$ cm
Incremental volume of material points: $\Delta V = 8 \times 10^{-6}\,\text{cm}^3$
Volume of boundary layer: $V_\Delta = (1 \times 100 \times 10) \times \Delta V = 8 \times 10^{-3}\text{cm}^3$
Volume of fictitious boundary layer: $V_\delta = (3 \times 100 \times 10) \times \Delta V = 24 \times 10^{-3}\text{cm}^3$
Horizon: $\delta = 3.015 \times \Delta$
Time step size: $\Delta t = 10^{-5}\text{s}$
Rate of heat generation per unit volume at $x = -L/2$ and $x = L/2$:
$h_s(\mathbf{x}, t) = \frac{1}{\Delta}h(\Theta_\infty - \Theta(\mathbf{x}, t)), \ \mathbf{x} \in \mathcal{R}_c$

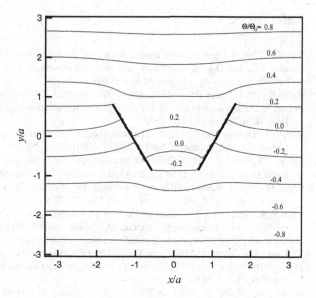

**Fig. 12.22** Three-dimensional peridynamic temperature predictions on the mid-plane with a normal in the $+z$ direction at $t = 0.45$ s for boundary conditions type-I ($\Theta_0 = 100°C$)

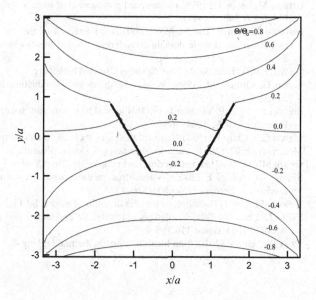

**Fig. 12.23** Three-dimensional peridynamic temperature predictions on the mid-plane with a normal in the $+z$ direction at $t = 0.45$ s for boundary conditions type-II ($\Theta_0 = 100°C$)

For the first type of boundary condition, the peridynamic prediction for the temperature field is shown in Fig. 12.22. They are in close agreement with the classical solution (Chang and Ma 2001; Chen and Chang 1994). For the second type of boundary condition, the peridynamic prediction for the temperature field is shown in Fig. 12.23. For this case, there exists no classical solution for comparison.

# References

Agwai A (2011) A peridynamic approach for coupled fields. Dissertation, University of Arizona

Alvarez FX, Jou D (2007) Memory and nonlocal effects in heat transport: from diffusive to ballistic regimes. Appl Phys Lett 90:083109

Bathe K (1996) Finite element procedures. Prentice Hall, Englewood Cliffs

Bobaru F, Duangpanya M (2010) The peridynamic formulation for transient heat conduction. Int J Heat Mass Transf 53:4047–4059

Bobaru F, Duangpanya M (2012) The peridynamic formulation for transient heat conduction in bodies with discontinuities. Int J Comp Phys 231:2764–2785

Chang CY, Ma CC (2001) Transient thermal conduction analysis of a rectangular plate with multiple insulated cracks by the alternating method. Int J Heat Mass Transf 44:2423–2437

Chen G (2002) Ballistic-diffusive equations for transient heat conduction from nano to macroscales. Trans ASME J Heat Transf 124:320–328

Chen WH, Chang CL (1994) Heat conduction analysis of a plate with multiple insulated cracks by the finite element alternating method. Int J Solids Struct 3:1343–1355

Dorfman AS (2004) Transient heat transfer between a semi-infinite hot plate and a flowing cooling liquid film. Trans ASME J Heat Transf 126:149–154

Duffey RB, Porthous D (1973) Physics of rewetting in water reactor emergency core cooling. Nucl Eng Des 25:379–394

Gerstle W, Silling S, Read D, Tewary V, Lehoucq R (2008) Peridynamic simulation of electromigration. Comput Mater Continua 8:75–92

Grmela M, Lebon G (1998) Finite-speed propagation of heat: a nonlocal and nonlinear approach. Physica A 248:428–441

Hosseini-Tehrani P, Eslami MR (2000) BEM analysis of thermal and mechanical shock in a two-dimensional finite domain considering coupled thermoelasticity. Eng Anal Bound Elem 24:249–257

Jiji ML (2009) Heat conduction. Springer, Berlin/Heidelberg

Lebon G, Grmela M (1996) Weakly nonlocal heat conduction in rigid solids. Phys Lett A 214:184–188

Luciani JF, Mora P, Virmont J (1983) Nonlocal heat-transport due to steep temperature-gradients. Phys Rev Lett 51:1664–1667

Mahan GD, Claro F (1988) Nonlocal theory of thermal-conductivity. Phys Rev B 38:1963–1969

Moiseiwitsch BL (2004) Variational principles. Dover, Mineola

Özişik MN (1980) Heat conduction, 2nd edn. Wiley, New York

Silling SA, Askari E (2005) A meshfree method based on the peridynamic model of solid mechanics. Comput Struct 83:1526–1535

Sobolev SL (1994) Equations of transfer in nonlocal media. Int J Heat Mass Transf 37:2175–2182

Tien CL, Chen G (1994) Challenges in microscale conductive and radiative heat-transfer. Trans ASME J Heat Transf 116:799–807

Tzou DY, Guo Z (2010) Non local behavior in thermal lagging. Int J Therm Sci 49:1133–1137

# Chapter 13
# Fully Coupled Peridynamic Thermomechanics

This chapter concerns the derivation of the coupled peridynamic (PD) thermomechanics equations based on thermodynamic considerations. The generalized peridynamic model for fully coupled thermomechanics is derived using the conservation of energy and the free-energy function. Subsequently, the bond-based coupled PD thermomechanics equations are obtained by reducing the generalized formulation. These equations are also cast into their nondimensional forms. After describing the numerical solution scheme, solutions to certain coupled thermomechanical problems with known previous solutions are presented.

Thermomechanics concerns the influence of the thermal state of a solid body on the deformation and the influence of the deformation on the thermal state. In many cases, the effect of the deformation field on the thermal state may be ignored. This leads to a decoupled or uncoupled thermomechanical analysis, for which only the effect of the temperature field on the deformation is present. However, the uncoupled thermomechanics may not be satisfactory for certain transient problems. Experimental verification of the influence of the deformation on the thermal state exists. It was shown that an adiabatic solid experiences a temperature drop when it is strained in tension (Chadwick 1960; Fung 1965). Also, elastic bodies under tensile loading experience cooling below the yield stress; however, beyond the yield stress the bodies heat up due to the irreversible nature of plasticity (Nowinski 1978).

Also, the temperature field induced by structural loading may not be uniform. For example, when a beam with an initially uniform temperature is under bending, part of the beam is in tension while the other part is in compression. Due to the thermomechanical coupling, the part of the beam that is in tension cools and the region that is in compression heats up, establishing a thermal gradient. This leads to the onset of heat diffusion. The heat flow is irreversible; thus, some of the mechanical energy supplied to bend the beam is dissipated through its conversion to heat energy. This phenomenon is called thermoelastic damping and it plays a critical role in vibrations and wave propagation.

It is well known that during fracture in metals a plastic region, in which the material has locally yielded, occurs ahead of the crack tip. As a result, the

E. Madenci and E. Oterkus, *Peridynamic Theory and Its Applications*, DOI 10.1007/978-1-4614-8465-3_13, © Springer Science+Business Media New York 2014

mechanical energy is dissipated as heat and the temperature rises in the local region ahead of the crack tip. A slightly different phenomenon is observed for fracture in polymers. During fracture in polymers, it was experimentally observed that thermoelastic cooling is followed by a temperature rise due to the plastic zone and/or fracture process itself, which exposes new surfaces (Rittel 1998). Consequently, in order to accurately model fracture, especially the crack tip, thermal consideration needs to be taken into account and a coupled thermomechanical analysis becomes necessary. The thermal and structural interaction becomes especially important for high-speed impact and penetration fracture problems (Brünig et al. 2011).

The derivation of the classical thermomechanics equation from a thermodynamic perspective did not occur till the mid 1950s (Biot 1956). Biot used generalized irreversible thermodynamics to formulate the classical thermomechanical laws in variational form, with the corresponding Euler equations representing the coupled momentum and energy equations.

The fully coupled thermomechanical equations based on the classical theory are well established. The classical equations of thermoelasticity are comprised of the deformation equation of motion with a thermoelastic constitutive law and the heat transfer equation with a structural (or deformational) heating and cooling term contributing to the thermal energy. For isotropic materials, the thermoelastic constitutive law includes the thermal stresses, which are related to the temperature gradient, while the structural heating and cooling are dependent on the thermal modulus and rate of dilatation. Depending on the structural idealization, the thermal modulus is defined as

$$\beta_{cl} = (3\lambda + 2\mu)\alpha = \frac{E\alpha}{1 - 2\nu} \text{ for three dimensions,} \qquad (13.1a)$$

$$\beta_{cl} = (2\lambda + 2\mu)\alpha = \frac{E\alpha}{(1 - \nu)} \text{ for two dimensions,} \qquad (13.1b)$$

$$\beta_{cl} = (2\mu)\alpha = E\alpha \text{ for one dimension,} \qquad (13.1c)$$

in which $E$ is the elastic modulus, $\alpha$ is the coefficient of thermal expansion, and $\nu$ is the Poisson's ratio. The parameters $\lambda$ and $\mu$ are Lamé's constants.

Typically, the strength of coupling is measured via the nondimensional quantity known as the coupling coefficient and defined as

$$\epsilon = \frac{\beta_{cl}^2 \Theta_0}{\rho \, c_v (\lambda + 2\mu)}, \qquad (13.2)$$

for which $\rho$ is the mass density, $c_v$ is the specific heat capacity, and $\Theta_0$ is the reference temperature at which the stress in the body is zero (Nowinski 1978). The coupling coefficients of metals are significantly lower than those of plastics. Steel, for example, has a coupling coefficient of about 0.011 while certain plastics have a value of $\epsilon = 0.43$.

## 13.1 Local Theory

Various researchers analytically examined plane waves in thermoelastic solids (Chadwick and Sneddon 1958; Deresiewicz 1957). In a one-dimensional formulation, they showed that the presence of thermal and elastic waves are dispersed and attenuated. They also studied the effect of frequency on the phase velocity, attenuation, and damping. Later, Chadwick (1962) extended the analysis to two dimensions and investigated the propagation of thermoelastic waves in thin plates. Paria (1958) determined the temperature and stress distribution of a two-dimensional half-space problem using Laplace and Hankel transforms. Laplace transforms have also been used by Boley and Hetnarski (1968) to characterize propagating discontinuities in various one-dimensional coupled thermoelastic problems. Fourier transforms were employed by Boley and Tolins (1962) to determine the mechanical and thermal response of a one-dimensional semi-infinite bar with transient boundary conditions. The major challenge with transform methods is in finding the analytical inverse transforms—in many cases this is not possible and numerical inversion is necessary. Other analytical solution methods have been adopted to solve coupled thermoelastic problems. Soler and Brull (1965) used perturbation techniques and more recently Lychev et al. (2010) determined a closed-form solution by an expansion of the eigenfunctions generated by the coupled equations of motion and heat conduction.

Numerical approximations to the classical thermoelastic equations have been very commonly found using the finite element (FE) method. A transient thermoelastic FE model was developed by Nickell and Sackman (1968) and Ting and Chen (1982). The approximations from their FE model were compared against analytical solutions for various one-dimensional semi-infinite problems. Oden (1969) and Givoli and Rand (1995) developed dynamic thermoelastic FE models. Additionally, Chen and Weng (1988, 1989a, b) modeled various thermoelastic problems such as the transient response of an axisymmetric infinite cylinder and an infinitely long plate using a finite element formulation in the Laplace transform domain. Hosseini-Tehrani and Eslami (2000) presented solutions for thermal and mechanical shocks in a finite domain based on the boundary element method (BEM) in conjunction with the Laplace-transform method in a time domain. They provided results for small time durations (early stages of the shock loads) using the numerical inversion of the Laplace-transform method.

Numerical solution schemes for thermomechanical problems are divided into two categories—monolithic schemes and staggered schemes. In monolithic schemes, the differential equations for different variables are solved simultaneously. On the other hand, for staggered or partitioned schemes, the solutions of the different variables are determined separately. In general, the staggered schemes have been favored over monolithic schemes, as the monolithic systems can be very large, making it unfeasible to solve practical problems. In addition, the mechanical and thermal parts of a thermomechanical problem may have very different time scales, hence requiring different time stepping schemes. However, the very nature of monolithic schemes renders this impossible.

One of the major issues associated with staggered numerical analysis of coupled thermomechanics is the concern of stability. When conditionally stable techniques are used to solve the coupled momentum and energy equations, a small time step size is required, which may be computationally impractical for certain problems. Even when various unconditionally stable methods are used to solve the equation of motion and heat transfer equation, the overall solution to the coupled problem may still be conditionally stable. A substantial amount of work has been done to combat this issue and to develop unconditionally stable staggered algorithms. Examples of such algorithms based on the finite element method include an adiabatic split scheme by Armero and Simo (1992) and various implicit-implicit and implicit-explicit schemes (Farhat et al. 1991; Liu and Zhang 1983; Liu and Chang 1985).

## 13.2   Nonlocal Theory

Research into nonlocal coupled thermomechanics is undoubtedly emerging. Classical nonlinear constitutive equations for nonlocal fully coupled thermoelasticity have been presented by Huang (1999). Ardito and Comi (2009) developed a fully nonlocal thermoelastic model that has an internal length scale. They analytically solved the nonlocal equations in order to determine the dissipation in microelectromechanical resonators. Comparison of their results with experimental observations revealed that the nonlocal model is able to capture the size effect that the standard local thermoelastic analysis is unable to capture. The work by Ardito and Comi (2009) illustrates the importance of nonlocality in small-scale problems. With the peridynamic thermomechanical model, not only are the problems that require nonlocality solvable, such as the microelectromechanical problems, but also the problems with discontinuities can be readily modeled. A crack that forms and propagates in a body with a varying temperature or temperature gradient is an example of such a problem. Therefore, the peridynamic approach to thermomechanics is advantageous as it not only accounts for nonlocality but also allows for coupled deformation and temperature fields to be determined in spite of cracks and other discontinuities. Uncoupled thermomechanics using the bond-based theory was developed within the realm of peridynamics by Kilic and Madenci (2010). However, no work has been published on fully coupled thermomechanics within the peridynamic framework.

## 13.3   Peridynamic Thermomechanics

Similar to the derivation of classical thermomechanical equations (Nowinski 1978), the generalized peridynamics for fully coupled thermomechanics is based on irreversible thermodynamics, i.e., the conservation of energy and the free-energy density function.

### 13.3.1 Peridynamic Thermal Diffusion with a Structural Coupling Term

The first law of thermodynamics based on peridynamic quantities, accounting for the conservation of mechanical and thermal energy, has been given by Silling and Lehoucq (2010) as

$$\dot{\varepsilon}_s = \mathbf{T} \cdot \dot{\underline{\mathbf{Y}}} + \bar{Q}_b + s_b, \tag{13.3}$$

where $\dot{\varepsilon}_s$ is the time rate of change of the internal energy storage density, and $s_b$ is the prescribed volumetric heat generation per unit mass. The term $\mathbf{T} \cdot \dot{\underline{\mathbf{Y}}}$ represents the absorbed power density; it is the dot product of the force state and the time rate of deformation state. The absorbed power density in peridynamics is analogous to the stress power, $\boldsymbol{\sigma} \cdot \dot{\mathbf{F}}$ in classical continuum mechanics, where $\boldsymbol{\sigma}$ is the Piola stress tensor and $\mathbf{F}$ is the deformation gradient tensor. The variable $\bar{Q}$ is the rate of heat energy exchange with other material points, and it is given by

$$\bar{Q} = \int_H (\underline{h}(\mathbf{x}, t)\langle \mathbf{x}' - \mathbf{x} \rangle - \underline{h}(\mathbf{x}', t)\langle \mathbf{x} - \mathbf{x}' \rangle)dV', \tag{13.4}$$

in which $\underline{h}$ is the heat flow *scalar* state. The quantity $\bar{Q}$ is related to $\bar{Q}_b$ as $\bar{Q} = \rho \bar{Q}_b$.

The free-energy density function is defined as (Silling and Lehoucq 2010)

$$\psi = \varepsilon_s - \Theta \eta, \tag{13.5}$$

where $\Theta$ is the absolute temperature and $\eta$ is the entropy density. The time derivative of Eq. 13.5 becomes

$$\dot{\psi} = \dot{\varepsilon}_s - \dot{\Theta} \eta - \Theta \dot{\eta}. \tag{13.6}$$

Substituting for $\dot{\varepsilon}_s$ in Eq. 13.6 from the conservation of energy given in Eq. 13.3 leads to the following expression:

$$\dot{\psi} = \mathbf{T} \cdot \dot{\underline{\mathbf{Y}}} + \bar{Q}_b + s_b - \dot{\Theta} \eta - \Theta \dot{\eta}. \tag{13.7}$$

The functional dependency of the free-energy density and the entropy density can be defined in terms of the deformation state, time rate of change of the deformation state, and the temperature in the form

$$\psi = \psi\left(\underline{\mathbf{Y}}, \dot{\underline{\mathbf{Y}}}, \Theta\right), \tag{13.8a}$$

$$\eta = \eta\left(\underline{\mathbf{Y}}, \dot{\underline{\mathbf{Y}}}, \Theta\right). \tag{13.8b}$$

In conjunction with the chain rule, the time rate of change of the free-energy density can be expressed as

$$\dot{\psi} = \psi_{,\underline{Y}} \cdot \dot{\underline{Y}} + \psi_{,\dot{\underline{Y}}} \cdot \ddot{\underline{Y}} + \psi_{,\Theta}\dot{\Theta}, \tag{13.9}$$

where the variable after the subscript comma indicates differentiation. If it is a state variable, its differentiation is known as the Fréchet derivative, as explained in the Appendix.

Substituting from Eq. 13.7 into Eq. 13.9 results in

$$\left(\Theta\dot{\eta} - \bar{Q}_b - s_b\right) + \left(\psi_{,\Theta} + \eta\right)\dot{\Theta} + \left(\psi_{,\underline{Y}} - \underline{T}\right) \cdot \dot{\underline{Y}} + \psi_{,\dot{\underline{Y}}} \cdot \ddot{\underline{Y}} = 0. \tag{13.10}$$

Adopting the assumption of Nowinski (1978) that $\dot{\underline{Y}}$, $\ddot{\underline{Y}}$, and $\dot{\Theta}$ vary independently, Eq. 13.10 leads to

$$\Theta\dot{\eta} - \bar{Q}_b - s_b = 0, \tag{13.11a}$$

$$\eta = -\psi_{,\Theta}, \tag{13.11b}$$

$$\underline{T} = \psi_{,\underline{Y}}, \tag{13.11c}$$

$$\psi_{,\dot{\underline{Y}}} = 0. \tag{13.11d}$$

By using the free-energy density, the first law of thermodynamics, and the Clausius-Duhem inequality, Silling and Lehoucq (2010) also determined Eqs. 13.11b and 13.11d. In addition, they obtained expressions for the *equilibrium*, $\underline{T}^e$, and *dissipative*, $\underline{T}^d$, parts of force vector state as

$$\underline{T}^e(\underline{Y},\Theta) = \psi_{,\underline{Y}}(\underline{Y},\Theta), \tag{13.12a}$$

$$\underline{T}^d(\underline{Y},\dot{\underline{Y}},\Theta) \cdot \dot{\underline{Y}} \geq 0. \tag{13.12b}$$

Using Eqs. 13.11b, 13.11d, and 13.8b in conjunction with the chain rule, the time derivative of the entropy density may be rewritten in the form

$$\dot{\eta} = -\psi_{,\Theta\underline{Y}} \cdot \dot{\underline{Y}} - \psi_{,\Theta\Theta}\dot{\Theta}. \tag{13.13}$$

Substituting from Eq. 13.13 into Eq. 13.11a and multiplying by $\rho$ leads to

$$\rho\Theta\psi_{,\Theta\underline{Y}} \cdot \dot{\underline{Y}} + \rho\Theta\psi_{,\Theta\Theta}\dot{\Theta} + \bar{Q} + \rho s_b = 0. \tag{13.14}$$

Based on the classical theory (Nowinski 1978), the specific heat capacity, $c_v$, can be related to the classical free-energy density, $\bar{\psi}$, as

$$\Theta\bar{\psi}_{,\Theta\Theta} = -c_v. \tag{13.15}$$

The assumption of the classical free-energy density at a point being equal to the peridynamic free-energy density, $\psi$ leads to

$$\Theta\psi_{,\Theta\Theta} = -c_v. \tag{13.16}$$

Based on this observation, it is evident that the specific heat capacity has a similar meaning in the peridynamic theory as in the classical theory. Therefore, the term $\Theta\psi_{,\Theta\Theta}$ in Eq. 13.14 can be replaced by $-c_v$.

Based on the classical theory (Fung 1965), the thermal modulus $\beta_{ij}$ can be related to the classical free-energy density $\bar{\psi}$ through

$$\beta_{clij} = \rho\frac{\partial^2\bar{\psi}}{\partial e_{ij}\partial\Theta}, \tag{13.17}$$

where $e_{ij}$ is the strain tensor. Note that $\beta_{clij} = \beta_{cl}\delta_{ij}$ for isotropic materials.

Analogus to the thermal modulus of the classical theory thermal modulus state, a *vector* state, $\underline{\mathbf{B}}$, can be defined as

$$\underline{\mathbf{B}} = \rho\psi_{,\Theta\underline{\mathbf{Y}}}. \tag{13.18}$$

Substituting from Eqs. 13.4, 13.16, and 13.18 into Eq. 13.14 and after rearranging some of terms results in

$$\rho c_v\dot{\Theta}(\mathbf{x}, t) = \int_H (\underline{h}(\mathbf{x}, t)\langle\mathbf{x}'-\mathbf{x}\rangle - \underline{h}(\mathbf{x}', t)\langle\mathbf{x} - \mathbf{x}'\rangle)dV' \tag{13.19}$$
$$+ \Theta(\mathbf{x}, t)\underline{\mathbf{B}}(\mathbf{x}, t)\cdot\underline{\dot{\mathbf{Y}}}(\mathbf{x}, t) + \rho s_b(\mathbf{x}, t).$$

Applying the definition of the vector state dot product (see Appendix) renders the equation

$$\rho c_v\dot{\Theta}(\mathbf{x}, t) = \int_H ((\underline{h}(\mathbf{x}, t)\langle\mathbf{x}' - \mathbf{x}\rangle - \underline{h}(\mathbf{x}', t)\langle\mathbf{x} - \mathbf{x}'\rangle) \tag{13.20}$$
$$+ \Theta(\mathbf{x}, t)\underline{\mathbf{B}}\langle\mathbf{x}' - \mathbf{x}\rangle\cdot\underline{\dot{\mathbf{Y}}}\langle\mathbf{x}' - \mathbf{x}\rangle)dV' + \rho s_b(\mathbf{x}, t),$$

in which the term $\underline{\mathbf{B}}\cdot\underline{\dot{\mathbf{Y}}}$ represents the effect of deformation on temperature. The final form of this equation can be obtained by defining $\underline{\dot{\mathbf{Y}}}$ and $\underline{\mathbf{B}}$ in terms of the time rate of change of the extension *scalar* state, $\underline{\dot{e}}$, and the thermal modulus scalar *state*, $\underline{\beta}$, as

$$\underline{\mathbf{B}}\langle \mathbf{x}' - \mathbf{x} \rangle = \underline{\beta}\langle \mathbf{x}' - \mathbf{x} \rangle \frac{\mathbf{y}' - \mathbf{y}}{|\mathbf{y}' - \mathbf{y}|}, \tag{13.21a}$$

$$\underline{\dot{\mathbf{Y}}}\langle \mathbf{x}' - \mathbf{x} \rangle = \underline{\dot{e}}\langle \mathbf{x}' - \mathbf{x} \rangle \frac{\mathbf{y}' - \mathbf{y}}{|\mathbf{y}' - \mathbf{y}|}, \tag{13.21b}$$

in which the extension *scalar* state, $\underline{e}$, and thermal modulus scalar *state*, $\underline{\beta}$, are defined as

$$\underline{e} = \underline{y} - \underline{x}, \tag{13.21c}$$

$$\underline{\beta} = \rho \psi_{,\Theta \underline{e}}, \tag{13.21d}$$

with $\underline{y} = |\underline{\mathbf{Y}}|$ and $\underline{x} = |\underline{\mathbf{X}}|$. Thus, Eq. 13.20 can be recast as

$$\rho c_v \dot{\Theta}(\mathbf{x}, t) = \int_H \big( (\underline{h}(\mathbf{x}, t)\langle \mathbf{x}' - \mathbf{x} \rangle - \underline{h}(\mathbf{x}', t)\langle \mathbf{x} - \mathbf{x}' \rangle)$$
$$+ \Theta(\mathbf{x}, t)\underline{\beta}\langle \mathbf{x}' - \mathbf{x} \rangle \underline{\dot{e}}\langle \mathbf{x}' - \mathbf{x} \rangle \big) dV' + \rho s_b(\mathbf{x}, t). \tag{13.22}$$

### 13.3.2  Peridynamic Deformation with a Thermal Coupling Term

Based on the classical linear theory of thermoelasticity (Nowinski 1978), the free-energy density is a potential function given by

$$\bar{\psi} = \bar{\psi}(e_{ij}, T) = \frac{1}{2} c_{ijkl} e_{ij} e_{kl} - \beta_{clij} e_{ij} T - \frac{c_v}{2\Theta_0} T^2, \tag{13.23}$$

where $c_{ijkl}$ is the elastic moduli of the material, $T = \Theta - \Theta_0$, and $\Theta_0$ is the reference temperature. A similar approach is adopted herein for the derivation of the peridynamic deformation equation with a thermal coupling term.

Silling (2009) developed a linearized form of the state-based peridynamics for small elastic deformation by introducing the force *vector* state, $\underline{\mathbf{T}}$, as

$$\underline{\mathbf{T}} = \underline{\mathbf{T}}(\underline{\mathbf{U}}), \tag{13.24}$$

where $\underline{U}$ is the displacement vector state. The free-energy density function is expressed in terms of $\underline{U}$ as

$$\psi(\underline{U}) = \psi(\underline{Y}^0) + \underline{T}^0 \cdot \underline{U} + \frac{1}{2}\underline{U} \cdot \underline{\mathbb{K}} \cdot \underline{U}, \tag{13.25}$$

where $\underline{Y}^0$ and $\underline{T}^0$ are defined as the equilibrated deformation and force states, respectively. The double state $\underline{\mathbb{K}}$ is called the modulus state, and it is given by Silling (2009) as

$$\underline{\mathbb{K}} = \underline{T}^0{}_{,\underline{Y}}. \tag{13.26}$$

For linear thermoelastic material response, in accordance with Eq. 13.23, this form of the free energy is modified by including $T$ and $\underline{U}$ as

$$\psi(\underline{U}, T) = \psi(\underline{Y}^0) + \underline{T}^0 \cdot \underline{U} + \frac{1}{2}\underline{U} \cdot \underline{\mathbb{K}} \cdot \underline{U} - \underline{B} \cdot \underline{U}T - \frac{c_v}{2\Theta_0}T^2. \tag{13.27}$$

Invoking this equation into Eq. 13.11c results in the explicit form of the force state as

$$\underline{T} = \underline{\mathbb{K}} \cdot \underline{U} - \underline{B}T. \tag{13.28}$$

It represents the state-based constitutive relation for a linearized peridynamic thermoelastic material. Substituting from Eq. 13.28 into the peridynamic equation of motion, Eq. 2.22a results in the following:

$$\rho\ddot{\mathbf{u}} = \int_H [(\underline{\mathbb{K}} \cdot \underline{U} - \underline{B}T)(\mathbf{x}, t)\langle \mathbf{x}' - \mathbf{x}\rangle - (\underline{\mathbb{K}} \cdot \underline{U} - \underline{B}T)(\mathbf{x}', t)\langle \mathbf{x} - \mathbf{x}'\rangle]dV'$$
$$+ \mathbf{b}(\mathbf{x}, t), \tag{13.29}$$

in which the term $\underline{B}\langle \mathbf{x}' - \mathbf{x}\rangle T$ represents the effect of the thermal state on deformation. For a nonlinear elastic material model, the free energy is composed of a thermal and a mechanical component. Therefore, one possible form of the force state can be

$$\underline{T} = \nabla W - \underline{B}T \tag{13.30}$$

in which $W$ is the deformational strain energy density and $\nabla W$ is its Fréchet derivative. The part of the force state, $\underline{T}_s$, that includes only the structural deformation can be defined as

$$\underline{T}_s = \nabla W. \tag{13.31}$$

Substituting from these equations into the peridynamic equation of motion, Eq. 2.22a can be recast as

$$\rho\ddot{\mathbf{u}}(\mathbf{x},t) = \int_H [(\underline{\mathbf{T}}_s\langle\mathbf{x}'-\mathbf{x}\rangle - \underline{\mathbf{B}}\,\langle\mathbf{x}'-\mathbf{x}\rangle T)]$$
$$- (\underline{\mathbf{T}}'_s\langle\mathbf{x}-\mathbf{x}'\rangle - \underline{\mathbf{B}}'\langle\mathbf{x}-\mathbf{x}'\rangle\,T')]dV' + \mathbf{b}(\mathbf{x},t), \tag{13.32}$$

where $\underline{\mathbf{T}}_s = \underline{\mathbf{T}}_s(\mathbf{x},t)$ and $\underline{\mathbf{T}}'_s = \underline{\mathbf{T}}_s(\mathbf{x}',t)$; similar notation is used for $\underline{\mathbf{B}}$ and $T$.

Substituting from Eq. 4.8 into Eq. 2.22b in conjunction with Eqs. 4.11 and 4.12 results in the bond-based PD equation for an isotropic material including the effect of temperature as

$$\rho\ddot{\mathbf{u}}(\mathbf{x},t) = \int_H \left\{ \left( \frac{c}{2}\frac{|\mathbf{y}'-\mathbf{y}|-|\mathbf{x}'-\mathbf{x}|}{|\mathbf{x}'-\mathbf{x}|} - \frac{c}{2}\alpha T \right) \frac{\mathbf{y}'-\mathbf{y}}{|\mathbf{y}'-\mathbf{y}|} \right.$$
$$\left. - \left( \frac{c}{2}\frac{|\mathbf{y}-\mathbf{y}'|-|\mathbf{x}-\mathbf{x}'|}{|\mathbf{x}-\mathbf{x}'|} - \frac{c}{2}\alpha T' \right) \frac{\mathbf{y}-\mathbf{y}'}{|\mathbf{y}-\mathbf{y}'|} \right\}dV' + \mathbf{b}(\mathbf{x},t)\ . \tag{13.33}$$

Comparison of this equation with Eq. 13.32 leads to the explicit forms of

$$\underline{\mathbf{T}}_s\langle\mathbf{x}'-\mathbf{x}\rangle = \frac{c}{2}\frac{|\mathbf{y}'-\mathbf{y}|-|\mathbf{x}'-\mathbf{x}|}{|\mathbf{x}'-\mathbf{x}|}\frac{\mathbf{y}'-\mathbf{y}}{|\mathbf{y}'-\mathbf{y}|} \tag{13.34a}$$

and

$$\underline{\mathbf{B}}\,\langle\mathbf{x}'-\mathbf{x}\rangle = \frac{c}{2}\alpha\frac{\mathbf{y}'-\mathbf{y}}{|\mathbf{y}'-\mathbf{y}|}. \tag{13.34b}$$

Comparison of Eq. 13.34b with Eq. 13.21a results in the expression for the thermal modulus scalar *state* $\beta$ as

$$\underline{\beta}\,\langle\mathbf{x}'-\mathbf{x}\rangle = \frac{c}{2}\alpha. \tag{13.35}$$

### 13.3.3   Bond-Based Peridynamic Thermomechanics

The difference between the generalized heat transfer equation, Eq. 12.51, and thermomechanical heat transfer equation for an isotropic material, Eq. 13.22, is due to the deformational heating and cooling term, $(\underline{\beta}\cdot\underline{\dot{e}})$. In light of this difference, the bond-based heat transfer equation, Eq. 12.51, can be modified to include the deformational heating and cooling term. Therefore, the bond-based coupled thermomechanical heat transfer equation can be cast as

$$\rho c_v \dot{\Theta}(\mathbf{x}, t) = \int_H \left( f_h - \Theta \frac{c}{2} \alpha \dot{e} \right) dV' + \rho s_b(\mathbf{x}, t), \tag{13.36}$$

where $\dot{e}$ is the time rate of change of the extension between the material points, and it is defined as

$$e = |\boldsymbol{\eta} + \boldsymbol{\xi}| - |\boldsymbol{\xi}|, \tag{13.37a}$$

with its time rate of change

$$\dot{e} = \frac{\boldsymbol{\eta} + \boldsymbol{\xi}}{|\boldsymbol{\eta} + \boldsymbol{\xi}|} \cdot \dot{\boldsymbol{\eta}}, \tag{13.37b}$$

where $\dot{\boldsymbol{\eta}}$ is the time rate of change of the relative displacement vector. Equation 13.36 can be rewritten in terms of the change in temperature, $T = \Theta - \Theta_0$, by replacing $\Theta$ with $T + \Theta_0$ and $\dot{\Theta}$ with $\dot{T}$ as

$$\rho c_v \dot{T}(\mathbf{x}, t) = \int_H \left( f_h - (T + \Theta_o) \frac{c}{2} \alpha \dot{e} \right) dV' + \rho s_b(\mathbf{x}, t), \tag{13.38a}$$

or

$$\rho c_v \dot{T}(\mathbf{x}, t) = \int_H \left( f_h - \Theta_o \left( \frac{T}{\Theta_o} + 1 \right) \frac{c}{2} \alpha \dot{e} \right) dV' + \rho s_b(\mathbf{x}, t). \tag{13.38b}$$

As suggested by Nowinski (1978), if the temperature change, $T$, is very small when compared with the reference temperature, $\Theta_o$, Eq. 13.38b can be approximated as

$$\rho c_v \dot{T}(\mathbf{x}, t) = \int_H \left( f_h - \Theta_o \frac{c}{2} \alpha \dot{e} \right) dV' + \rho s_b(\mathbf{x}, t). \tag{13.39}$$

Substituting for the thermal response (heat flow density) function from Eq. 12.55 leads to its final form as

$$\rho c_v \dot{T}(\mathbf{x}, t) = \int_H \left( \kappa \frac{\tau}{|\boldsymbol{\xi}|} - \Theta_o \frac{c}{2} \alpha \dot{e} \right) dV' + \rho s_b(\mathbf{x}, t). \tag{13.40}$$

From Eq. 13.33, the bond-based PD equation of motion including the effect of temperature can be rewritten as

$$\rho \ddot{\mathbf{u}}(\mathbf{x}, t) = \int_H \frac{\boldsymbol{\xi} + \boldsymbol{\eta}}{|\boldsymbol{\xi} + \boldsymbol{\eta}|} (cs - c\alpha T_{avg}) dV' + \mathbf{b}(\mathbf{x}, t), \tag{13.41}$$

in which $c$ is the peridynamic material parameter. The initial relative position and relative displacement vectors are defined as $\boldsymbol{\xi} = \mathbf{x}' - \mathbf{x}$ and $\boldsymbol{\eta} = \mathbf{u}' - \mathbf{u}$, respectively. The parameter $s$ represents the stretch between material points $\mathbf{x}'$ and $\mathbf{x}$, and $T_{avg}$ is the mean value of the change in temperatures at material points $\mathbf{x}'$ and $\mathbf{x}$ defined as

$$T_{avg} = \frac{T + T'}{2}. \tag{13.42}$$

Introducing $\beta$ as the bond-based peridynamic thermal modulus, the final form of the fully coupled bond-based thermomechanical equations becomes

$$\rho c_v \dot{T}(\mathbf{x}, t) = \int_H \left( \kappa \frac{\tau}{|\boldsymbol{\xi}|} - \Theta_o \frac{\beta}{2} \dot{e} \right) dV' + h_s(\mathbf{x}, t), \tag{13.43a}$$

with $h_s = \rho s_b$ representing the heat source due to volumetric heat generation, and

$$\rho \ddot{\mathbf{u}}(\mathbf{x}, t) = \int_H \frac{\boldsymbol{\xi} + \boldsymbol{\eta}}{|\boldsymbol{\xi} + \boldsymbol{\eta}|} (cs - \beta T_{avg}) dV' + \mathbf{b}(\mathbf{x}, t), \tag{13.43b}$$

with

$$\beta = c\alpha. \tag{13.43c}$$

The first equation is the conservation of thermal energy (i.e., the heat transfer equation) with a contribution from deformational heating and cooling, and the second equation is the conservation of linear momentum (i.e., the equation of motion) with a thermoelastic constitutive relation.

## 13.4   Nondimensional Form of Thermomechanical Equations

The nondimensional form of an equation or system of equations involves eliminating the units associated with the variables and parameters. For coupled systems, various parameters may differ in size and the effects of certain parameters may not be apparent. The nondimensional form of equations can permit the effects of the different parameters to become more evident. The appropriate scaling, relative measure of quantities, and characteristic properties of the system, such as time constants, length scales, and resonance frequencies, can be revealed through nondimensionalization.

### 13.4.1   Characteristic Length and Time Scales

The characteristic length/time quantity for heat conduction is the diffusivity defined as

$$\gamma = \frac{k}{\rho c_v} = \frac{\ell^{*2}}{t^*}, \tag{13.44}$$

where $\ell^*$ and $t^*$ represent the characteristic length and time, respectively. For the equation of motion, the characteristic length/time is the elastic wave speed. The square of the elastic wave speed, $\tilde{a}$, is

$$\tilde{a}^2 = \frac{(\lambda + 2\mu)}{\rho} = \frac{\ell^{*2}}{t^{*2}}, \tag{13.45}$$

where $\lambda$ and $\mu$ are Lamé's constants. Combining the characteristic length/time scale from Eqs. 13.44 and 13.45 leads to the characteristic length and time for thermomechanics as

$$\ell^* = \frac{\gamma}{\tilde{a}} \quad \text{and} \quad t^* = \frac{\gamma}{\tilde{a}^2}. \tag{13.46}$$

The characteristic length and time are typically employed in the non dimensionalization of the thermomechanical equations.

### 13.4.2   Nondimensional Parameters

The nondimensional form of Eq. 13.43a can be achieved by adopting the approach by Nickell and Sackman (1968) using Eqs. 13.44 and 13.45 for thermal diffusivity and the square of the elastic wave speed. The nondimensional variables are denoted with an overscore. The nondimensionalization procedure for length-related variables, i.e., $x$, $\delta$, $A$, and $V$ (the volume), employs the characteristic length, and they are defined as

$$x = \left(\frac{\gamma}{\tilde{a}}\right)\bar{x}, \quad \delta = \left(\frac{\gamma}{\tilde{a}}\right)\bar{\delta}, \quad A = \left(\frac{\gamma}{\tilde{a}}\right)^2 \bar{A} \quad \text{and} \quad V = \left(\frac{\gamma}{\tilde{a}}\right)^3 \bar{V}. \tag{13.47}$$

The displacement is nondimensionalized as

$$u = \left(\frac{\gamma}{\tilde{a}}\right)\frac{\beta_{cl}\Theta_o}{(\lambda + 2\mu)}\bar{u}. \tag{13.48}$$

The stretch is nondimensionalized as

$$s = \frac{\beta_{cl}\Theta_o}{(\lambda + 2\mu)}\bar{s}. \tag{13.49}$$

The time is scaled using the characteristic length as

$$t = \left(\frac{\gamma}{\bar{a}^2}\right)\bar{t}. \tag{13.50}$$

The nondimensionalization for the velocity-related variables is achieved by

$$v = \frac{\beta_{cl}\Theta_o}{(\lambda + 2\mu)}\bar{a}\,\bar{v} \quad \text{and} \quad \dot{e} = \frac{\beta_{cl}\Theta_o}{(\lambda + 2\mu)}\bar{a}\,\bar{\dot{e}}. \tag{13.51}$$

Finally, the temperature and temperature difference are nondimensionalized as

$$T = \Theta_o\bar{T} \quad \text{and} \quad \tau = \Theta_o\bar{\tau}. \tag{13.52}$$

It is worth noting that the definitions of thermal modulus, bulk modulus, Lamé constants, shear modulus, peridynamic parameters, and microconductivity depend on the structural idealization. Their definitions for one-, two-, and three-dimensional analysis are summarized as:

*One-dimensional analysis*

$$\lambda = 0, \mu = \frac{E}{2}, \alpha = \frac{\beta_{cl}}{2\mu}, c = \frac{2E}{A\delta^2}, \kappa = \frac{2k}{A\delta^2}. \tag{13.53a}$$

*Two-dimensional analysis*

$$\lambda = \frac{E\nu}{(1-\nu)(1+\nu)}, \mu = \frac{E}{2(1+\nu)}, \alpha = \frac{\beta_{cl}}{2\lambda + 2\mu},$$
$$c = \frac{9E}{\pi h\delta^3}, \kappa = \frac{6k}{\pi h\delta^3}. \tag{13.53b}$$

*Three-dimensional analysis*

$$\lambda = \frac{E\nu}{(1+\nu)(1-2\nu)}, \mu = \frac{E}{2(1+\nu)}, \alpha = \frac{\beta_{cl}}{3\lambda + 2\mu},$$
$$c = \frac{12E}{\pi\delta^4}, \kappa = \frac{6k}{\pi\delta^4}. \tag{13.53c}$$

Equating the coefficient of thermal expansion from Eqs. 13.1 and 13.43c leads to the thermal modulus as

$$\beta = \frac{\beta_{cl}}{3\lambda + 2\mu}c \text{ for three dimensions,} \tag{13.54a}$$

$$\beta = \frac{\beta_{cl}}{2\lambda + 2\mu}c \text{ for two dimensions,} \tag{13.54b}$$

$$\beta = \frac{\beta_{cl}}{2\mu}c \text{ for one dimension.} \tag{13.54c}$$

Substituting from Eqs. 13.47, 13.48, 13.49, 13.50, 13.51, and 13.52 with the dimensional considerations from Eq. 13.53, the fully coupled bond-based thermomechanical equations in the absence of body force and heat source can be cast into their nondimensional forms:

*One-dimensional analysis*

$$\frac{\partial^2 \bar{u}}{\partial \bar{t}^2} = \frac{2}{\bar{\delta}^2 \bar{A}} \int_H \frac{\xi + \eta}{|\xi + \eta|} (\bar{s} - \bar{T}_{avg}) d\bar{V}_{x'} + \bar{b}, \tag{13.55a}$$

$$\frac{\partial \bar{T}}{\partial \bar{t}} = \frac{2}{\bar{\delta}^2 \bar{A}} \int_H \left( \frac{\bar{\tau}}{|\bar{\xi}|} - \epsilon \frac{\dot{\bar{e}}}{2} \right) d\bar{V}_{x'} + \bar{h}_s, \tag{13.55b}$$

*Two-dimensional analysis*

$$\frac{\partial^2 \bar{u}}{\partial \bar{t}^2} = \frac{9(1 - \nu)}{\pi \bar{\delta}^3 \bar{h}} \int_H \frac{\xi + \eta}{|\xi + \eta|} ((1 + \nu)\bar{s} - \bar{T}_{avg}) d\bar{V}_{x'} + \bar{b}, \tag{13.56a}$$

$$\frac{\partial \bar{T}}{\partial \bar{t}} = \frac{6}{\pi \bar{\delta}^3 \bar{h}} \int_H \left( \frac{\bar{\tau}}{|\bar{\xi}|} - \frac{3}{4}(1 - \nu)\epsilon\dot{\bar{e}} \right) d\bar{V}_{x'} + \bar{h}_s, \tag{13.56b}$$

*Three-dimensional analysis*

$$\frac{\partial^2 \bar{u}}{\partial \bar{t}^2} = \frac{6}{\pi \bar{\delta}^4} \int_H \frac{\xi + \eta}{|\xi + \eta|} \left( \frac{1 + \nu}{1 - \nu}\bar{s} - \bar{T}_{avg} \right) d\bar{V}_{x'} + \bar{b}, \tag{13.57a}$$

$$\frac{\partial \bar{T}}{\partial \bar{t}} = \frac{6}{\pi \bar{\delta}^4} \int_H \left( \frac{\bar{\tau}}{|\bar{\xi}|} - \frac{1}{2}\epsilon\dot{\bar{e}} \right) d\bar{V}_{x'} + \bar{h}_s, \tag{13.57b}$$

in which the nondimensional coupling coefficient, $\epsilon$, body force density, $\bar{\mathbf{b}}$, and heat source due to volumetric heat generation, $\bar{h}_s$, are defined as

$$\epsilon = \frac{\beta_{cl}^2 \Theta_0}{\rho c_v (\lambda + 2\mu)},$$  (13.58a)

$$\bar{\mathbf{b}} = \frac{\gamma(\lambda + 2\mu)}{\rho \tilde{a}^3 \beta_{cl} \Theta_0} \mathbf{b},$$  (13.58b)

$$\bar{h}_s = \frac{\gamma}{\rho c_v \tilde{a}^2 \Theta_0} h_s.$$  (13.58c)

The coupling coefficient $\epsilon$ measures the strength of thermal and deformation coupling and it appears in the nondimensional thermomechanical equations associated with the heating and cooling term due to deformation. The coupling coefficient transpires out of the nondimensional form of these peridynamic equations in a similar manner as it does out of the classical thermomechanical equations, as illustrated by Nickell and Sackman (1968). The nondimensional equation represents decoupled thermomechanics for $\epsilon = 0$. It is worth noting that the equation of motion still contains the effect of temperature even if $\epsilon = 0$.

## 13.5  Numerical Procedure

For numerically approximating the solution to the classical fully coupled equations for thermoelasticity, one of two different time stepping strategies is generally employed by researchers. The monolithic or simultaneous scheme is one time stepping strategy. For a monolithic algorithm, the time stepping scheme is applied simultaneously to the full system of equations and the unknown variables are solved for at the same time. If the time stepping scheme for the monolithic algorithm is implicit, unconditional stability is usually achieved. However, monolithic algorithms can result in practical large systems, in spite of their unconditional stability. For the staggered or partitioned scheme, the coupled system of equations are split, typically according to two different fields, the displacement and temperature fields. Each field is then individually treated with a different time stepping algorithm. Staggered algorithms generally circumvent the shortcomings of their monolithic counterparts; however, this is often accomplished at the expense of the unconditional stability. In many scenarios, even when unconditionally stable time stepping schemes are used to solve each partitioned equation, the overall stability of the thermomechanical system of equations is only conditional (Wood 1990). As a result, a good deal of work has been performed to successfully develop unconditionally stable staggered algorithms for thermoelasticity (Armero and Simo 1992; Farhat et al. 1991; Liu and Chang 1985).

**Fig. 13.1** Discretization
of one-dimensional domain
with collocation points

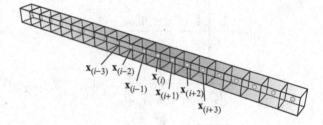

For the numerical treatment of the fully coupled thermoelastic peridynamic system of equations, a staggered strategy is adopted. The system is partitioned naturally according to the structural and thermal fields; thus, the equation of motion is solved for the displacement field and the heat transfer equation is solved for the temperature field. Explicit time stepping schemes are utilized to approximate the solutions to both equations.

In order to illustrate the numerical implementation, one-dimensional peridynamic thermoelastic equations, Eq. 13.55a,b, are considered, and they can be discretized in the forms

$$\ddot{\bar{u}}^n_{(i)} = \frac{2}{\bar{\delta}^2 \bar{A}} \sum_{j=1}^{N} \frac{\bar{\xi}^n_{(i)(j)} + \bar{\eta}^n_{(i)(j)}}{\left|\bar{\xi}^n_{(i)(j)} + \bar{\eta}^n_{(i)(j)}\right|} \left(\bar{s}^n_{(i)(j)} - \bar{T}^n_{(i)(j)}\right)\bar{V}_{(j)} \qquad (13.59a)$$

and

$$\dot{\bar{T}}^n_{(i)} = \frac{2}{\bar{\delta}^2 \bar{A}} \sum_{j=1}^{N} \left(\frac{\bar{\tau}^n_{(i)(j)}}{\left|\bar{\xi}^n_{(i)(j)}\right|} - \epsilon \frac{\dot{\bar{e}}^n_{(i)(j)}}{2}\right)\bar{V}_{(j)}, \qquad (13.59b)$$

in which the term $2/(\bar{\delta}^2 \bar{A})$ is assumed to be constant throughout the domain, $n$ represents the time step number, $i$ is the collocation point that is being solved for, and $j$ represents the collocation points within the horizon of $i$. The nondimensional volume of the subdomain represented by the collocation point $j$ is denoted by $\bar{V}_{(j)}$.

The discretization of a one-dimensional domain is illustrated in Fig. 13.1. The one-dimensional domain is discretized into subdomains, with the collocation points at the center of each subdomain.

The horizon is $\bar{\delta} = 3\bar{\Delta}$, where $\bar{\Delta}$ is the nondimensional spacing between material points. The material point of interest is denoted by $i$ and it interacts with the three points to its left and right. Thus, points $j$ within the horizon of $i$ are $i-3, i-2, i-1$, $i+1, i+2$, and $i+3$, as shown in Fig. 13.1

The nondimensional displacement, velocity, and temperature of all the collocation points are known at the $n^{th}$ time step, i.e., the current time step. Based on Fig. 13.1, Eq. 13.55a can be discretized as

$$
\begin{aligned}
\bar{\bar{u}}_{(i)}^{n} = \frac{2}{\delta^{2}\bar{A}} \Bigg[ & \frac{\bar{\boldsymbol{\xi}}_{(i)(i+3)}^{n} + \bar{\boldsymbol{\eta}}_{(i)(i+3)}^{n}}{\left|\bar{\boldsymbol{\xi}}_{(i)(i+3)}^{n} + \bar{\boldsymbol{\eta}}_{(i)(i+3)}^{n}\right|} \left(\bar{s}_{(i)(i+3)}^{n} - \bar{T}_{(i)(i+3)}^{n}\right)\bar{V}_{(i+3)} \\
& + \frac{\bar{\boldsymbol{\xi}}_{(i)(i+2)}^{n} + \bar{\boldsymbol{\eta}}_{(i)(i+2)}^{n}}{\left|\bar{\boldsymbol{\xi}}_{(i)(i+2)}^{n} + \bar{\boldsymbol{\eta}}_{(i)(i+2)}^{n}\right|} \left(\bar{s}_{(i)(i+2)}^{n} - \bar{T}_{(i)(i+2)}^{n}\right)\bar{V}_{(i+2)} \\
& + \frac{\bar{\boldsymbol{\xi}}_{(i)(i+1)}^{n} + \bar{\boldsymbol{\eta}}_{(i)(i+1)}^{n}}{\left|\bar{\boldsymbol{\xi}}_{(i)(i+1)}^{n} + \bar{\boldsymbol{\eta}}_{(i)(i+1)}^{n}\right|} \left(\bar{s}_{(i)(i+1)}^{n} - \bar{T}_{(i)(i+1)}^{n}\right)\bar{V}_{(i+1)} \\
& + \frac{\bar{\boldsymbol{\xi}}_{(i)(i-3)}^{n} + \bar{\boldsymbol{\eta}}_{(i)(i-3)}^{n}}{\left|\bar{\boldsymbol{\xi}}_{(i)(i-3)}^{n} + \bar{\boldsymbol{\eta}}_{(i)(i-3)}^{n}\right|} \left(\bar{s}_{(i)(i-3)}^{n} - \bar{T}_{(i)(i-3)}^{n}\right)\bar{V}_{(i-3)} \\
& + \frac{\bar{\boldsymbol{\xi}}_{(i)(i-2)}^{n} + \bar{\boldsymbol{\eta}}_{(i)(i-2)}^{n}}{\left|\bar{\boldsymbol{\xi}}_{(i)(i-2)}^{n} + \bar{\boldsymbol{\eta}}_{(i)(i-2)}^{n}\right|} \left(\bar{s}_{(i)(i-2)}^{n} - \bar{T}_{(i)(i-2)}^{n}\right)\bar{V}_{(i-2)} \\
& + \frac{\bar{\boldsymbol{\xi}}_{(i)(i-1)}^{n} + \bar{\boldsymbol{\eta}}_{(i)(i-1)}^{n}}{\left|\bar{\boldsymbol{\xi}}_{(i)(i-1)}^{n} + \bar{\boldsymbol{\eta}}_{(i)(i-1)}^{n}\right|} \left(\bar{s}_{(i)(i-1)}^{n} - \bar{T}_{(i)(i-1)}^{n}\right)\bar{V}_{(i-1)} \Bigg],
\end{aligned}
\tag{13.60}
$$

where the nondimensional stretch is denoted by $\bar{s}_{(i)(j)}^{n}$, and it is defined as

$$
\bar{s}_{(i)(j)}^{n} = \frac{\left|\bar{\boldsymbol{\xi}}_{(i)(j)}^{n} + \bar{\boldsymbol{\eta}}_{(i)(j)}^{n}\right| - \left|\bar{\boldsymbol{\xi}}_{(i)(j)}^{n}\right|}{\left|\bar{\boldsymbol{\xi}}_{(i)(j)}^{n}\right|}.
\tag{13.61}
$$

The position of the $i$th and $j$th collocation points are given by $\bar{\mathbf{x}}_{(i)}$ and $\bar{\mathbf{x}}_{(j)}$, respectively, and, as such, the nondimensional relative initial position is defined as

$$
\bar{\boldsymbol{\xi}}_{(i)(j)}^{n} = \bar{\mathbf{x}}_{(j)} - \bar{\mathbf{x}}_{(i)}.
\tag{13.62}
$$

The nondimensional displacements of the $i$th and $j$th collocation points are given by $\bar{\mathbf{u}}_{(i)}^{n}$ and $\bar{\mathbf{u}}_{(j)}^{n}$, respectively. Therefore, the nondimensional relative displacement becomes

$$
\bar{\boldsymbol{\eta}}_{(i)(j)}^{n} = \bar{\mathbf{u}}_{(j)}^{n} - \bar{\mathbf{u}}_{(i)}^{n},
\tag{13.63a}
$$

and the term $\bar{T}_{(i)(j)}^{n}$ is defined as

$$
\bar{T}_{(i)(j)}^{n} = \frac{\bar{T}_{(j)}^{n} + \bar{T}_{(i)}^{n}}{2}.
\tag{13.63b}
$$

Based on Fig. 13.1, Eq. 13.55b can be discretized as

$$
\bar{\bar{T}}^n_{(i)} = \frac{2}{\bar{\delta}^2 \bar{A}} \left[ \left( \frac{\bar{\tau}^n_{(i)(i+3)}}{\left| \bar{\xi}^n_{(i)(i+3)} \right|} - \epsilon \frac{\bar{\bar{e}}^n_{(i)(i+3)}}{2} \right) \bar{V}_{(i+3)} + \left( \frac{\bar{\tau}^n_{(i)(i+2)}}{\left| \bar{\xi}^n_{(i)(i+2)} \right|} - \epsilon \frac{\bar{\bar{e}}^n_{(i)(i+2)}}{2} \right) \bar{V}_{(i+2)} \right.
$$

$$
+ \left( \frac{\bar{\tau}^n_{(i)(i+1)}}{\left| \bar{\xi}^n_{(i)(i+1)} \right|} - \epsilon \frac{\bar{\bar{e}}^n_{(i)(i+1)}}{2} \right) \bar{V}_{(i+1)} + \left( \frac{\bar{\tau}^n_{(i)(i-3)}}{\left| \bar{\xi}^n_{(i)(i-3)} \right|} - \epsilon \frac{\bar{\bar{e}}^n_{(i)(i-3)}}{2} \right) \bar{V}_{(i-3)}
$$

$$
\left. + \left( \frac{\bar{\tau}^n_{(i)(i-2)}}{\left| \bar{\xi}^n_{(i)(i-2)} \right|} - \epsilon \frac{\bar{\bar{e}}^n_{(i)(i-2)}}{2} \right) \bar{V}_{(i-2)} + \left( \frac{\bar{\tau}^n_{(i)(i-1)}}{\left| \bar{\xi}^n_{(i)(i-1)} \right|} - \epsilon \frac{\bar{\bar{e}}^n_{(i)(i-1)}}{2} \right) \bar{V}_{(i-1)} \right],
$$

$$
\text{(13.64)}
$$

where

$$
\bar{\tau}^n_{(i)(j)} = \bar{T}^n_{(j)} - \bar{T}^n_{(i)}, \tag{13.65a}
$$

and the nondimensional rate of extension between the material points is given by

$$
\bar{\bar{e}}^n_{(i)(j)} = \frac{\bar{\boldsymbol{\eta}}^n_{(i)(j)} + \bar{\boldsymbol{\xi}}^n_{(i)(j)}}{\left| \bar{\boldsymbol{\eta}}^n_{(i)(j)} + \bar{\boldsymbol{\xi}}^n_{(i)(j)} \right|} \cdot \left( \bar{\bar{\mathbf{u}}}^n_{(j)} - \bar{\bar{\mathbf{u}}}^n_{(i)} \right). \tag{13.65b}
$$

As explained in Sects. 7.3 and 12.9, the time integration of Eq. 13.60 can be performed by using explicit forward and backward difference techniques and Eq. 13.64 by forward difference time integration scheme.

## 13.6   Validation

The validity of the fully coupled PD thermomechanical equations is established by constructing PD solutions to previously considered problems. The first problem is a semi-infinite bar subjected to a transient thermal boundary condition. The second problem concerns the dynamic response of a thermoelastic bar with an initial sinusoidal velocity. The solutions to these problems are obtained by constructing one-dimensional PD models.

The third problem is a finite plate subjected to either a pressure shock or a thermal shock, and their combination. The solutions to these problems are obtained by constructing two-dimensional PD models. The fourth is a block of material subjected to a transient thermal boundary condition. The solution to this problem is obtained by constructing a three-dimensional PD model.

**Fig. 13.2** Peridynamic
model of the fields in the
one-dimensional bar: (a)
thermal, (b) deformation

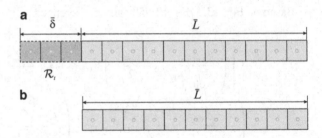

## 13.6.1   A Semi-infinite Bar Under Thermal Loading

A semi-infinite bar is subjected to the temperature boundary condition on the
bounding end. The bounding end is stress free and is gradually heated. The
stress-free condition on the bounding end is represented by not specifying any
displacement or velocity conditions. The peridynamic discretization of the bar for
thermal and deformational fields is shown in Fig. 13.2.

The peridynamic predictions for the nondimensional temperature and displacement for the three different coupling scenarios are compared against the classical
solution reported by Nickell and Sackman (1968). The coupling coefficient values of
$\epsilon = 0, 0.36, 1$ are used to depict the decoupled, moderate, and strong coupling
situations, respectively. The temperature boundary condition is imposed through the
fictitious region $\mathcal{R}_t$, as explained in Chap. 12. The solution is obtained by specifying
the geometric parameters, material properties, initial and boundary conditions, as well
as the peridynamic discretization and time integration parameters as:

Geometric Parameters

Length of bar: $\bar{L} = 5$
Area of cross section: $\bar{A} = 6.25 \times 10^{-4}$

Boundary Conditions

$\bar{T}(0, \bar{t}) = (\bar{t}/\bar{t}_0)H(\bar{t}_0 - \bar{t}) + H(\bar{t} - \bar{t}_0)$   , with $\bar{t}_o = 0.25$

Initial Conditions

$\bar{u}(\bar{x}, 0) = \partial \bar{u}(\bar{x}, 0)/\partial \bar{t} = \bar{T}(\bar{x}, 0) = 0$

PD Discretization Parameters

Total number of material points in the $\bar{x}$- direction: 200
Spacing between material points: $\bar{\Delta} = 0.025$
Incremental volume of material points: $\Delta \bar{V} = 1.5625 \times 10^{-5}$
Volume of fictitious boundary layer: $\bar{V}_{\bar{\delta}} = (3) \times \Delta \bar{V} = 4.6875 \times 10^{-5}$
Horizon: $\bar{\delta} = 3.015\bar{\Delta}$
Time step size: $\Delta \bar{t} = 0.5 \times 10^{-3}$

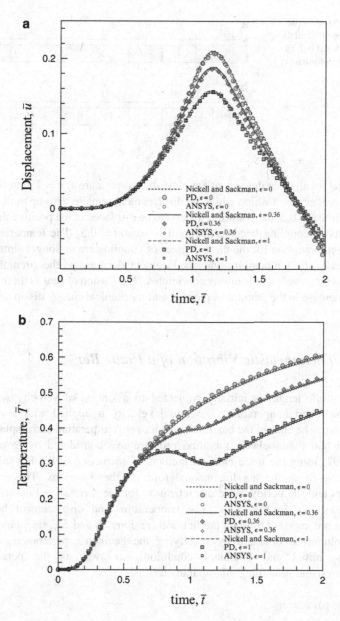

**Fig. 13.3** For different coupling coefficients: (**a**) Displacement and (**b**) temperative predictions at $\bar{x} = 1$

*Numerical Results:* Figure 13.3 provides a comparison of the temperature and displacement distribution predicted by the peridynamic simulation against the finite element predictions using ANSYS at $\bar{x} = 1$ for $\epsilon = 0, \ 0.36, \ 1$. These results also agree extremely well with those reported by Nickell and Sackman (1968). It is

**Fig. 13.4** Peridynamic model of the fields in the one-dimensional bar: (a) thermal, (b) deformation

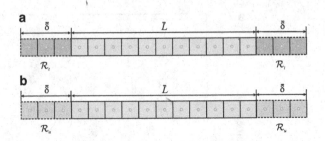

evident that for all three degrees of coupling the temperature at $\bar{x} = 1$ increases with time in a very similar fashion while the displacement remains zero up until $\bar{t} = 0.5$. At about time $\bar{t} = 0.5$, the point $\bar{x} = 1$ starts to be displaced in the positive direction. The effects of coupling become apparent beyond $\bar{t} = 0.5$. The temperature and displacement variation for the three degrees of coupling are no longer similar. The amplitudes of the temperature and displacement decrease as the strength of the coupling is increased. The coupling accelerates the diffusion of heat as there appears to be an increase in the amount of thermal and mechanical energy dissipated.

## 13.6.2  Thermoelastic Vibration of a Finite Bar

A bar of finite length is initially subjected to a sinusoidal velocity with zero displacement and temperature. The initial velocity is applied with a specified wavenumber. The ends of the bar are fixed with zero temperature and displacement. This particular thermoelastic vibration problem was considered by Armero and Simo (1992) using the finite element method. Construction of the PD solution is achieved by using the nondimensional form of the equations. The geometric parameters and the peridynamic discretization for the thermal and deformational fields are shown in Fig. 13.4. The temperature and displacement boundary conditions are imposed through the fictitious regions $\mathcal{R}_t$ and $\mathcal{R}_u$, respectively.

The solution is obtained by specifying the geometric parameters, material properties, initial and boundary conditions, as well as the peridynamic discretization and time integration parameters as:

Geometric parameters

Length of bar: $\bar{L} = 100$

Boundary Conditions

$$\bar{T}(0,\bar{t}) = \bar{T}(\bar{L},\bar{t}) = 0$$
$$\bar{u}(0,\bar{t}) = \bar{u}(\bar{L},\bar{t}) = 0$$

Initial Conditions

$$\bar{u}(\bar{x}, 0) = \bar{T}(\bar{x}, 0) = 0$$
$$\partial \bar{u}(\bar{x}, 0)/\partial t = \sin(\pi \bar{x}/\bar{L})$$

PD Discretization Parameters

Total number of material points in the $\bar{x}$- direction: 5000
Spacing between material points: $\bar{\Delta} = 0.02$
Incremental volume of material points: $\Delta \bar{V} = 8 \times 10^{-6}$
Volume of fictitious boundary layer: $\bar{V}_{\bar{\delta}} = (3) \times \Delta \bar{V} = 24 \times 10^{-6}$
Horizon: $\bar{\delta} = 3.015 \times \bar{\Delta}$
Time step size: $\Delta \bar{t} = 1 \times 10^{-4}$

*Numerical Results:* The resulting elastic waves are progressive traveling waves. In the case of a fully coupled thermoelastic problem, there exist two types of waves: elastic and thermal. Both types of waves have been modified from their uncoupled forms. The modified elastic waves are attenuated, compared to the uncoupled elastic waves, and are subjected to dispersion and damping in time. The modified thermal waves also exhibit dispersion and damping in time. The peridynamic predictions for the temporal distribution of displacement and temperature at $\bar{x} = 50$ and $\bar{x} = 25$, respectively, are shown in Fig. 13.5 for coupling coefficients of $\epsilon = 0$ and $\epsilon = 1$. The peridynamic predictions are also compared with the classical finite element approximations given by Armero and Simo (1992).

## 13.6.3  Plate Subjected to a Shock of Pressure and Temperature, and Their Combination

The fully coupled nondimensional PD thermomechanical equations are further verified by solving a problem previously considered by Hosseini-Tehrani and Eslami (2000) using the Boundary Element Method. It concerns a square plate of isotropic material under either a pressure shock or a thermal shock, and their combination on the free edge in the positive $\bar{x}$-direction. As shown in Fig. 13.6, it is clamped at the other edge and the insulated horizontal edges are free of any loading. The thermomechanical equations are solved for both uncoupled and coupled cases.

Geometric Parameters

Length: $\bar{L} = 10$
Width: $\bar{W} = 10$
Thickness: $\bar{H} = 1$

Initial Conditions

$$\bar{T}(\bar{x}, \bar{y}, 0) = 0$$
$$\bar{u}_{\bar{x}}(\bar{x}, \bar{y}, 0) = \bar{u}_{\bar{y}}(\bar{x}, \bar{y}, 0) = 0$$

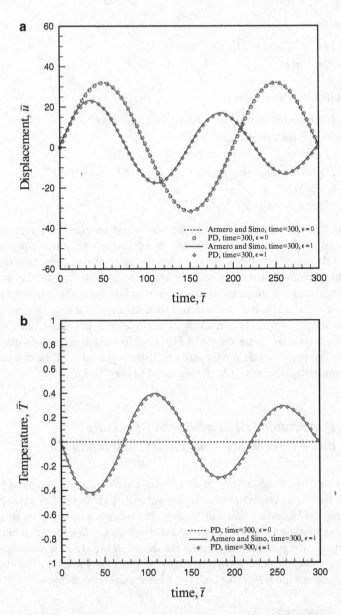

**Fig. 13.5** Variation of (**a**) displacement at $\bar{x} = 50$ and (**b**) temperature at $\bar{x} = 25$

## Boundary Conditions

$$\bar{T}_{,\bar{x}}(\bar{x} = 10, \bar{y}, \bar{t}) = 0$$
$$\bar{T}_{,\bar{y}}(\bar{x}, \bar{y} = \pm 5, \bar{t}) = 0$$
$$\bar{u}_{\bar{x}}(\bar{x} = 10, \bar{y}, \bar{t}) = \bar{u}_{\bar{y}}(\bar{x} = 10, \bar{y}, \bar{t}) = 0$$

**Fig. 13.6** Geometry and boundary conditions of the plate under pressure or thermal shock

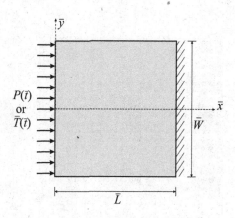

$\sigma_{\bar{y}\bar{y}}(\bar{x}, \bar{y} = \pm 5, \bar{t}) = \sigma_{\bar{x}\bar{y}}(\bar{x}, \bar{y} = \pm 5, \bar{t}) = 0$
where $\bar{t}$ is the nondimensional time.

Pressure Shock

$\bar{T}(\bar{x} = 0, \bar{y}, \bar{t}) = 0$
$\sigma_{\bar{x}\bar{x}}(\bar{x} = 0, \bar{y}, \bar{t}) = -P(\bar{t}) = -5\bar{t}e^{-2\bar{t}}$

Thermal Shock

$\bar{T}(\bar{x} = 0, \bar{y}, \bar{t}) = 5\bar{t}e^{-2\bar{t}}$
$\sigma_{\bar{x}\bar{x}}(\bar{x} = 0, \bar{y}, \bar{t}) = 0$

Combined Pressure and Thermal Shock

$\bar{T}(\bar{x} = 0, \bar{y}, \bar{t}) = 5\bar{t}e^{-2\bar{t}}$
$\sigma_{\bar{x}\bar{x}}(\bar{x} = 0, \bar{y}, \bar{t}) = -P(\bar{t}) = -5\bar{t}e^{-2\bar{t}}$

PD Discretization Parameters

Total number of material points in the $\bar{x}$- direction: 200
Total number of material points in the $\bar{y}$- direction: 200
Spacing between material points: $\bar{\Delta} = 0.05$
Incremental volume of material points: $\Delta\bar{V} = 1.25 \times 10^{-4}$
Volume of fictitious boundary layer: $\bar{V}_{\bar{\delta}} = (3 \times 200) \times \Delta\bar{V} = 0.075$
Volume of boundary layer: $\bar{V}_{\bar{\Delta}} = (1 \times 200) \times \Delta\bar{V} = 0.025$
Horizon: $\bar{\delta} = 3.015 \times \bar{\Delta}$
Time step size: $\Delta\bar{t} = 0.5 \times 10^{-3}$

The peridynamic discretization for the thermal field is shown in Fig. 13.7. The temperature boundary condition is imposed in fictitious region $\mathcal{R}_t$. The peridynamic discretization for the deformational field is shown in Fig. 13.8. The displacement

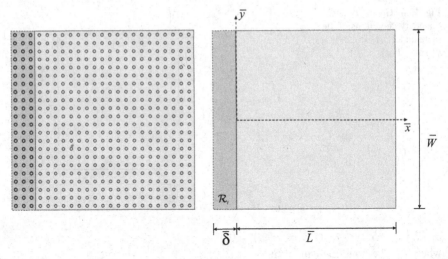

**Fig. 13.7** Peridynamic model of the thermal field in a plate

boundary condition is imposed in fictitious region $\mathcal{R}_u$. The pressure is applied through boundary layer region $\mathcal{R}_p$.

*Numerical results:* Figure 13.9 shows the temperature and displacement variations at $\bar{y} = 0$ due to the pressure shock at times $\bar{t} = 3$ and $\bar{t} = 6$. When the coupling coefficient is zero, no temperature change is expected. However, when the coupled effect is included, even though mechanical loading is applied, temperature change is expected. The compressive stress along the boundary causes a temperature rise. As observed in this figure, the peak of the temperature distribution moves to the right as time progresses. Figure 13.9 also shows the axial displacement along the $\bar{x}$-axis. The PD results are also in close agreement with the BEM results (Hosseini-Tehrani and Eslami 2000). Figure 13.10 shows the temperature and displacement variations at $\bar{y} = 0$ due to thermal shock at times $\bar{t} = 3$ and $\bar{t} = 6$. As observed, the coupling term in the thermal field causes a temperature drop, and the peridynamic predictions are in close agreement with the BEM solution published by Hosseini-Tehrani and Eslami (2000). Figure 13.11 shows the temperature and displacement variations at $\bar{y} = 0$ due to combined pressure and thermal shock at times $\bar{t} = 3$ and $\bar{t} = 6$. The PD predictions are in close agreement with the BEM results by Hosseini-Tehrani and Eslami (2000).

### 13.6.4   A Block of Material Under Thermal Loading

A three-dimensional finite block of material is gradually heated at one end, and the remaining surfaces are insulated. As shown in Fig. 13.12, it is clamped at the other end without any other type of loading. The PD discretization of the thermal and deformational fields is shown in Fig. 13.13

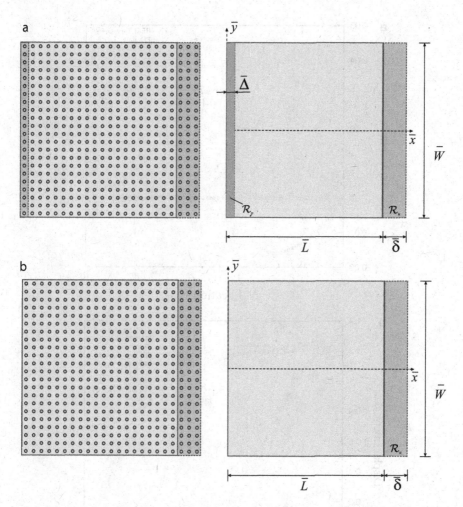

**Fig. 13.8** Peridynamic model of deformational field in a plate: (**a**) pressure shock, (**b**) thermal shock

The solution is obtained by specifying the geometric parameters, initial and boundary conditions, as well as the peridynamic discretization and time integration parameters as:

**Geometric Parameters**

Length: $\bar{L} = 5$
Width: $\bar{W} = 0.15$
Thickness: $\bar{H} = 0.15$

**Initial Conditions**

$$\bar{u}(\bar{x}, \bar{y}, \bar{z}, 0) = \partial \bar{u}(\bar{x}, \bar{y}, \bar{z}, 0)/\partial \bar{t} = \bar{T}(\bar{x}, \bar{y}, \bar{z}, 0) = 0$$

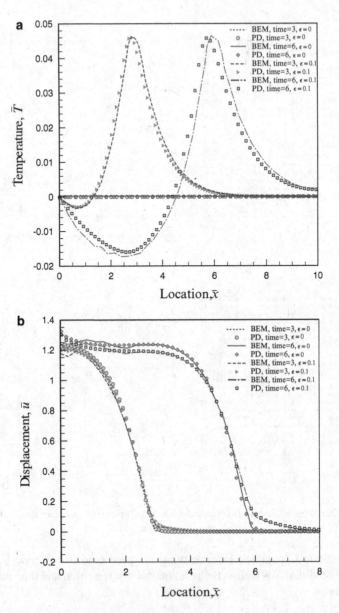

**Fig. 13.9** Variations along the centerline in the plate for uncoupled ($\epsilon = 0$) and coupled ($\epsilon \neq 0$) cases under pressure shock loading: (**a**) temperature, and (**b**) displacement

Boundary Conditions

$$\bar{T}(0, \bar{y}, \bar{z}, \bar{t}) = (\bar{t}/\bar{t}_0)H(\bar{t}_0 - \bar{t}) + H(\bar{t} - \bar{t}_0)$$
$$\bar{T}_{,\bar{x}}(\bar{x} = \bar{L}, \bar{y}, \bar{z}, \bar{t}) = 0$$
$$\bar{T}_{,\bar{y}}(\bar{x}, \bar{y} = 0, \bar{z}, \bar{t}) = 0$$

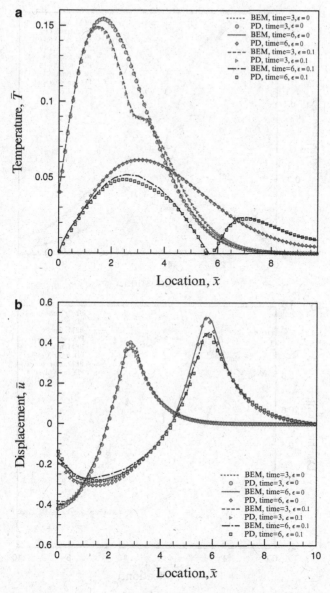

**Fig. 13.10** Variations along the centerline in the plate for uncoupled ($\epsilon = 0$) and coupled ($\epsilon \neq 0$) cases under thermal shock loading: (**a**) temperature, and (**b**) displacement

$$\bar{T}_{,\bar{y}}(\bar{x}, \bar{y} = \bar{W}, \bar{z}, \bar{t}) = 0$$
$$\bar{T}_{,\bar{z}}(\bar{x}, \bar{y}, \bar{z} = 0, \bar{t}) = 0$$
$$\bar{T}_{,\bar{z}}(\bar{x}, \bar{y}, \bar{z} = \bar{H}, \bar{t}) = 0$$
$$\bar{u}_{\bar{x}}(\bar{x} = \bar{L}, \bar{y}, \bar{z}, \bar{t}) = \bar{u}_{\bar{y}}(\bar{x} = \bar{L}, \bar{y}, \bar{z}, \bar{t}) = \bar{u}_{\bar{z}}(\bar{x} = \bar{L}, \bar{y}, \bar{z}, \bar{t}) = 0$$

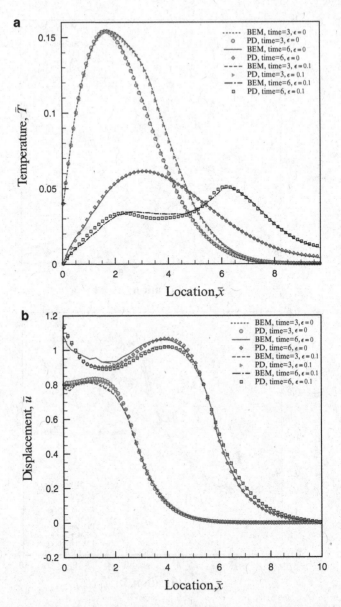

**Fig. 13.11** Variations along the centerline in the plate for uncoupled ($\epsilon = 0$) and coupled ($\epsilon \neq 0$) cases under combined thermal and pressure shock loading: (**a**) temperature, and (**b**) displacement

$$\sigma_{\bar{x}\bar{x}}(\bar{x} = 0, \bar{y}, \bar{z}, \bar{t}) = \sigma_{\bar{x}\bar{y}}(\bar{x} = 0, \bar{y}, \bar{z}, \bar{t}) = \sigma_{\bar{x}\bar{z}}(\bar{x} = 0, \bar{y}, \bar{z}, \bar{t}) = 0$$
$$\sigma_{\bar{y}\bar{y}}(\bar{x}, \bar{y} = 0, \bar{z}, \bar{t}) = \sigma_{\bar{x}\bar{y}}(\bar{x}, \bar{y} = 0, \bar{z}, \bar{t}) = \sigma_{\bar{y}\bar{z}}(\bar{x}, \bar{y} = 0, \bar{z}, \bar{t}) = 0$$
$$\sigma_{\bar{y}\bar{y}}(\bar{x}, \bar{y} = \bar{W}, \bar{z}, \bar{t}) = \sigma_{\bar{x}\bar{y}}(\bar{x}, \bar{y} = \bar{W}, \bar{z}, \bar{t}) = \sigma_{\bar{y}\bar{z}}(\bar{x}, \bar{y} = \bar{W}, \bar{z}, \bar{t}) = 0$$
$$\sigma_{\bar{z}\bar{z}}(\bar{x}, \bar{y}, \bar{z} = 0, \bar{t}) = \sigma_{\bar{x}\bar{z}}(\bar{x}, \bar{y}, \bar{z} = 0, \bar{t}) = \sigma_{\bar{y}\bar{z}}(\bar{x}, \bar{y}, \bar{z} = 0, \bar{t}) = 0$$
$$\sigma_{\bar{z}\bar{z}}(\bar{x}, \bar{y}, \bar{z} = \bar{H}, \bar{t}) = \sigma_{\bar{x}\bar{z}}(\bar{x}, \bar{y}, \bar{z} = \bar{H}, \bar{t}) = \sigma_{\bar{y}\bar{z}}(\bar{x}, \bar{y}, \bar{z} = \bar{H}, \bar{t}) = 0$$

**Fig. 13.12** Geometry
and boundary conditions
of the block under
thermal loading

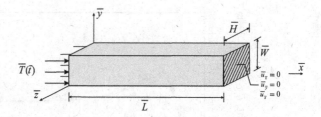

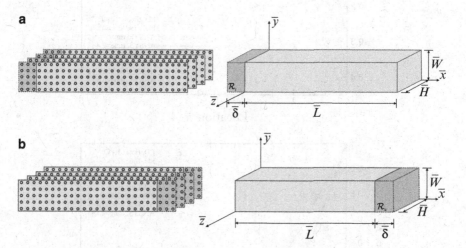

**Fig. 13.13** Three-dimensional peridynamic model of the fields: (**a**) thermal, (**b**) deformation

PD Discretization Parameters

Total number of material points in the $\bar{x}$- direction: 200
Total number of material points in the $\bar{y}$- direction: 6
Total number of material points in the $\bar{z}$- direction: 6
Spacing between material points: $\bar{\Delta} = 0.025$
Incremental volume of material points: $\Delta \bar{V} = 1.5625 \times 10^{-5}$
Volume of fictitious boundary layer: $\bar{V}_{\bar{\delta}} = (3 \times 6 \times 6) \times \Delta \bar{V} = 1.6875 \times 10^{-3}$
Horizon: $\bar{\delta} = 3.015 \times \bar{\Delta}$
Time step size: $\Delta \bar{t} = 1 \times 10^{-4}$

*Numerical Results:* As shown in Fig. 13.14, the PD predictions for temperature and
displacement variations along the length of the block are compared with the FEA
results from ANSYS at $\bar{t} = 1$ and 2 for $\epsilon = 0$ and $\epsilon = 1$. The comparison indicates
excellent agreement.

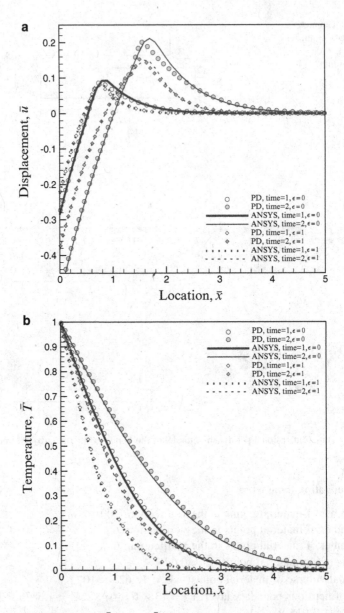

**Fig. 13.14** Predictions at $(\bar{y} = \bar{W}/2,\ \bar{z} = \bar{H}/2)$ for coupled and uncoupled cases: (**a**) displacement, and (**b**) temperature

# References

Ardito R, Comi C (2009) Nonlocal thermoelastic damping in microelectromechanical resonators. J Eng Mech-ASCE 135:214–220

Armero F, Simo JC (1992) A new unconditionally stable fractional step method for nonlinear coupled thermomechanical problems. Int J Numer Meth Eng 35:737–766

Biot MA (1956) Thermoelasticity and irreversible thermodynamics. J Appl Phys 27:240–253

Boley BA, Hetnarski RB (1968) Propagation of discontinuities in coupled thermoelastic problems. ASME J Appl Mech 35:489–494

Boley BA, Tolins IS (1962) Transient coupled thermoelastic boundary value problems in the half-space. J Appl Mech 29:637–646

Brünig M, Albrecht D, Gerke S (2011) Numerical analyses of stress-triaxiality-dependent inelastic deformation behavior of aluminum alloys. Int J Damage Mech 20:299–317

Chadwick P (1960) Thermoelasticity: the dynamical theory. In: Sneddon IN, Hill R (eds) Progress in solid mechanics, vol I. North-Holland, Amsterdam, pp 263–328

Chadwick P (1962) On the propagation of thermoelastic disturbances in thin plates and rods. J Mech Phys Solids 10:99–109

Chadwick P, Sneddon IN (1958) Plane waves in an elastic solid conducting heat. J Mech Phys Solids 6:223–230

Chen TC, Weng CI (1988) Generalized coupled transient thermoelastic plane problems by Laplace transform finite-element method. ASME J Appl Mech 55:377–382

Chen TC, Weng CI (1989a) Coupled transient thermoelastic response in an axi-symmetric circular-cylinder by Laplace transform finite-element method. Comput Struct 33:533–542

Chen TC, Weng CI (1989b) Generalized coupled transient thermoelastic problem of a square cylinder with elliptical hole by Laplace transform finite-element method. J Therm Stresses 12:305–320

Deresiewicz H (1957) Plane waves in a thermoelastic solid. J Acoust Soc Am 29:204–209

Farhat C, Park KC, Duboispelerin Y (1991) An unconditionally stable staggered algorithm for transient finite-element analysis of coupled thermoelastic problems. Comput Methods Appl Mech Eng 85:349–365

Fung YC (1965) Foundations of solid mechanics. Prentice-Hall, Englewood Cliffs

Givoli D, Rand O (1995) Dynamic thermoelastic coupling effects in a rod. AIAA J 33:776–778

Hosseini-Tehrani P, Eslami MR (2000) BEM analysis of thermal and mechanical shock in a two-dimensional finite domain considering coupled thermoelasticity. Eng Anal Bound Elem 24:249–257

Huang ZX (1999) Points of view on the nonlocal field theory and their applications to the fracture mechanics (II)—re-discuss nonlinear constitutive equations of nonlocal thermoelastic bodies. Appl Math Mech 20:764–772

Kilic B, Madenci E (2010) An adaptive dynamic relaxation method for quasi-static simulations using the peridynamic theory. Theor Appl Fract Mech 53:194–204

Liu WK, Chang HG (1985) A note on numerical-analysis of dynamic coupled thermoelasticity. ASME J Appl Mech 52:483–485

Liu WK, Zhang YF (1983) Unconditionally stable implicit explicit algorithms for coupled thermal-stress waves. Comput Struct 17:371–374

Lychev SA, Manzhirov AV, Joubert SV (2010) Closed solutions of boundary-value problems of coupled thermoelasticity. Mech Solids 45:610–623

Nickell RE, Sackman JL (1968) Approximate solutions in linear coupled thermoelasticity. J Appl Mech 35:255–266

Nowinski JL (1978) Theory of thermoelasticity with applications. Sijthoff & Noordhoff International Publishers, Alphen aan den Rijn

Oden JT (1969) Finite element analysis of nonlinear problems in dynamical theory of coupled thermoelasticity. Nucl Eng Des 10:465–475

Paria G (1958) Coupling of elastic and thermal deformations. Appl Sci Res 7:463–475

Rittel D (1998) Experimental investigation of transient thermoelastic effects in dynamic fracture. Int J Solids Struct 35:2959–2973

Silling SA (2009) Linearized theory of peridynamic states. Report SAND2009-2458

Silling SA, Lehoucq RB (2010) Peridynamic theory of solid mechanics. In: Aref H, van der Giessen H (eds) Advances in applied mechanics. Elsevier, San Diego, pp 73–168

Soler AI, Brull MA (1965) On solution to transient coupled thermoelastic problems by perturbation techniques. J Appl Mech 32:389–399

Ting EC, Chen HC (1982) A unified numerical approach for thermal-stress waves. Comput Struct 15:165–175

Wood WL (1990) Practical time-stepping schemes. Clarendon Press, Oxford

# Appendix

## A.1 Concept of State

A continuous function $g(x)$ for $-\infty < x < \infty$ can be $g(x_i)$ considered as a combination of an infinite number of discrete function values, for $i = 1, ...., \infty$. These discrete function values can be stored in an infinite-dimensional array, or a "state," $\underline{g}$ as

$$\underline{g} = \left\{ \begin{array}{c} g(x_1) \\ \vdots \\ g(x_i) \\ \vdots \\ g(x_\infty) \end{array} \right\}. \tag{A.1}$$

For notation purposes, all states are denoted with an underscore.

The states of order 2 (double state) are written in an upper case font, $\underline{\mathbb{A}}$. States of order 1 (vector state) are written in a bold upper case font, $\underline{\mathbf{A}}$. States of order 0 (scalar state) are written in a lower case (nonbold) font, $\underline{a}$. When the double state $\underline{\mathbb{A}}$ operates on an angle bracket $\langle \bullet \rangle$, the result is the second-order tensor $\underline{\mathbb{A}}\langle \bullet \rangle$; when the vector state $\underline{\mathbf{A}}$ operates on an angle bracket $\langle \bullet \rangle$, the result is the vector $\underline{\mathbf{A}}\langle \bullet \rangle$; and when the scalar state $\underline{a}$ operates on $\langle \bullet \rangle$, the result is the scalar $\underline{a}\langle \bullet \rangle$.

The "state" concept is not restricted to continuous functions. It is also applicable to discontinuous functions. As explained by Silling et al. (2007), "states" can also be described as a general form of tensors. It is possible to convert states to tensors or vice versa. The process of converting a tensor to a state is referred to as "*expansion*" and the process of converting a state to a tensor as "*reduction*."

If a second-order tensor, $\mathbf{F}$, operates on a vector $(\mathbf{x}_{(j)} - \mathbf{x}_{(k)})$, the corresponding vector $(\mathbf{y}_{(j)} - \mathbf{y}_{(k)})$ is obtained as

$$\left( \mathbf{y}_{(j)} - \mathbf{y}_{(k)} \right) = \mathbf{F} \left( \mathbf{x}_{(j)} - \mathbf{x}_{(k)} \right), \tag{A.2}$$

E. Madenci and E. Oterkus, *Peridynamic Theory and Its Applications*,
DOI 10.1007/978-1-4614-8465-3, © Springer Science+Business Media New York 2014

**Fig. A.1** The "*expansion*"
of the second-order tensor **F**

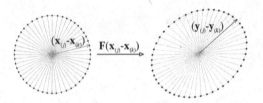

where $j = 1, \ldots, \infty$. All the $(\mathbf{y}_{(j)} - \mathbf{y}_{(k)})$ vectors can be stored in an infinite-dimensional array, or a vector state, $\underline{\mathbf{Y}}$:

$$\underline{\mathbf{Y}} = \left\{ \begin{array}{c} \left(\mathbf{y}_{(1)} - \mathbf{y}_{(k)}\right) \\ \vdots \\ \left(\mathbf{y}_{(\infty)} - \mathbf{y}_{(k)}\right) \end{array} \right\} \quad \text{or} \quad \underline{\mathbf{Y}} = \left\{ \begin{array}{c} \mathbf{F}\left(\mathbf{x}_{(1)} - \mathbf{x}_{(k)}\right) \\ \vdots \\ \mathbf{F}\left(\mathbf{x}_{(\infty)} - \mathbf{x}_{(k)}\right) \end{array} \right\}. \tag{A.3}$$

In this equation, there is a direct relationship between the vector state $\underline{\mathbf{Y}}$ and the second-order tensor **F**. This relationship can be expressed as the "*expansion*" of the second-order tensor **F**. The "*expansion*" process can be visualized as shown in Fig. A.1. In this figure, the second-order tensor **F** operates on an infinite number of vectors, forming a circle, $(\mathbf{x}_{(j)} - \mathbf{x}_{(k)})$ with $j = 1, \ldots, \infty$, and the resulting vectors, $(\mathbf{y}_{(j)} - \mathbf{y}_{(k)})$, form an ellipse.

Therefore, the "state" can be viewed as a data bank to extract information about the state of material points. For example, the vector states of reference position $\underline{\mathbf{X}}$ and deformation $\underline{\mathbf{Y}}$ provide information about the relative position of material points in the reference and deformed configurations. The mathematical operations for such extraction of information are denoted as

$$\underline{\mathbf{X}}\langle\mathbf{x}' - \mathbf{x}\rangle = \mathbf{x}' - \mathbf{x} \tag{A.4a}$$

and

$$\underline{\mathbf{Y}}\langle\mathbf{x}' - \mathbf{x}\rangle = \mathbf{y}' - \mathbf{y}, \tag{A.4b}$$

in which $\mathbf{x}' - \mathbf{x}$ and $\mathbf{y}' - \mathbf{y}$ represent the relative position of the points $\mathbf{x}'$ and $\mathbf{x}$ in the reference and deformed configurations. Similarly, a temperature scalar state, $\underline{\tau}$, can provide information about the temperatures, $T'$ and $T$, at these two material points in the form

$$\underline{\tau}\langle\mathbf{x}' - \mathbf{x}\rangle = T' - T. \tag{A.4c}$$

As presented by Silling et al. (2007), the dot product of two vector states, $\underline{\mathbf{A}}$ and $\underline{\mathbf{D}}$, and two scalar states, $\underline{a}$ and $\underline{d}$, can be cast as

$$\underline{\mathbf{A}} \bullet \underline{\mathbf{D}} = \int_H \underline{\mathbf{A}} \langle \mathbf{x}' - \mathbf{x} \rangle \bullet \underline{\mathbf{D}} \langle \mathbf{x}' - \mathbf{x} \rangle dH \tag{A.5a}$$

and

$$\underline{a} \bullet \underline{d} = \int_H \underline{a} \langle \mathbf{x}' - \mathbf{x} \rangle \underline{d} \langle \mathbf{x}' - \mathbf{x} \rangle dH. \tag{A.5b}$$

Their point products are expressed as

$$(\underline{\mathbf{A}}\underline{\mathbf{D}}) \langle \mathbf{x}' - \mathbf{x} \rangle = \underline{\mathbf{A}} \langle \mathbf{x}' - \mathbf{x} \rangle \bullet \underline{\mathbf{D}} \langle \mathbf{x}' - \mathbf{x} \rangle \tag{A.6a}$$

and

$$(\underline{a}\,\underline{d}) \langle \mathbf{x}' - \mathbf{x} \rangle = \underline{a} \langle \mathbf{x}' - \mathbf{x} \rangle \underline{d} \langle \mathbf{x}' - \mathbf{x} \rangle. \tag{A.6b}$$

The tensor product of vector states $\underline{\mathbf{A}}$ and $\underline{\mathbf{D}}$ is defined as

$$\underline{\mathbf{A}} * \underline{\mathbf{D}} = \int_H \underline{\quad}w \langle \mathbf{x}' - \mathbf{x} \rangle \underline{\mathbf{A}} \langle \mathbf{x}' - \mathbf{x} \rangle \otimes \underline{\mathbf{D}} \langle \mathbf{x}' - \mathbf{x} \rangle dH, \tag{A.7}$$

where $\underline{\quad}w$ is the influence function, a scalar state, and $\otimes$ represents the dyadic product of two vectors i.e., $\mathbf{C} = \mathbf{a} \otimes \mathbf{b}$ or $C_{ij} = a_i b_j$.

The reverse transformation from a vector state to a second-order tensor, which is called the "*reduction*" process, can be approximated by the expression given by Silling et al. (2007). The tensor $\mathcal{R}\{\underline{\mathbf{Y}}\}$ is the vector state reduction of the vector state $\underline{\mathbf{Y}}$ and is defined as

$$\mathcal{R}\{\underline{\mathbf{Y}}\} = (\underline{\mathbf{Y}} * \underline{\mathbf{X}})\mathbf{K}^{-1}. \tag{A.8}$$

Hence, a vector state $\underline{\mathbf{Y}}$ can be reduced to a second-order tensor $\mathbf{F}$

$$\mathbf{F} = \mathcal{R}\{\underline{\mathbf{Y}}\}. \tag{A.9}$$

The shape tensor, $\mathbf{K}$, is defined as

$$\mathbf{K} = \underline{\mathbf{X}} * \underline{\mathbf{X}}. \tag{A.10}$$

Therefore, the shape tensor, $\mathbf{K}$, can be obtained as

$$\mathbf{K} = \int_H \underline{w} \langle \mathbf{x}' - \mathbf{x} \rangle \underline{\mathbf{X}} \langle \mathbf{x}' - \mathbf{x} \rangle \otimes \underline{\mathbf{X}} \langle \mathbf{x}' - \mathbf{x} \rangle \, dH. \tag{A.11}$$

**Fig. A.2** Components
of the position vector, ξ,
between material points
at **x** and **x**′

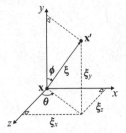

The influence function, $\underline{w}\langle \mathbf{x}' - \mathbf{x} \rangle$, as discussed in Chap. 4, can be defined as

$$\underline{w}\langle \mathbf{x}' - \mathbf{x} \rangle = \frac{\delta}{|\mathbf{x}' - \mathbf{x}|}, \qquad (A.12)$$

with $\delta$ defining the radius of the horizon, $H$. The shape tensor, **K**, has a direct relationship with the volume of the horizon. Defining the position vector in the form $\boldsymbol{\xi} = \mathbf{x}' - \mathbf{x}$, the shape tensor can be rewritten as

$$\mathbf{K} = \int_H \underline{\omega}\langle \boldsymbol{\xi} \rangle \boldsymbol{\xi} \otimes \boldsymbol{\xi} \, dH \qquad (A.13a)$$

or

$$K_{ij} = \int_H \underline{\omega}\langle \boldsymbol{\xi} \rangle \, \xi_i \, \xi_j \, dH, \quad \text{with } i,j = 1,2,3. \qquad (A.13b)$$

The components $(\xi_x, \xi_y, \xi_z)$ of the position vector $\boldsymbol{\xi}$ in reference to a Cartesian coordinate system $(x, y, z)$, whose origin is located at **x**, between material points at **x** and **x**′ can be expressed as

$$\xi_1 = \xi_x = \xi \sin(\phi) \sin(\theta), \qquad (A.14a)$$

$$\xi_2 = \xi_y = \xi \cos(\phi), \qquad (A.14b)$$

$$\xi_3 = \xi_z = \xi \sin(\phi) \cos(\theta), \qquad (A.14c)$$

where $\xi = |\boldsymbol{\xi}|$ is the length of the position vector; definitions of angles $\phi \in (0, \pi)$ and $\theta \in (0, 2\pi)$ are shown in Fig. A.2.

The components of the shape tensor, **K**, become

$$K_{ij} = \int_H \frac{\delta}{\xi} \xi_i \, \xi_j \, dH, \quad i,j = 1,2,3 \qquad (A.15a)$$

or

$$K_{ij} = \int\limits_0^\delta \int\limits_0^{2\pi} \int\limits_0^\pi \frac{\delta}{\xi}\, \xi_i\, \xi_j\, \xi^2\, \sin(\phi)\, d\phi\, d\theta\, d\xi. \tag{A.15b}$$

After performing the integration in Eq. A.15b, the components of the shape tensor, $\mathbf{K}$, are obtained as

$$K_{ij} = \frac{\pi \delta^5}{3}\delta_{ij}, \tag{A.16}$$

where $\delta_{ij}$ is the Kronecker delta with $i,j = 1,2,3$. By defining the volume of the horizon, $V = 4/3\,\pi\delta^3$, the shape tensor, $\mathbf{K}$, can be expressed as

$$\mathbf{K} = \frac{V\delta^2}{4}\mathbf{I}, \tag{A.17}$$

with $\mathbf{I}$ representing the identity matrix. Therefore, the shape tensor can be viewed as a quantity that serves as volume averaging of the tensor product of vector states, $(\underline{\mathbf{Y}} * \underline{\mathbf{X}})$.

Based on the definition of reduction, Eq. A.8, a scalar state, $\underline{a}$, can be reduced to a vector, $\mathcal{R}\{\underline{a}\}$, as

$$\mathcal{R}\{\underline{a}\} = (\underline{a} * \underline{x})m^{-1}. \tag{A.18}$$

Hence, a vector state, $\underline{a}$, can be reduced to a vector, $\mathbf{f}$, as

$$\mathbf{f} = \mathcal{R}\{\underline{a}\}. \tag{A.19}$$

The scalar weighted volume, $m$, is defined as

$$m = \int\limits_H -w\langle\mathbf{x}'-\mathbf{x}\rangle|\underline{\mathbf{X}}|\langle\mathbf{x}'-\mathbf{x}\rangle \otimes |\underline{\mathbf{X}}|\langle\mathbf{x}'-\mathbf{x}\rangle dH. \tag{A.20}$$

The dyadic, $\otimes$, operation annuls because both $\underline{a}\langle\mathbf{x}' - \mathbf{x}\rangle$ and $|\underline{\mathbf{X}}|\langle\mathbf{x}'-\mathbf{x}\rangle = |\mathbf{x}'-\mathbf{x}|$ are scalar; thus, the reduction expression can be rewritten as

$$\mathbf{f} = \frac{1}{m}\int\limits_H \underline{w}\langle\mathbf{x}' - \mathbf{x}\rangle\underline{\mathbf{X}}\langle\mathbf{x}' - \mathbf{x}\rangle\underline{a}\langle\mathbf{x}' - \mathbf{x}\rangle\, dH, \tag{A.21a}$$

with

$$m = \int\limits_H \underline{w}\langle\mathbf{x}'-\mathbf{x}\rangle|\mathbf{x}'-\mathbf{x}||\mathbf{x}'-\mathbf{x}|dH. \tag{A.21b}$$

Substituting for the influence function, $\underline{w}\langle \mathbf{x}' - \mathbf{x} \rangle$, from Eq. A.12, the scalar weighted volume can be evaluated as

$$m = \delta \int_H |\mathbf{x}' - x| dH. \tag{A.22}$$

In light of Fig. A.2, it can be explicitly evaluated as

$$m = \delta \int_0^\delta \int_0^{2\pi} \int_0^\pi \xi \xi^2 \, \sin(\phi) \, d\phi \, d\theta \, d\xi = \frac{3}{4} V \delta^2. \tag{A.23}$$

The scalar weighted volume can be viewed as a quantity that serves as volume averaging of the product of a scalar and vector states, $\underline{a} * \underline{\mathbf{X}}$.

## A.2   The Fréchet Derivative

Let a scalar function, $\Psi$, be dependent on a state, $\underline{\mathbf{A}}$, i.e., $\Psi = \Psi(\underline{\mathbf{A}})$. Its variation is defined as

$$d\Psi = \Psi(\underline{\mathbf{A}} + d\underline{\mathbf{A}}) - \Psi(\underline{\mathbf{A}}), \tag{A.24}$$

in which $d\underline{\mathbf{A}}$ is the differential of $\underline{\mathbf{A}}$. Silling et al. (2007) notes that if $\Psi$ is differentiable then the variation of $\Psi$ can be defined as

$$d\Psi = \nabla\Psi(\underline{\mathbf{A}}) \cdot d\underline{\mathbf{A}} \tag{A.25a}$$

or

$$d\Psi = \Psi,_{\underline{\mathbf{A}}}(\underline{\mathbf{A}}) \cdot d\underline{\mathbf{A}}, \tag{A.25b}$$

and the term $\nabla\Psi(\underline{\mathbf{A}}) = \Psi,_{\underline{\mathbf{A}}}(\underline{\mathbf{A}})$ is called the Fréchet derivative of $\Psi$ at $\underline{\mathbf{A}}$. Since $\Psi$ is a scalar value function, $\Psi,_{\underline{\mathbf{A}}}(\underline{\mathbf{A}})$ is a state of the same order as $\underline{\mathbf{A}}$. Fréchet derivatives of various functions of states are given by Silling et al. (2007) and Silling and Lehoucq (2010).

## References

Silling SA, Lehoucq RB (2010) Peridynamic theory of solid mechanics. Adv Appl Mech 44:73–168

# Index

Printed in the United States
By Bookmasters